Edizioni PensareDiverso.
Collana Cenacolo Jung Pauli..

Hinduism and quantum theory.

Bruno Del Medico.

Hinduism and Quantum Theory

All the similarities between quantum principles and the concepts proper to Hindu philosophy: Brahma, Atman, Karma, Moksha, Dharma.

Bruno Del Medico Editor.
ThinkingDiverse Editions
edizioni@delmedico.it
www.giardinoquantico.it

On the cover: at the entrance to the European Laboratory for Particle Physics, (CERN in Geneva) stands a statue of Shiva Nataraja, the dancing god of Hinduism. This statue represents the perpetual cycle of creation, preservation and destruction, a central concept in Indian cosmology.

Summary.

Introduction.

Comparing Eastern philosophies and quantum physics reveals fascinating similarities that invite us to reconsider our understanding of reality.

The intent of the book is to explore the surprising connection between the ancient philosophical traditions of Hinduism and the principles of quantum physics. An initial part explains in a totally understandable way the basic principles of quantum physics and the philosophy related to this new science. In the sequel, an evocative journey, leads the reader to discover how these two seemingly distant realities can interact and influence each other.

Hinduism, with its profound metaphysical principles, offers a framework that seems to anticipate in many respects the discoveries of quantum physics. The book compares some of the major themes of Hinduism with their corresponding quantum notions.

Brahman represents ultimate reality, an interconnected whole that permeates everything. This aligns with the concept of quantum entanglement, in which particles can remain connected regardless of the distance between them, and suggests that separation at the fundamental level may just be an illusion. One of the Upanishads reads, "All this is Brahman," highlighting universal interconnectedness, and recalls the Higgs field, which gives mass and wave nature to particles.

Atman, the individual soul that is a manifestation of Brahman, can be viewed through the lens of quantum superposition, in which one particle exists in multiple states simultaneously. This reflects the idea that the true essence of the self exists on multiple planes of reality.

The concept of Karma is based on the law of cause and effect, parallel to Heisenberg's indeterminacy, where the precision of one measure affects that of another. Actions in the present can thus influence future outcomes in ways that are not always predictable.

Moksha, liberation from the cycle of death and rebirth, finds a parallel in quantum decoherence, the process by which a quantum system loses its quantum properties. This transition represents the transformation of consciousness from one state to another, analogous to the quest for spiritual liberation.

Dharma, or each person's ethical duty, is reflected in correlations between particles, where interactions influence the behavior of a complex system. Any action taken in accordance with Dharma can have long-term effects; similarly, quantum relationships shape our universe.

It should be pointed out that this book deals with Hindu philosophy, not Hindu religion.

This distinction takes on particular relevance in the context of quantum physics. While Hindu religion deals with devotion and the cosmic order established by deities, Hindu philosophy offers a framework for understanding a complex, interconnected and constantly changing reality. For example, the idea of Maya-the illusory perception of the material world-has parallels with the uncertainty principle in quantum physics, according to which we cannot simultaneously know precisely the position and velocity of a particle.

Finally, religion is a collective and ritual path, while philosophy is a more individual and contemplative path. Both enrich Hinduism, but with different perspectives and goals. In religion one seeks union with the divine. In philosophy one seeks an understanding of being. The two dimensions, therefore, coexist but offer different tools for exploring the same reality. As Swami Vivekananda told the World Parliament of Religions in 1893, "Hinduism is not a religion, but an infinite wealth of human experience." A phrase that perfectly sums up this richness and complexity

Foreword.

Hindu philosophy and Hindu religion. Differences.

When approaching the topic of Hinduism, it is easy to confuse philosophy and religion, partly because their boundaries are often blurred and intertwined. However, to understand the connections between Hindu philosophy and areas such as quantum physics, it is essential to distinguish the two territories.

Hindu religion is the practical and devotional dimension of Hinduism. It includes rituals, ceremonies, folk beliefs and the veneration of a multiplicity of deities. This religious form, also called Sanatana Dharma (the "*eternal law*"), is expressed in temples, the recitation of mantras, the celebration of festivals such as Diwali or Holi, and the worship of deities such as Vishnu, Shiva, Durga or Krishna. The Hindu religion is linked to community, tradition and daily practices. Famous in this context is the Jagannath temple in Puri, a major religious destination in India.

Hindu philosophy, on the other hand, represents a more abstract and speculative quest. It focuses on fundamental questions: Who are we really? What is reality? What is the meaning of existence? This quest is developed through the texts of the Vedas, the Upanishads, and the six major philosophical schools, including Samkhya, Yoga, and Vedanta. The philosophy often has a universal tone, transcending the boundaries of religion and opening the door to individual reflection. For example, the concept of Brahman in the Upanishads is metaphysical in nature and unrelated to the worship of a specific god. It is described as the absolute, the fundamental essence of all that exists.

A practical example of this double face can be seen in the "Om." Religiously, Hindus use this symbol in rituals to call the sacred. Philosophically, the Om represents the primordial vibration of the universe, an idea that parallels the modern understanding of waves and frequencies in quantum laws.

One of the most relevant episodes for understanding the difference between Hindu religion and philosophy is found in the Bhagavad Gita. This ancient text is part of the epic poem Mahabharata and is presented as a dialogue between Krishna and Arjuna on the battlefield of Kurukshetra. Religiously, it is a sacred text recited in Hindu temples and homes. Philosophically, however, it offers profound reflections on the nature of action, duty (Dharma), and ultimate reality (Moksha). Krishna explains that each individual should act without attachment to results, a teaching that echoes the theory of ego renunciation.

German philosopher Arthur Schopenhauer was fascinated by the Upanishads, which he described as "*the noblest reading in the world.*" This enthusiasm shows how Hindu philosophy, as a path of universal inquiry, can find echoes outside its religious matrix.

Another key concept that separates religion and philosophy is that of Karma. Religiously, Karma is often understood as a moral code: every action has consequences, whether blessings or punishments. Philosophically, however, Karma is the law of cause and effect that governs the universe, a rational and impersonal principle. In this light, Karma transcends any religious context and approaches modern physical theories of the interconnectedness of events.

In addition to thematic difference, a central element of Hindu philosophy is doubt. Texts such as the dialogues in the Upanishad often encourage critical reflection and the overcoming of dogmatic beliefs. For example, the dialogue between the young Nachiketa and the god of death, Yama, in the Katha Upanishad focuses on questions regarding the nature of the soul and what happens after death. Religion, in contrast, tends to provide definite answers and codified rituals.

These distinctions take on particular relevance in the context of quantum physics. While Hindu religion deals with devotion and the cosmic order established by deities, Hindu philosophy offers a framework for understanding a complex, interconnected and constantly changing reality. For example, the idea of Maya-the illusory perception of the material world-has parallels with the uncertainty principle in quantum physics, according to which we cannot simultaneously know precisely the position and velocity of a particle.

Finally, religion is a collective and ritual path, while philosophy is a more individual and contemplative path. Both enrich Hinduism, but with different perspectives and goals. In religion one seeks union with the divine. In philosophy one seeks understanding of being.

The two dimensions, then, coexist but offer different tools for exploring the same reality. As Swami Vivekananda told the World Parliament of Religions in 1893:

> *"Hinduism is not a religion, but an infinite heritage of human experience."*

This sentence perfectly sums up this richness and complexity. This book refers to Hindu philosophy.

Hindu doctrine.

There is no single, standard doctrine in Hinduism, as it is an extremely complex and diverse religious and philosophical tradition. Hinduism is not a centralized theological system like some other religions, but rather a collection of traditions, texts, rituals, beliefs and practices that have evolved over thousands of years.

Hindu doctrines are based primarily on a vast collection of sacred and philosophical texts, which can be divided into two broad categories:

"*Shruti*" (that which is "heard"), includes the Vedas and Upanishads, which represent revealed scriptures and explore concepts such as Brahman (the Absolute) and Atman (the Self).

"*Smriti*" (that which is "remembered"), includes texts such as the Puranas, the Mahabharata (which includes the Bhagavad Gita), the Ramayana and the Sutras. These texts offer more accessible teachings, mythological stories and moralistic guidelines.

The teachings are based on central spiritual concepts such as Dharma (duty), Karma (action and its consequences), Moksha (spiritual liberation) and meditation on Brahman, the universal essence.

Hindu teaching is often transmitted by spiritual masters, called *gurus*, who interpret and teach the scriptures in a personal way based on both tradition and their own inner realization. Each master may belong to a particular school of thought or tradition (sampradāya) and follow a particular spiritual path such as yoga, bhakti, jnana, or tantra.

The teachings of one master may differ from those of another, depending on the school or lineage to which they belong and the spiritual path they promote. For example:

Advaita Vedanta (non-dualism), represented by masters such as Adi Shankaracharya, focuses on the unity between Atman and Brahman.

Bhakti (devotion), promoted by figures such as Ramanuja or Chaitanya, privileges the worship of a personal deity such as Vishnu or Krishna.

Yoga and meditation, taught by Patanjali or modern masters such as Swami Vivekananda, emphasize spiritual discipline of body and mind.

In this book I try to provide a generic and inclusive understanding based on academic and historical knowledge of Hinduism. I do not rely on a specific teacher, but rather draw inspiration from the fundamental texts of Hinduism (such as the Vedas, the Upanishads, the Bhagavad Gita, and the Puranas). I can also refer to the philosophy of great teachers such as Adi Shankaracharya, Ramanuja, Swami Vivekananda, or some modern approaches that link Hinduism to science or universal spirituality. In this sense, my references on the subject of Hinduism are always indicative and never exclusive.

Chapter I. Quantum physics, a new paradigm.

From Sanskrit to science: a physics that speaks the language of Hinduism.

Faced with the enigma of reality, two seemingly distant worlds converge: quantum physics and ancient Hindu philosophy. Together, they tell a vision of the cosmos in which everything is connected, where the observer influences what he observes, and reality itself, as Advaita Vedanta suggests, can be illusory. But this is not just a poetic suggestion. It is a deep interweaving of science and millennia-old wisdom.

Maya and the veil of "what it looks like".

In the Upanishads, the sapiential texts of Hinduism, the material world is described as Maya, an illusory reality, a veil hiding the ultimate truth. According to Advaita Vedanta, this ultimate truth is Brahman, the universal, indivisible and all-pervading principle. This concept is strikingly echoed in modern quantum physics, where the observable seems only a partial manifestation of a deeper underlying unity.

The example of wave-particle dualism is a symbol of this. The electron, which sometimes behaves like a particle and sometimes like a wave, highlights a paradox that defies all traditional intuition. The electron, like the cosmos described in the Upanishads, seems to have no intrinsic existence without interaction with an observer. It is our act of observing that "collapses" the possibilities, turning potential into reality.

In the Chandogya Upanishad, it states:

Tat tvam asi." (You are That).

This teaching invites the recognition that the individual self (Atman) is not separate from the cosmos (Brahman). Similarly, quantum physics suggests that what we call "objects" are not really separate from each other, but emerge from a unified quantum field that underlies everything.

The role of the observer.

One of the most intriguing principles of quantum physics is the role of the observer. The double-slit experiment illustrates this clearly. When an electron is observed, it behaves like a particle but, when the same electron is not observed, it behaves like a wave. The presence of the observer, inexplicably, changes the behavior of matter....

Here we can draw a parallel with a central concept of Advaita Vedanta. This school of thought emphasizes that perceived reality is the result of a limitation of the human mind. The world, as we see it, is a product of our "observation," filtered by the senses and the mind. Without this filter, all that remains is Brahman, the nondual reality. This philosophical echo emerges as a metaphysical explanation for the oddities of the quantum paradigm, where the boundary between observer and observable dissolves.

The quantum field is one of the fundamental concepts of modern physics. It is a state of pure potential energy, an invisible sea from which all particles and forces emerge. Physicist David Bohm described it as the "implicit order" of reality, from which the visible "explicit order" emerges.

Similarly, Advaita Vedanta describes Brahman as the ultimate and undivided reality, the source of all that exists. The phenomenal world, called Jagat, emerges and dissolves in Brahman, just as particles are born and exist temporarily in the quantum field.

This parallelism is not just a strange coincidence. It shows that both science and Hindu philosophy seek, each through its own language, to understand the invisible unity from which the observable universe is derived.

The amazement of Feynman and the *rishis*.

Richard Feynman, one of the fathers of quantum physics, declared:

"No one really understands quantum physics."

To describe the oddities of particle behavior, Feynman often resorted to poetic metaphors to fill the void left by logic. Quantum mystery, for him, was a source of profound awe.

This awe is reminiscent of that expressed by many Indian sages (*rishis*) who, more than three thousand years ago, meditated on the nature of the universe in the forests of ancient India. Like Feynman, the rishis perceived that reality was more complex and mysterious than human senses could grasp. Their answer, however, was not experimentation but introspection, deep immersion in consciousness to discover universal truths.

An example from the Brhadaranyaka Upanishad is illuminating:

"What is infinitely small is also infinitely large."

This seemingly mystical principle finds remarkable resonance in the quantum world, where the laws governing subatomic particles seem to influence the macrocosm.

A bridge between cultures.

The encounter between quantum physics and Hindu philosophy is not just a curious parallel. It is a cultural and intellectual bridge that invites us to rethink the way we see the world. Science and spirituality are not antagonistic, but complementary perspectives that explore the same mystery: what is beyond appearance?

Advaita Vedanta urges us to recognize our oneness with the cosmos. Quantum physics shows us that the world is fundamentally interconnected, and that our role as observers is inextricably linked to the reality around us. Like two mirrors reflecting each other, these two disciplines offer a symphony of ideas that, while expressing themselves in different forms, speak the same language.

Paradoxes and philosophies of quantum mechanics.

Quantum mechanics represents one of the greatest conceptual leaps in the history of science. Born in the first decades of the 20th century, it revolutionized the understanding of physical reality, unveiling an astonishing universe on the borderline between science, philosophy and, for some, metaphysics.

"Schrödinger's Cat": the physics of apparent contradiction.

In 1935, Erwin Schrödinger, an Austrian physicist known for his work on quantum theory, formulated an immortal mind experiment in the history of science. He imagined a cat locked in a box along with a mechanism that included a radioactive substance and a vial of poison. If the radioactive substance emitted a particle, a mechanism would detect the event, break the vial and kill the cat. According to quantum mechanics, until an observer opened the box, the cat would be in a state of quantum superposition: alive and dead at the same time.

This paradox was not just a provocation. Schrödinger wanted to point out an intrinsic oddity of quantum formalism, expressed through the equation of state he co-created. According to this theory, a particle can exist simultaneously in multiple states (e.g., "emitted" and "not emitted") until an observation is made. The cat, therefore, participated in the phenomenon, metaphorically embodying the problem of the interaction between the quantum world and the macroscopic world..

This paradox was not just a provocation. Schrödinger wanted to point out an intrinsic oddity of quantum formalism, expressed through the equation of state he co-created. According to this theory, a particle can exist simultaneously in multiple states (e.g., "emitted" and "not emitted") until an observation is made. The cat, therefore, participated in the phenomenon, metaphorically embodying the problem of the interaction between the quantum world and the macroscopic world.

Everett and parallel worlds.

Twenty years later, Hugh Everett proposed the "*Many-Worlds Interpretation*" (1957), which provided a radical solution to the "Schrödinger's Cat" paradox. According to Everett, whenever a quantum event such as measuring the state of a particle occurs, the universe splits. In one universe, the cat is alive. In another universe, it is dead. This is not a superposition in the same reality, but it is the emergence of parallel branches.

Everett, a young Princeton researcher, was initially ignored by the academic community. However, his work gained interest in the following decades, mainly due to the popularization of the theory by physicist Bryce DeWitt. The idea of parallel worlds not only offered a more direct approach to the paradoxes of quantum mechanics, but opened new perspectives in philosophy and popular culture, fueling works of fiction, films and modern discussions of the multiverse.

The observer and the nature of reality.

The paradoxes of quantum mechanics do not stop with cats or multiple worlds. From a philosophical point of view, the real core of the issue concerns the role of the observer. Max Planck, considered the father of quantum theory, stated:

"We cannot gain any knowledge of the objective nature of the world without referring to human consciousness."

This perspective, later developed by scientists such as Niels Bohr and Werner Heisenberg, posed a fundamental question: does reality exist independently of the subject who observes it?

Bohr advocated the principle of complementarity, according to which observation inevitably changes what is observed. Heisenberg, with his famous uncertainty principle, demonstrated that it is not possible to know with absolute precision simultaneously the position and momentum of a particle. The presence of the observer is, therefore, an integral part of the measured phenomenon, a theme that resonates with some Eastern philosophical traditions.

In this context, the relationships between quantum mechanics and Eastern philosophies, such as Hinduism, are not coincidental. A mainstay of Vedantic philosophy is the idea that perceived reality is Maya, an illusion created by the interaction between subject and object. Similar concepts are found in quantum mechanics, where the microscopic world does not "exist" in a definite way without interaction. The analogy even extends to the concept of Brahman, the ultimate undifferentiated reality, reminiscent of the "*unified quantum field*" hypothesized by some more speculative interpretations.

Places and historical anecdotes help contextualize these ideas. The Copenhagen laboratory, site of conversations between Bohr and Heisenberg in the 1920s, is a kind of "cradle" of quantum mechanics. At the same time, scholars such as David Bohm have sought to link modern physics with Eastern philosophical concepts, fostering a dialogue between the two traditions.

The paradoxes of quantum mechanics are not mere abstract curiosities. They challenge the foundations of human thinking, prompting us to rethink concepts such as reality, observation and consciousness.

History and development of quantum physics.

A chronology of ideas and discoveries.

Mankind has always sought to understand the universe, starting with the great celestial bodies and descending to the infinitely small. It was this second path that opened wide, in the early 20th century, the door to an unseen and surprising world: quantum physics. Its origins go back to a time when the laws of classical physics, enshrined by Newton, seemed to offer sufficient answers to all observable phenomena. However, something was changing, and the upheaval began with a very specific name: Max Planck.

The idea of a universe governed by predictable, mechanical laws has dominated physics since the 17th century thanks to Isaac Newton. Natural reality, in the eyes of classical physicists, was a perfect machine. Every phenomenon followed a deterministic order. However, in the early 20th century, this view encountered an unexpected obstacle: the behavior of subatomic particles. The infinitely small seemed to violate every known rule. Thus was born quantum physics, a new paradigm that would change not only science but also the way humanity conceives of reality.

The turning point came in 1900, when Max Planck, a German physicist, proposed a revolutionary idea to explain a seemingly technical problem. The conundrum concerned blackbody radiation, that is, how objects emit energy in the form of light. Planck was faced with an anomaly: classical physics could not accurately explain the energy emitted.

Planck introduced a bold hypothesis. He suggested that energy was not emitted continuously, as imagined by classical physics, but in discrete packets, which he called "quanta." In retrospect, Planck confessed that he had found a purely technical solution, without imagining that it would lay the groundwork for a philosophical and scientific revolution.

A few years later, Albert Einstein extended the concept of "quantum" to light itself. In 1905, in his famous series of papers that changed science forever, Einstein suggested that light was composed of discrete particles called photons. This explains the photoelectric effect, a phenomenon in which some metals release electrons when struck by light.

Einstein's insight into the photoelectric effect earned him the Nobel Prize in 1921. But what is surprising is that even Einstein did not fully accept the deeper implications of quantum physics. Famous is his statement, "God does not play dice with the universe," an expression of his discomfort with the indeterminacy introduced by the new physics...

In the following years, the implications of quantum theory also revolutionized the understanding of atoms. In 1913, Danish physicist Niels Bohr proposed a new atomic model. He imagined electrons moving around the atomic nucleus in specific orbits, like planets around the sun. However, unlike a classical orbit, electrons could "jump" from one level to another by emitting or absorbing energy.

Bohr integrated Planck's quantum theory into his model to explain the emission and absorption of light by atoms. His work laid the foundation for further developments and made him one of the leading figures of the new physics.

The year 1925 marked another decisive turning point thanks to Werner Heisenberg, a brilliant young German physicist. Heisenberg introduced the uncertainty principle, which establishes a fundamental limit to our ability to know the subatomic world. It states that we cannot simultaneously know precisely the position and velocity of a particle.

This discovery is not only technical, but philosophical. The uncertainty principle challenges the very idea of a fully knowable universe. According to Heisenberg, observing a system inevitably changes the system itself. Subatomic reality seemed to escape the control of classical determinism.

In 1926, Erwin Schrödinger, an Austrian physicist, introduced the wave equation, another pillar of quantum physics. This equation describes the behavior of subatomic particles in terms of probability. However, the puzzling implications led Schrödinger to devise a famous thought experiment, the "Schrödinger cat."

In this hypothetical situation, a cat locked in a box can be simultaneously alive and dead, depending on the observation. The experiment highlights the paradox of quantum superposition, a condition in which a particle can be in multiple states at once until the moment it is observed.

The cultural context: a paradigm shift.

Quantum physics has not only had a scientific impact. It has also deeply influenced philosophy, art and even spirituality. This discipline challenges the concepts of objective reality and causality, opening dialogues with millennia-old philosophical traditions, such as those in the East, which have been reflecting on the illusory nature of the world and the interconnectedness of subject and object for centuries.

During the 20th century, figures such as J. Robert Oppenheimer and Fritjof Capra explored the links between quantum physics and Eastern thought. Oppenheimer, for example, found analogies between quantum physics and the verses of the Bhagavad Gita, while Capra, in his famous book "*The Tao of Physics*," paralleled quantum theory with Eastern philosophy.

Quantum physics represents one of humanity's greatest achievements. It not only describes the subatomic world but redefines our relationship with reality. Its history, punctuated by revolutionary discoveries and extraordinary figures, continues to stimulate profound questions concerning both science and the nature of existence. It is a journey of exploration that, like the thinkers of ancient India, invites us to reflect on the invisible and the infinite.

Max Planck and the birth of "quantum."

In 1900, German physicist Max Planck asked a question that proved revolutionary. He was trying to explain a specific problem related to electromagnetic radiation: the so-called "blackbody problem." His experiments showed that the radiation emitted by a black body, an ideal object capable of absorbing and radiating energy perfectly, could not be explained by the laws of physics then known.

Planck formulated a bold hypothesis: energy is not transferred continuously, as had been believed, but in small discrete packets that he called "quanta." This changed everything. Planck himself initially regarded his idea as a mathematical gimmick, not as a new view of reality. Yet his work laid the foundation of a scientific revolution. In 1918 he received the Nobel Prize for this insight. To this day, many consider 1900 to be the true birth year of quantum physics.

In December 1900, in Berlin, Max Planck presented his idea. The occasion was a meeting at the prestigious *German Physical Society*.

The problem Planck was addressing was known as the "ultraviolet catastrophe," a seemingly insoluble headache related to the study of the so-called black body.

A blackbody, in theoretical physics, is an ideal object that absorbs and radiates energy perfectly. Experiments conducted in the 19th century had shown that when analyzing the emission spectrum of the black body, predictions based on the traditional laws of classical physics simply did not add up. Simply put, it was expected that as the frequency of the emitted radiation increased, the energy would grow without limit. But the experimental data showed something very different: at high frequencies, the energy decreased dramatically. This discrepancy was one of the main puzzles of physics at the time, and no one seemed able to solve it.

Max Planck set to work and proposed a solution he called revolutionary: to describe the energy emitted by a black body, it was necessary to imagine that the energy itself was not continuous, but divided into "quanta." Planck hypothesized that energy traveled in the form of discrete units, each proportional to the frequency of the radiation according to a constant known today as Planck's "constant," denoted by the symbol (h).

> *"My idea did not originate as a philosophical conviction. It was a technical necessity."*

said Planck several years later, admitting that initially the concept of how much seemed to him little more than a mathematical trick. But that "trick" turned out to be the first crack in the solid but now antiquated structure of classical physics. The rest, as they say, is history.

Berlin, 1900: the epicenter of a revolution.

Let us imagine for a moment the Berlin of the early 20th century. The city, Germany's scientific and economic capital, was a vibrant intellectual center. Universities and academies teemed with brilliant minds who challenged the boundaries of knowledge. Max Planck, distinguished for his brilliance in the theoretical sciences, taught at the University of Berlin. Officially, he was a thermodynamics scholar and studied, among other things, the fundamental principles governing energy transfer.

His dedication to science was almost religious. Planck, in fact, was a man deeply attached to a vision of the universe in which order and harmony were fundamental elements.

It took a few years, and numerous subsequent theoretical developments, for the significance of Planck's ideas to be fully understood. Around 1913, Planck's quantum theory also found important applications in the atomic model of Niels Bohr, another giant of physics, who used "quanta" to explain how electrons jumped between discrete orbits while emitting or absorbing light. The seed planted by Planck was growing, and the entire scientific community was beginning to realize that it was facing a new era of knowledge.

In 1918, Max Planck received the Nobel Prize in Physics "*for his services in the progress of physics through the discovery of quanta.*" During the ceremony, Planck reiterated one of his most famous maxims:

> *"A new scientific principle does not triumph because it convinces its opponents. Rather, it triumphs because new generations accept it."*

A prophetic phrase that summed up the difficulties and resistance that every major cultural and scientific revolution encounters.

The scientific scandal of "how much.

When Planck introduced the concept of quantum in 1900, the world of classical physics, which until then had been dominated by its laws, was shaken to its core. The idea that energy was discrete had no place in the vision of a universe governed by continuous, predictable lines. Even Planck, known for his intellectual prudence, seemed hesitant. Later, a new generation of physicists, including Albert Einstein, further developed the concept of quantum.

Einstein, in fact, built on Planck's work one of his best-known discoveries: the photoelectric effect. It was 1905 when the young scientist explained that light, just like blackbody energy, travels in the form of "light quanta," photons. Einstein's theory not only provided practical confirmation for Planck's model, but opened the door to modern physics.

Max Planck was not just a physicist. He was a man of culture, fascinated by philosophy and spirituality. It is not surprising, then, that many reflections on "quanta" later found resonance in philosophical thought and even in the spiritual traditions of the East. Planck insistently sought an underlying order, a "hidden truth." And it is precisely this tension between apparent chaos and deep order, so typical of quantum physics, that is one of the subtle links that bring the Western world closer to ancient Eastern philosophies, such as Hinduism.

However, this is a story we will tell in future chapters.

Max Planck's First Step.

In the apparent calm of Dec. 14, 1900, a shy man, mathematician by training and physicist by vocation, uttered a sentence that would set 20th-century science on fire. Max Planck, a professor in Berlin, presented a radical idea to the German Physical Society. Speaking softly, Planck proposed that energy did not flow as a continuous river, but behaved as discrete drops. These drops were called "quanta." No one, not even Planck himself, suspected that that simple insight would lay the foundation for quantum physics, a paradigm destined to subvert centuries of scientific thought.

Physicists in the 19th century were plagued by a seemingly simple conundrum: How to accurately describe the radiation emitted by a hot object, the so-called "black body"? Experiments showed that hot objects emitted light waves with an intensity that varied with wavelength. But, according to the laws of classical electromagnetism and thermodynamics, the light intensity should have become infinite at short wavelengths. This was an absurdity, known as the "ultraviolet catastrophe." This dissonance challenged the entire theoretical framework of nineteenth-century physics.

Planck approached this problem with the method he was most familiar with: pure mathematics. He devoted years to this challenge, but could not find a solution. In the end, as he himself later admitted, he was forced to perform an act of extreme pragmatism. He proposed that energy should not flow continuously, but only in discrete packets proportional to the frequency of the light wave.

At first, Planck did not attribute any physical significance to this theory. He called it a "mathematical trick." Planck was a conservative of classical physics and did not imagine that such an insight would demolish the foundations of the very physics he so revered.

Initial acceptance of Planck's work was more lukewarm than enthusiastic. His ideas seemed bizarre even to his colleagues.

However, before long, experimental data confirmed the validity of his model.

Planck's work was not just a scientific discovery. It was a paradigm shift that transformed the way human beings viewed the universe. Quantizations broke the illusion of continuity, revealing a universe made up of discontinuities and probabilities. This principle challenged not only classical physics but also many of the philosophical insights of the West.

Interestingly, the concept of discontinuity was echoed in some Eastern philosophical traditions, such as Hinduism. In Vedanta philosophy, the universe is seen as a discontinuous manifestation of Brahman, the universal principle. This parallelism between science and spirituality would be explored decades later when quantum physics met Eastern thought.

Max Planck was a man of great humility. He did not seek fame or revolution. Yet from his hands a new science was born. Quantum physics, with its enigmatic language, unveiled a universe profoundly different from what scientists had known up to that time.

Albert Einstein and the photoelectric effect.

In 1905, Albert Einstein was twenty-six years old, living in Bern and working at the Patent Office as a third-class clerk. It was in that extraordinary year, later called "*Annus Mirabilis*," that the young scientist published four papers destined to change physics forever. Prominent among them was a short but intense publication: his study of the photoelectric effect, work that would lay the foundation for quantum physics as we know it today.

Until the early twentieth century, light had been considered essentially as a wave. The work of James Clerk Maxwell had shown that electromagnetic waves propagate through space following well-defined mathematical laws. However, some observed phenomena did not follow those patterns. One of these was, precisely, the photoelectric effect: when light of a certain frequency strikes the surface of a metal, the metal emits electrons. Experiments conducted on this effect had left physicists without answers. Why was the intensity of light not enough to explain the process? Why was it not enough to increase the energy of the light with greater brilliance to generate electrons? The key seemed to be the frequency of light, but no one could explain why.

Einstein approached the problem from a different perspective, inspired by the work of Max Planck. In 1900, Planck, in an attempt to explain blackbody radiation, had introduced a revolutionary idea: energy is emitted in "discrete quanta," not continuously. This insight, initially conceived as a simple mathematical artifice, opened an unexpected path. Einstein, going against the prevailing conception, suggested that light itself was composed of particles, which would later be called photons. Each photon carries energy proportional to its frequency.

Einstein explained the photoelectric effect with disarming simplicity. When a photon hits an electron in a metal atom, it can transfer all its energy to the electron. If the photon's energy is sufficient to overcome the electron's binding energy (*called the work function or extraction potential*), it is ejected from the metal surface. The higher the frequency of the light, the greater the energy of the photon and, therefore, the greater the probability of the electron being emitted. It was not the intensity of the light that counted, but the quality of the light itself, that is, its frequency.

Einstein's theory had an enormous impact. Not only did it explain the photoelectric effect, but it overturned the traditional conception of light. Until that time, many scientists were certain that light was a wave. Now, with Einstein's work, it was shown that light also had corpuscular behavior, introducing the concept of "*wave-particle dualism.*" It was the beginning of a revolution that would redefine our understanding of the very nature of reality.

Einstein published his article on the photoelectric effect in a relatively unknown journal, "*Annalen der Physik*," in March 1905. Despite its importance, the idea of photons was initially greeted with skepticism. Even prominent physicists of the time, such as Max Planck and Niels Bohr, considered the concept of light particles too radical an idea to be accepted easily.

Planck himself, who had unwittingly paved the way for the quantum idea, hesitated to embrace Einstein's vision. Planck would say privately:

> *"Radiation cannot be like that. This is too bold an interpretation."*

However, experiments conducted in the following years confirmed Einstein's view. American physicist Robert Millikan, who initially tried to prove Einstein wrong, ended up experimentally verifying his theory instead, in turn winning a Nobel Prize in 1923.

Einstein himself received the Nobel Prize in 1921, but not for special relativity, as is often believed, but for work on the photoelectric effect. The Nobel committee chose to reward a quantifiable contribution in experimental physics rather than a still controversial theory such as relativity.

The story of the photoelectric effect embodies the tension between scientific intuition and established beliefs. Einstein's work not only opened new horizons, but posed a larger question about the foundations of classical physics.

How could light be both wave and particle? This mystery, would haunt Einstein for the rest of his life, especially after the development of quantum mechanics. Although he was one of the founders of the new paradigm, Einstein remained critical of some of its more extreme implications.

Einstein's view of the photoelectric effect also offers an important cultural connection. Science, like art or philosophy, never develops in a vacuum. The first half of the twentieth century was a time of radical change in every field of knowledge. Art movements such as Cubism and Futurism redefined space and time, while in philosophy thinkers such as Henri Bergson challenged the concept of linear time. Einstein's scientific ideas reflected the spirit of an age that was not afraid to break with the past.

The photoelectric effect, in its apparent simplicity, was one of the first leaps toward a new way of thinking. It showed that nature does not conform to our intuitions, requiring instead that we adapt our thinking to reality, however paradoxical it may seem.

A spark of light changes history

In 1905, Albert Einstein revolutionized physics with an insight as bold as it was simple: light does not just behave like a wave, but under certain circumstances acts as if it were composed of tiny discrete packets of energy, called photons. This concept would shake the foundations of classical physics and open the door to the era of quantum physics.

Until then, many scientists believed that light was just a wave, following the theories of James Clerk Maxwell, who had described light as an electromagnetic wave in the 19th century. Einstein, however, building on the insight of Max Planck, who in the early twentieth century had introduced the idea of quantized energy to explain blackbody radiation, proposed something radical. Starting from the observation that light could "rip" electrons from a material such as metal (the so-called photoelectric effect), Einstein showed that only by considering light as composed of particles-future photons-could this phenomenon be explained.

To understand the importance of this result, one must consider the scientific context of the time. The wave theory of light seemed to be confirmed by a great many experiments, such as interference and diffraction. But Einstein realized that these observations were not sufficient to explain all the properties of light. In his landmark paper published in June 1905, Einstein borrowed Planck's idea about the quantization of energy and applied it to a new level. He wrote:

"The energy of a beam of light is spatially distributed not continuously, but in small packets of energy."

It was the beginning of a revolutionary idea: light had a dual aspect, simultaneously wave and particle.

This insight found extraordinary practical confirmation in laboratories. Metals, when exposed to light of a certain frequency, released electrons, but only if the energy "quanta"-photons-possessed enough energy to overcome the barrier that kept the electrons bound to the metal. It did not matter how intense the light was: if the photons did not have enough energy (i.e., if their frequency was too low), the photoelectric effect simply did not occur.

The importance of this discovery was enormous. Einstein received the Nobel Prize in Physics in 1921 precisely for his work on the photoelectric effect. Interestingly, the prize was not for his Theory of Relativity, which was probably less understood and accepted at the time, but for an insight that had an immediate practical impact on emerging technologies in the 20th century. Today, the photoelectric effect underlies fundamental devices, from solar-powered devices to cameras.

An interesting anecdote comes from Einstein's own relationship with the concept of quantization. Although Einstein had used the idea to explain the photoelectric effect, he himself was later reluctant to accept some of the more extreme implications of quantum mechanics, such as uncertainty and probability.

However, his work on the photoelectric effect remains central to quantum thinking and represents an extraordinary bridge between the classical and new worlds of physics. Whenever we use a photovoltaic cell to capture sunlight or an image sensor in a camera, we are applying the fundamental principle discovered by Einstein in 1905.

Niels Bohr and the atomic model (1913).

Meanwhile, the internal structure of the atom, that mysterious little world, was beginning to be explored with new clarity.

In 1913, a young Danish physicist named Niels Bohr forever changed the way we observe the universe. The son of a physiology professor in Copenhagen, Bohr grew up in an environment permeated with philosophical and scientific questions. He was a bridge between two worlds: the classical Newtonian tradition and new insights into the mysterious quantum realm that was beginning to be explored. With his revolutionary atomic model, Bohr demonstrated that the hidden order in the universe was far stranger than we could imagine.

Bohr started from an open problem. Ernest Rutherford, in 1911, had proposed that the atom was like a small solar system: a positive nucleus at the center and electrons orbiting around it. But the model had a fatal flaw. The classical laws of electromagnetism, Maxwell's, said that a moving electron would emit energy in the form of radiation. Result? In a short time the electron would fall toward the nucleus, collapsing the entire atom in an explosion of instability. The atom could not have existed in the form in which Rutherford envisioned it.

Bohr was fascinated by the problem. Studying in England with Rutherford, he made a bold proposal. Making use of the concept of "quantum" introduced by Max Planck in 1900 and further developed in the following years by Albert Einstein, Bohr decided to apply quantum ideas directly to the atom. In 1913 he published an article in the scientific "*Philosophical Magazine*" in which he outlined his model.

In Bohr's model, electrons orbited the nucleus on defined energy levels, called "*quantized orbits.*" These levels were not continuous, as classical physics imagined, but discrete. Electrons could occupy only certain orbits, like lines on a musical staff on which only precise notes could be placed.

Here was the breakthrough that shocked the scientific world: Bohr suggested that an electron could "*jump*" from one orbit to another *without traveling through intermediate space*. When an electron descended to a lower energy level, it emitted a quantum of energy in the form of light. This "quantum" was a photon. This confirmed the quantum miracle discovered a few years earlier by Einstein to explain the photoelectric effect.

Bohr's model immediately found a concrete application in the case of the hydrogen spectrum. Scientists had long known that hydrogen atoms, subjected to an electrical discharge, emitted light at specific wavelengths. No one, until then, had been able to explain why. Bohr's new model matched those spectral lines perfectly, as if the universe had finally revealed one of its most hidden secrets.

Reactions to Bohr's work were varied. Many were enthusiastic, others more skeptical. Rutherford himself was impressed by the boldness of the model. But details, such as evidence for the concept of "quantum leaps," raised doubts. A collaborator close to Bohr, German physicist Arnold Sommerfeld, improved the model by introducing the ideas of elliptical orbits. Sommerfeld noted that, in some cases, electrons were not limited to circular orbits, but the development of these extensions still led to a more complex and precise view.

However, Bohr's work marked only the beginning. In the following years, new scientists entered the scene. Werner Heisenberg, from Germany, would become one of its main heirs and friends. Heisenberg would later say:

> *"Bohr taught to look not only at the atom, but at reality itself as a process of relationships, more mysterious than we had ever imagined."*

The 1913 atomic model is not perfect. It was superseded in the 1920s with the arrival of Schrödinger's quantum mechanics and Heisenberg's uncertainty principle. However, Bohr's insight was fundamental, not only because it explains concrete phenomena such as the hydrogen spectrum, but because it opens the door to a new way of thinking: the universe is not a predictable machine, but a quantized mystery made up of jumps, discretions and dualities.

Niels Bohr, in his approach to science, was also influenced by philosophical ideas. He studied Eastern thought and was fond of quoting an ancient Chinese saying:

> *"Words open the door, but it is silence that goes through it."*

It is no accident that his insights are intertwined with philosophical themes reminiscent, in some ways, of Hinduism's Brahman, that absolute that eludes linear understanding. Bohr further asserted:

> *"Truth is learned through complementary opposites."*

We could already see these opposites in the atomic model: stability and instability, order and quantum leap. His work reflected not only a scientific revolution, but also a profound cultural transformation. Bohr, with his atom, was leading science toward a new paradigm. A world where nothing was continuous and predictable anymore, but where every phenomenon-even the smallest-carried with it the possibility of a greater mystery.

Niels Bohr and the atom that does not explode.

In 1913, a young Danish physicist named Niels Bohr published a revolutionary theory that would forever change the way we view matter. Bohr was born and raised in Copenhagen, a city overlooking the Baltic Sea, and even as a boy he displayed a brilliant intelligence. His mother came from a family of Jewish intellectuals, while his father was a renowned physiologist, and there was a fertile atmosphere for innovative ideas in the Bohr household. Not surprisingly, it was Niels who became one of the leading figures in the nascent quantum physics.

At the beginning of the 20th century, physics was experiencing a crisis. The classical models of Newton and Maxwell, though mighty, could not explain phenomena at the scale of the atom. Why were atoms, these tiny basic units of matter, not collapsing in on themselves? Why did the electrons, circling the positive nucleus, not lose energy and fall into the nucleus, as would have been expected according to classical theory?

The answer came with Bohr and his atomic model of quantized orbits. Based on the hypotheses proposed by Max Planck and the work of Einstein, Bohr introduced an extraordinary concept: electrons can orbit the nucleus only in specific, or quantized, orbits. In these defined states, electrons do not lose energy.

This insight not only explained the stability of matter, but also solved another mystery: the emission spectrum of the hydrogen atom. When the atom is excited, electrons move from a higher to a lower orbit, emitting light with a precise frequency. Experimental observations matched Bohr's theory perfectly.

An atom that does not explode and the link to Eastern philosophy.

Bohr's model was extraordinary in its implications. It showed an atomic world governed by laws that did not follow intuitive logic but imposed a duality between continuity and discontinuity. In a sense, matter seemed to dance following a rule that was reminiscent of concepts expressed in some Eastern philosophies, such as Hinduism.

In Indian philosophical texts such as the Upanishad, the material world is described as "*Maya*" (illusion), where appearance hides a deep and indescribable reality. Similarly, in the modern view of quantum mechanics, matter is not fully "solid"; electrons behave as both particles and waves, and their position is 'fuzzy' until observed. This paradoxical nature links, at least on a conceptual level, Bohr's scientific insights with ancient Eastern wisdom.

Today, visitors to Copenhagen can go to Bohr*'s House of Honor*, a center dedicated to the memory of one of the 20th century's greatest scientists. It is a place that chronicles Bohr's life and thought from his early studies to the application of his vision to the emerging quantum theory.

But there is another aspect that makes Bohr unique: his philosophy of science. He constantly sought to combine "truth and clarity." This commitment was not just a methodological approach. It was an ethical idea, based on the belief that science should bring us closer to understanding reality without sacrificing simplicity and transparency.

Bohr was not only a physicist but also a philosophical thinker. He often said

> *"We cannot separate the world of physics from the way we observe it."*

This statement anticipates many modern reflections on the nature of reality and the role of consciousness, topics dear to both quantum mechanics and Eastern philosophies.

Looking back, we can glimpse a thread linking the quantum revolution to ancient wisdom. Bohr, with his atomic model, showed that matter is neither rigid nor absolute, but dynamic, relational and governed by laws beyond ordinary understanding. This concept, so counterintuitive, finds surprising resonance in Indian philosophy, which invites us to look beyond the phenomenal world to discover the ultimate reality, Brahman.

Werner Heisenberg and the uncertainty principle.

In the 1920s, quantum physics finally took shape thanks to a wave of discoveries made by brilliant young minds, including Werner Heisenberg and Erwin Schrödinger. In 1927, Heisenberg enunciated the uncertainty principle, one of the pillars of quantum mechanics. This principle states that it is not possible to simultaneously know with absolute precision two complementary quantities, such as the position and velocity of a particle.

This concept overturned the idea that the scientific experiment was neutral. Microscopic reality, physicists discovered, was deeply connected to the experimenter.

In the 1920s, physics was experiencing a revolution. The clear and predictable concepts of classical science, which had been the masters of the scene from the dawn of Galileo until Newton, were faltering. The subatomic world, revealed thanks to quantum mechanics, had disturbing features that were seemingly contrary to common sense. It was in this context that in 1927, a young German physicist, Werner Heisenberg, at the age of just 26, formulated one of the most shocking principles of modern science: the uncertainty principle.

According to this principle, two complementary quantities, such as the position and velocity (or momentum) of a particle, cannot be known precisely at the same time. The greater the precision in measuring one, the less precisely the other can be known. This was not an instrumental limitation, but an intrinsic feature of nature. Classical science, which described the universe as a machine governed by deterministic laws, was faced with a different reality: that of uncertainty.

The idea of the uncertainty principle was born during one of the most fruitful summers of Heisenberg's life. In 1925 Heisenberg sought refuge from hay fever on the North Sea island of Heligoland to concentrate without distraction on the dilemmas of quantum mechanics. He was tormented by the difficulty in understanding the orbits of electrons in atoms and decided to abandon the spatial representation of particles. This insight prompted him to develop the matrix formulation of quantum mechanics, which formed the basis for the principle he would enunciate two years later.

In 1927, during a conference in Copenhagen, Heisenberg formalized this idea. The uncertainty principle profoundly influenced the philosophy of science. It was a paradigm shift. Not only was classical physics no longer sufficient, but now the observer himself became an integral part of the observed system. Heisenberg explained this point with a quote that became famous:

"What we observe is not nature itself, but nature exposed to our method of inquiry."

Heisenberg's principle not only transformed science, but had a cultural impact far beyond the boundaries of physics. The concept of indeterminacy inspired philosophers, artists and even writers. In Germany, Thomas Mann reflected on quantum mechanics in his work "Doktor Faustus." Martin Heidegger, Heisenberg's contemporary philosopher, grappled with the idea of a reality that eluded objective certainty.

Heisenberg's formulation led to a heated debate with Albert Einstein. Einstein, who advocated a universe governed by deterministic laws, rejected the idea that chance and indeterminacy could govern nature. Einstein's phrase remains famous:

"God does not play dice."

In response, Niels Bohr, Heisenberg's colleague and mentor, retorted:

"Don't tell God what to do with his dice."

Werner Heisenberg was more than a brilliant physicist. His exceptional intelligence was accompanied by a strong interest in philosophy. Heisenberg admired Plato and often questioned the theoretical basis of knowledge. Perhaps it was this philosophical spirit that enabled him to accept uncertainty as a fundamental part of the natural world.

The interaction between science and humanism was also reflected in his personal ties. Werner Heisenberg befriended poet Rainer Maria Rilke and cultivated the dialogue between art and science. In his famous book "*Physics and Philosophy*" published in 1958, Heisenberg emphasized how quantum mechanics not only redefined physics but offered a new worldview, intertwined with the subjectivity of human beings.

Nearly a century after its formulation, Heisenberg's principle continues to be a cornerstone of modern physics. It underlies technologies such as electron microscopes and has applications ranging from quantum computing to cryptography. But its value is not just technical. The uncertainty principle reminds us of the limits of human knowledge and the complexity of the natural world. As Hinduism and Eastern philosophies have suggested for millennia, knowledge is not a pure description of reality, but an interaction between subject and object, observer and observed.

Heisenberg, armed with his brilliant mind and the courage to defy convention, changed our perception of the world forever. And in doing so, he left an eternal message: the universe is not only to be explained, but also-and above all-to be contemplated.

Werner Heisenberg and the mysterious side of nature.

In 1927 Werner Heisenberg revolutionized the way we understand the physical world. A gifted young German physicist, Heisenberg was working at the "*Institute for Theoretical Physics*" in Leipzig, a vibrant environment where the brightest minds of the time were redesigning the foundations of modern science. It was there that he formulated the famous Principle of Indeterminacy, opening up a new perspective on how nature works.

The principle is based on a simple but at the same time disarming idea: it is impossible to know with absolute precision both the position and the velocity of a subatomic particle. The more one tries to measure exactly one of these aspects, the more the other escapes. Heisenberg initially explained this with a mathematical formulation, but the philosophical significance of the principle was immediately clear. Among scientists and thinkers, this discovery opened the door to profound questions about reality itself.

In the 1920s, the Leipzig Institute was a hub of the scientific vanguard. At that time, physicists such as Niels Bohr and Max Born were questioning the essence of the microscopic world. Heisenberg, then in his early thirties and a student of Bohr, took up these challenges with great audacity. His idea arose not only from a theoretical study but also from a long reflection on the way observations were being made in the domain of quantum physics.

A famous anecdote tells that Heisenberg, tormented by the problem, found inspiration during a walk in the meadows of the German countryside. It is said that the immensity of nature, with its invisible but perceptible motion, made him realize how elusive subatomic particles were. This insight enabled him to formalize the idea that "observation" itself altered the observed object.

The metaphor of light and the microscope.

To explain the principle, Heisenberg often used the example of a microscope. Imagine we want to observe an electron using a beam of light. The light, which is necessary to see the electron, pushes it and alters its position or velocity. This means that the simple act of observing alters the microscopic reality.

This is a technical issue, but also a deeply philosophical one. The very idea of measurement becomes ambiguous. What does it mean to know an object if our "gaze" ends up changing it?

The Principle of Indeterminacy did not remain confined to physics laboratories. The scope of this discovery transformed the philosophy of science, finding echoes in literature, the arts and even religions. Thinkers like Albert Einstein, initially skeptical, wondered if the universe could really be so elusive.

The differences between two giants of physics like Bohr and Einstein shows how much the Principle of Indeterminacy had touched the heart of the great mysteries of existence. For Einstein, the universe was supposed to be orderly and predictable. For Heisenberg and Bohr, on the other hand, indeterminacy was a fundamental feature of reality.

In the Western world, the Indeterminacy Principle was greeted with a mixture of astonishment and anguish. It disrupted the traditional view of the world as a "*perfectly predictable machine*," a paradigm that dated back to Newton. The universe was no longer an ordered clock, but an entity in which chance, uncertainty and mystery played a central role.

Interestingly, Eastern philosophies such as Hinduism and Buddhism had long embraced a similar view. In Vedic thought, the material world is considered *Maya*, an elusive and transitory illusion. Similarly, quantum mechanics suggested a deep and hidden idea of reality, accessible only through intuitions rather than measurements.

In the Upanishads, ancient Hindu texts, we read:

> *"What the spirit sees, it cannot be fixed with the eyes, for it is without form."*

These words seem to anticipate the challenge Heisenberg himself faced: to describe what cannot be seen, and to do so with the language of mathematics. No wonder that, years later, the same physicist declared that he felt a special affinity with Eastern thought.

A legacy that goes beyond science.

The Principle of Indeterminacy is not only about particles. It invites us to reflect on the relationship between knowledge and reality. How much can we really know about the world around us? And at what point does our observation become interaction, transforming what we seek to understand?

To this day, Heisenberg's insights remain a pillar of modern physics, influencing disciplines ranging from quantum computing to cosmology. But his real impact goes beyond the boundaries of science, touching the heart of the human condition: the desire to know, mixed with the mystery of the unknowable.

In 1927, Heisenberg could not have foreseen how much his work would forever change our worldview. Yet, with his Principle of Indeterminacy, he gave us a glimpse of the gulf between what we can measure and what we can only imagine. Today, that mystery continues to inspire scientists and philosophers, reminding us that knowledge, like particles, is always in motion.

Erwin Schrödinger and the cat.

In 1935, Erwin Schrödinger, known for his fundamental contributions to quantum physics, raised one of the most famous and debated questions of this new scientific frontier. The Austrian physicist, born in 1887 in Vienna, is recognized for the wave equation that bears his name. This equation is still a central tool for describing the probabilistic behavior of particles on a subatomic scale. But Schrödinger was also a man of critical spirit, capable of questioning the implications of the theories he helped develop.

Schrödinger's cat paradox was born out of this intellectual vein. It was a provocative way to explore the limits of the Copenhagen interpretation, formulated by Niels Bohr and Werner Heisenberg. According to the Copenhagen interpretation, the state of a quantum particle, such as position or velocity, remains undefined until observed. Schrödinger decided to make evident the paradoxical aspect of this idea with a (mental) experiment involving a cat.

He imagined a closed system: a box containing a cat, a bottle of poison, a Geiger counter and a radioactive source. The source has a 50 percent chance of emitting a particle in a given amount of time. If this particle is detected by the Geiger counter, a mechanism is activated, breaking the bottle and releasing the poison, which kills the cat. Until someone opens the box to observe the system, the cat would be in a "hybrid" quantum state, that is, it would be simultaneously alive and dead.

Why did Schrödinger propose this experiment? He was certainly not an advocate of the "zombie" cat. His intent was to demonstrate how strange and counterintuitive the implications of quantum physics were when applied to macroscopic reality. As a classically trained physicist, Schrödinger found it difficult to accept that reality could exist in a state of indefiniteness. In a letter to his friend Albert Einstein, he wrote:

> *"I don't like this vague and elusive interpretation of physical reality."*

The cat experiment had a huge impact in scientific culture. While Niels Bohr argued that the Copenhagen interpretation was the best description of quantum behavior, many others, including Einstein and Schrödinger himself, remained skeptical. Einstein, a well-known critic of quantum determinism, welcomed Schrödinger's paradox with interest.

The historical context of Schrödinger's proposal is equally significant. In the 1930s, Europe was a melting pot of scientific minds that, despite growing geopolitical tensions, continued to collaborate and push the boundaries of knowledge. Schrödinger, along with figures such as Bohr, Heisenberg, Dirac and Born, was redefining the foundations of physics while grappling with philosophical questions that seemed unsolvable.

Interestingly, Schrödinger never chose the cat as an example arbitrarily. He loved animals, and his thought experiment was not meant to be cruel, but rather ironic and provocative. Over the years, this image of the "dead and alive cat" has become a symbol not only of quantum physics but also of modern philosophical reflection on what it means to observe and understand reality.

The cultural echo of Schrödinger's paradox has propagated far beyond its original scientific context. It has inspired novels, comic books, movies and even popular expressions. In the literary sphere, for example, British writer Douglas Adams ironically referred to the cat in his work "*The Hitchhiker's Guide to the Galaxy*." In pop culture, the phrase "*like Schrödinger's cat*" is often used to refer to situations of uncertainty or ambiguity.

Schrödinger, with this paradox, left a legacy that goes far beyond the scientific world. In a sense, the famous "cat" embodies our being suspended between two worlds: that of classical determinism, which shaped pre-quantum physics, and that of indeterminism, which defines the modern quantum view. It is another example of how science is never just technical analysis or experimentation, but also deep philosophical exploration. And in this, Schrödinger, the physicist and thinker, remains a central figure, capable of stimulating questions that still echo today.

Waves, equations and paradoxes of nature

By the 1920s, classical theories could no longer describe the behavior of atomic and subatomic particles. Scientists, therefore, were looking for new paradigms to understand the deeper nature of matter. It was in this context that Erwin Schrödinger, an Austrian physicist with a deep philosophical sensibility, helped rewrite the way we observe the universe.

In 1926, while in the snowy mountains of Arosa, a Swiss valley far from the chaos of big cities, Schrödinger formulated one of the most famous equations in the history of science: the wave equation. Forced to move for health reasons and immersed in a period of creative isolation, Schrödinger found in this mountain quiet the ideal environment to ponder the questions that nagged at him. Light filtered sharply through the alpine landscapes, and this clarity seemed to be reflected in his work.

Schrödinger's wave equation describes the probabilistic nature of particles at the quantum level. No longer a point-like corpuscle, but a fuzzy entity distributed in space through a wave function. The idea displaced many physicists of the time. Schrödinger proposed that this function not only calculated the probability of finding a particle at a specific location, but also revealed its intrinsic duality between particle and wave.

The wave function.

The wave function is a fundamental concept in quantum mechanics. It is a mathematical function that describes the quantum state of a particle or system of particles.

The peculiarity of the wave function is that it has no direct physical meaning, but represents the probability density of finding a particle at a certain position and at a certain instant of time. This comes from the probabilistic interpretation of quantum mechanics provided by Max Born.

Some key points of the wave function:

It is a complex function (that is, it can have both real and imaginary parts).

It must satisfy the Schrödinger equation, which governs the time evolution of the system.

Contains all accessible information about the quantum system under investigation.

Simply put, the wave function is a mathematical tool that allows us to make predictions about the possible measurements (position, momentum, energy...) associated with a particle in a given quantum system.

Schrödinger's vision drew inspiration from both science and philosophy. Like many Austrian intellectuals of the time, Schrödinger studied the works of Arthur Schopenhauer, who was fascinated by Eastern philosophies. The deep connection between observer and reality at the heart of quantum mechanics recalled the principles of Advaita Vedānta, a Hindu philosophical current that he himself studied. The latter held that the phenomenal world is the result of a unique consciousness that permeates everything. Schrödinger was not a religious practitioner, but he recognized the philosophical intuition of the fundamental unity of existence.

The genius of his work was accompanied, however, by a provocative and skeptical streak. In 1935, to challenge the Copenhagen interpretation of quantum physics-which implied an "indeterminate" reality until it was observed-Schrödinger devised one of the most famous mental experiments ever: Schrödinger's cat.

While critical, Schrödinger knew that his work represented a revolution. Far from the deterministic explanations of Newtonian physics, quantum mechanics was revealing a counterintuitive universe, where certainty gave way to probability and potential. Understanding the "cat" involved confronting a dualism that seemed to oppose logic, but which found deep echoes not only in Western philosophy but also in Eastern philosophy.

Schrödinger's figure cannot be separated from the cultural and philosophical context of his time. He frequented intellectual and literary circles, and was known for his interest in topics that transcended the boundaries of physics. He loved to quote poets, thinkers and mystical traditions.

> *"What we observe as body, mind and world are manifestations of one."*

he wrote, revealing the unified vision that inspired him.

The stay in Arosa symbolically represents the spirit of the physicist. Like a sage retreating to a mountain to meditate, Schrödinger used solitude to observe the world from a new perspective. The silence of the Alps enabled him to formulate concepts that, decades later, would influence not only physics, but also technology, philosophy and even more poetic territories of thought.

Schrödinger's wave equation and cat perfectly embody the heart of quantum mechanics: a world suspended between waves of probability and observations that change reality itself. Schrödinger, with his synthesis of science, philosophy and imagination, gave us a lens with which to look at the universe. Not only by turning our gaze to the microcosm of atoms, but also by questioning the place of the human mind in this cosmic mystery.

Copenhagen 1927. The conference that changed the fate of physics.

In the landscape of modern science, few events have had an impact comparable to the 1927 Solvay Conference held in Copenhagen.

The Solvay Congresses (also called Solvay Conferences), founded by Belgian industrialist Ernest Solvay, are a series of scientific conferences devoted to important open problems involving physics and chemistry, held in Brussels every three years, starting in 1911, at the "*International Solvay Institutes for Physics and Chemistry*."

The 1927 meeting, also known as the Fifth Solvay Conference, represents a pivotal moment in the history of physics and human thought. On this occasion, the greatest scientists of the 20th century gathered to discuss a topic that would redefine our understanding of the world: the nature of quantum reality. Among them were people of extraordinary prestige, names that still evoke the depth and complexity of scientific thought: Max Planck, Albert Einstein, Niels Bohr, Werner Heisenberg and Erwin Schrödinger.

At the beginning of the 20th century, classical physics, based on Newton's laws, dominated every aspect of scientific understanding. However, during the first decades of the 20th century, phenomena emerged that the classical laws could not explain. The nature of light, the behavior of atoms and the relationship between energy and matter paved the way for new theories. In 1900, Max Planck formulated the quantum theory, a revolution that laid the foundation for a new physics. Later, Albert Einstein, in 1905, explained the photoelectric effect using the idea that light was composed of particles, photons, further strengthening the quantum paradigm.

In the 1920s, theories such as Heisenberg's uncertainty principle and Schrödinger's wave equation gave mathematical structure to the world of the infinitely small, where the unpredictable behavior of subatomic particles reigns. These discoveries marked the transition from a deterministic view of classical physics to a probabilistic and inherently uncertain interpretation of reality.

In this context, 1927 became the crucial year to compare these views and to begin to resolve one of the most intriguing dilemmas in science: is there an objective reality independent of the observer?

The conference was organized by the "*Solvay Foundation*" and was devoted to the theme "*Electrons and Photons*," but soon the debate shifted to deeper philosophical issues. Twenty-nine of the brightest physicists of the time attended the event, and almost all of them had contributed significantly to the birth of quantum physics.

Albert Einstein was already famous for his theory of relativity and his contribution to the quantum paradigm. However, Einstein was deeply troubled by the interpretation emerging from the new probabilistic models. He was firmly opposed to the idea that reality was governed in a fundamental way by chance. The Einsteinian view was opposed by Niels Bohr, the Danish physicist recognized as one of the founding fathers of quantum mechanics. Bohr developed the Copenhagen interpretation, according to which the behavior of particles cannot be described independently of the observer.

Other participants included Werner Heisenberg, who had proposed the uncertainty principle in 1927, and Erwin Schrödinger, who instead believed that waves could represent a more fundamental reality. Max Planck, the founder of quantum theory, presided over the gathering, representing the figure of the wise patriarch.

The debate between Einstein and Bohr.

One of the most famous moments of the conference was the philosophical clash between Einstein and Bohr. Einstein believed that quantum mechanics, with its probabilistic nature, was incomplete. His most iconic statement in this regard was, "God does not play dice." Einstein did not deny the successes of quantum theory in describing phenomena on a subatomic scale, but he argued that there was a deeper explanation, not yet understood, that would reconcile the quantum world with the principles of determinism.

Bohr, on the other hand, firmly reiterated that reality itself was inherently probabilistic. His approach was based on complementarity, the principle that particles such as the electron can behave as both waves and particles depending on the context in which they are observed. Bohr's position was clear: reality cannot be separated from the interaction between observer and observed object. During a heated confrontation, Bohr replied to Einstein with an observation that encapsulates the essence of modern physics:

"The universe is not obliged to be comprehensible to man."

A memorable anecdote.

Among the many fascinating pages in the history of science, the one written at the Solvay Congresses is undoubtedly one of the most emblematic. Here an extraordinary intellectual clash took place between two giants of physics, Albert Einstein and Niels Bohr, around an issue that would shape the fate of 20th-century science: Werner Heisenberg's uncertainty principle. The affair is so significant not only for the depth of the arguments developed, but also for the human and cultural entanglement that characterized the encounter.

We are in the late 1920s, an extraordinarily fertile period for theoretical physics. Quantum theories, developed since the early twentieth century with Max Planck and refined by contributions such as those of Einstein, Bohr, Heisenberg and Schrödinger, had opened up new perspectives on the microscopic world. However, these new discoveries brought with them profound questions about the nature of reality, determinism and the limits of human knowledge. Albert Einstein, winner of the 1921 Nobel Prize in Physics thanks to his explanation of the photoelectric effect and father of the theory of relativity, was one of the fiercest critics of the new quantum mechanics. He did not accept the idea that physics was governed by fundamental indeterminacy. As he stated in one of his famous phrases, "God does not play dice."

On the other hand, Niels Bohr, a Danish physicist and one of the founders of quantum mechanics, strenuously defended the indeterminacy introduced by the new theories. For Bohr, this feature was not a flaw in the theory, but an essential description of the deep nature of the world. Their philosophical differences were obvious: Einstein clung to a deterministic conception of reality, while Bohr embraced the idea that the phenomena of nature could only be described in probabilistic terms.

The Solvay Congress was the scene of a heated confrontation between these two giants of physics. Einstein brought to the assembly's attention a thought experiment designed to criticize Heisenberg's uncertainty principle.

Einstein, to challenge Bohr, proposed the following experiment: he imagined a box equipped with a small opening controlled by a clock. The box contained light or electromagnetic energy and allowed a single photon to pass through the opening in a well-defined time interval. Thanks to the clock, one could theoretically record the exact moment when the photon exited. Similarly, by measuring the change in the mass of the box before and after the photon exited (according to Einstein's relation (E = mc^2)), the photon's energy could also have been accurately calculated. This, according to Einstein, contradicted the uncertainty principle, allowing time and energy to be known simultaneously with absolute precision.

Faced with Einstein's argument, Niels Bohr immediately understood the magnitude of the challenge, but the solution was far from simple. That night, Bohr withdrew to reflect with great intensity; he apparently did not sleep a wink, meditative and immersed in the details of the problem. The following day, Bohr responded to Einstein's provocation with an intellectual twist that not only preserved the uncertainty principle but reinforced it, using the very theory of general relativity developed by Einstein himself.

Bohr pointed out that in order to determine the change in the mass of the box (needed to calculate the photon's energy), it had to be suspended on a balance. However, any measurement of mass implies that the box is subject to the influence of gravity. At this point, Bohr highlighted a crucial point: according to general relativity, the gravitational field affects the passage of time. The position of the box within the gravitational field caused a slight but unavoidable indeterminacy in the time at which the photon left the box, because of the relationship between gravity and clocks. This indeterminacy was of the same magnitude as the energy-time indeterminacy predicted by Heisenberg's principle. In other words, Einstein's attempt to circumvent the uncertainty principle failed because the phenomenon itself introduced an uncertainty compatible with quantum mechanics.

This memorable clash was not just about formulas or physical laws, but questioned the deeper meaning of reality. Bohr, in arguing with Einstein, stated:

> *"We cannot separate quantum phenomena from measuring instruments. Any description of reality must take into account both the observed behavior and the context in which it is observed."*

This view, known as the Copenhagen interpretation, would form the philosophical basis of quantum mechanics.

Einstein, despite Bohr's refutation, never gave in to the view that quantum mechanics was not a final theory. He continued for the rest of his life to search for a deeper, unifying theory capable of explaining reality without indeterminacy. However, this particular confrontation marked one of the rare occasions when Einstein was publicly defeated in debate.

The story goes that at the end of the Solvay Congress, Einstein acknowledged, albeit reluctantly, the ingenuity of Bohr's arguments, although he was never fully convinced. The confrontation became a true legend in the scientific world, a symbol of how the progress of knowledge is fueled by heated but intellectually honest debates.

In a broader sense, the debate between Bohr and Einstein became the paradigm of a persistent tension in science: that between the desire for certainty and the recognition of the inherent limits in human knowledge. That Solvay Congress represents not only a moment of glorious scientific confrontation, but also an illustration of the profound dialogue between philosophy and modern science.

Despite the differences between the two giants of thought, the Copenhagen Conference did not end with a winner. The discussions and criticisms raised in that room continued to profoundly influence science and philosophy in the following decades. Quantum mechanics quickly became a very powerful tool for describing and predicting natural phenomena, and most of the scientific community accepted Bohr's probabilistic interpretation. However, Einstein's doubts were not forgotten. His criticisms gave rise to a whole strand of research, culminating years later in Bell's theorems and experiments on quantum entanglement.

The 1927 Solvay Conference was not only a scientific event, but a symbol of the confrontation between radically opposed worldviews. On the one hand, the idea of an ordered universe, governed by deterministic laws; on the other, an elusive universe, resisting all human attempts to pigeonhole it into absolute categories. This tension, between certainty and uncertainty, between reality and observation, continues to resonate not only in science, but also in philosophy and spirituality.

Richard Feynman and the American contribution (1940s-60s).

In the years that followed, quantum physics expanded, finding technological and theoretical applications. Prominent among the great protagonists of the 20th century was Richard Feynman, an American physicist known for his talent for making complicated concepts understandable. Feynman developed the theory of quantum electrodynamics (QED), a milestone that described how light and matter interact. His famous "Feynman diagrams" offered a visual way to represent subatomic interactions, making the chaos of quantum reality more understandable.

Feynman loved to make science fun, but he never failed to emphasize its complexity.

Feynman was born in 1918 in Far Rockaway, a suburb of New York City. From a young age, he displayed a brilliant and creative mind. It is said that as a boy he took radios apart to understand how they worked. That curiosity remained alive throughout his career. During the years of World War II, Feynman was involved in the Manhattan Project, which led to the creation of the atomic bomb. This experience marked his entry into the American scientific elite.

After the war, Feynman devoted himself to one of the most complex problems in modern physics: understanding how light and matter interact. In theoretical terms, the problem involved quantum *electrodynamics*, or QED (*Quantum Electrodynamics*). Together with Julian Schwinger and Shin'ichirō Tomonaga, Feynman was one of the great architects of this revolutionary theory. In 1965, he received the Nobel Prize for this work, shared with the other two scientists.

Feynman's contribution was not limited to theoretical mathematics. He invented a new visual language for representing subatomic processes: the Feynman diagrams. These diagrams, simple in appearance, contained extraordinary explanatory power. They showed how particles exchanged photons or interacted with other particles in a way that could be precisely calculated. For millions of students, those diagrams became a stepping stone to understanding quantum mechanics.

Feynman's genius lay in his ability to simplify without ever trivializing. During his lectures at the "*California Institute of Technology*" (Caltech), he managed to explain even the most abstract concepts with practical examples and anecdotes. In 1964, he gave a famous series of lectures at *Cornell University*, documented in the BBC television series "*The Character of Physical Law*." In one of these lectures, he uttered a phrase that has become famous:

> *"I think I can say that no one really understands quantum mechanics."*

Feynman himself admitted that he found quantum physics puzzling, despite being one of its founding fathers.

An often-remembered episode about Feynman is his involvement in solving the "*Space Shuttle Challenger*" disaster in 1986. Although this was not an issue directly related to quantum physics, he once again demonstrated his fresh and direct approach to the problem: by immersing a simple gasket ring in cold water, he proved that the material lost flexibility, contributing to the identification of the cause of the disaster.

In the 1960s, quantum physics began to show its practical implications. The theories of Feynman and others went far beyond academic laboratories. They were applied in the design of transistors and semiconductors, pillars of modern electronics. Many years later, the principles of quantum physics would also find application in quantum computing and cryptography.

But Feynman was not only a scientist. He was also an observer of culture. He was fascinated by alternative paths of human thought, such as the languages of native peoples or Eastern philosophies. This interest was evident in his travels to Brazil, where he learned to play the bongo, and to Japan, where he observed with interest different customs and traditions.

Richard Feynman died in 1988, leaving an immense legacy. His work on quantum physics, and QED in particular, transformed the way we look at the nature of reality. But perhaps his greatest lesson lies in the words he often repeated to his students:

> *"Never be afraid of not understanding completely. The important thing is to ask questions."*

Chaos diagrams.

In the 1940s and 1950s, quantum physics was undergoing a revolution. It was a period of great scientific ferment, in which the classical laws seemed to be losing ground in the face of an increasingly complex and mysterious universe. The nature of light and matter, the interaction between subatomic particles, and probability instead of certainty-these issues dominated the scientific debate. In this chaotic scenario, a young American physicist emerged, full of imagination and with an out-of-the-box approach, destined to change the understanding of the universe forever. His name was Richard Feynman.

Richard Feynman was already shining during his undergraduate years. After graduating from MIT, he moved to Princeton, where he began working on problems that would shape his career. It was during that time that he began to focus on quantum electrodynamics (QED) theory. This theory aimed to describe the interaction between light and matter, a field where classical laws could no longer provide adequate answers.

QED, at the time, was a real puzzle. Mathematical calculations often led to infinite, contradictory, and completely unserviceable results. It was like trying to bring order to a chaos with no solution. This is where Feynman's genius came in. His insight was not only mathematical, but also visual. He decided to turn the complexity of subatomic interactions into something more "human," accessible to even the most skeptical colleagues.

Between 1948 and 1950, Feynman developed his famous diagrams, and presented them during a seminar at Princeton University, leaving colleagues stunned. Those seemingly simple drawings actually represented a revolutionary language for describing interactions between elementary particles. Straight and wavy lines, meeting points and angles-the diagrams translated years of complex calculations into understandable symbols capable of telling stories of subatomic processes... Those seemingly simple drawings actually represented a revolutionary language for describing the interactions between elementary particles. Straight and wavy lines, meeting points and angles-the diagrams translated years of complex calculations into understandable symbols capable of telling stories of subatomic processes.

A classic example is the contents of a typical diagram illustrating the exchange of a photon (the particle of light) between two electrons. The straight lines represent the electrons; the wavy one, the photon. In an instant, everyone could see what might previously have required pages of complex formulas. It was not just a mathematical tool: it was a visual lens to see the hidden chaos of the universe.

These diagrams not only simplified calculations but changed the way physicists thought about reality itself. The Nobel Prize in Physics, which Feynman received in 1965 along with Julian Schwinger and Sin-Itiro Tomonaga, emphasized precisely the importance of his contribution to QED.

A genius outside the box.

Richard Feynman, besides being an extraordinary thinker, was also an eccentric and fascinating character. His inquisitive wit and ability to explain complex concepts with simple metaphors made him an iconic figure not only on college campuses but also in popular culture. At Caltech (*California Institute of Technology*), where he spent much of his career, he was known for his brilliant courses and his humble, straightforward approach to science. He often said:

"What I can't build, I don't understand."

This phrase reflected his intellectual approach: everything he learned had to be rooted in concrete understanding, made clear and tangible.

One iconic anecdote features him at MIT. Feynman, called by students "*the genius with the bongos*," enjoyed playing improvised musical instruments and discussing science in the most informal ways. His passion for popularization also led him to write books that are now considered landmarks, such as "*Six Easy Pieces*."

What makes Feynman's work so fascinating is not just the precision of his science, but the way it is linked to a philosophical worldview. For Feynman, chaos was not something to be feared, but rather an opportunity to discover new hidden rules. In a sense, his genius was manifested precisely in his ability to discover an order in disorder. This view reflects, in part, Eastern thought and the idea that chaos and harmony can coexist.

In a later lecture, reflecting on the meaning of scientific research, he said:

"It is not important to know the answers. It is important to experience the joy of asking questions."

This boundless curiosity, combined with a natural talent for simplicity, made Feynman not only one of the greatest physicists of the 20th century, but also a bridge between the mathematician and the philosopher, the scientist and the artist.

Today, Feynman diagrams are a fundamental pillar of modern physics. They are used not only in QED, but also in more advanced fields such as string theory and theories of everything. Yet, they have never been just a tool. They are a symbol of his bold approach, his rejection of complexity for its own sake, and his extraordinary ability to see beauty in chaos.

Feynman left an imprint that goes beyond equations and scientific awards. He showed that there is often a flavor of innate simplicity behind the most complex discoveries. And that, in the end, even chaos can be told with an intersecting wavy line and a straight line.

A new understanding of reality.

At the beginning of the twentieth century, physics was experiencing a golden age. Scientists believed they understood almost everything there was to know about the physical universe. Sir Isaac Newton had provided the laws of motion and universal gravitation. James Clerk Maxwell had elegantly described electromagnetic phenomena. Only a few details seemed to be missing to perfect the immense machine of knowledge.

But in 1900, a German physicist, Max Planck, took an unintentionally revolutionary step. Studying the problem of blackbody radiation, he discovered something surprising. Electromagnetic radiation, like light, did not flow continuously, as expected, but behaved as if it consisted of discrete packets of energy. Planck called these mysterious packets "quanta." The idea was initially a mathematical gimmick, nothing more. But it would prove to be the beginning of an earthquake in the way we understand reality.

In the following years, these oddities began to multiply. Albert Einstein, in 1905, used Planck quanta to explain the photoelectric effect, showing that light itself could behave like a particle. Einstein won the Nobel Prize for this discovery, but a more shocking truth soon emerged: matter also exhibited apparently dual behavior. Experiments by scientists such as Louis de Broglie and Davisson-Germer proved that particles, such as electrons, behave sometimes as waves and sometimes as corpuscles, depending on how they are observed.

This ambiguity was not just a negligible detail. It implied a completely new view of reality: in the quantum world, it is impossible to know with certainty both the position and the velocity of a particle. Werner Heisenberg formalized this idea in 1927 with his uncertainty principle. Quantum physics no longer describes a deterministic universe, like Newton's, but a universe governed by probability.

Niels Bohr, another pioneer of quantum physics, expressed this reality clearly:

> *"If it doesn't confuse you, you don't really understand quantum physics."*

According to Bohr, the microscopic world is governed by bizarre rules. Particles can be here and there at the same time, and they can interact instantaneously over great distances, in a phenomenon that Einstein himself called "*a spectral action at a distance*," known today as *quantum entanglement*.

This new physics was not the result of one mind, but of an extraordinary era of collaborations and disagreements between some of the most brilliant minds in history. Chief among them was Einstein, who, while contributing greatly to the birth of quantum physics, remained skeptical of it for a long time.

The technological revolution of quantum physics.

Quantum physics is one of the most extraordinary intellectual achievements of the 20th century. It began in the early twentieth century as a countercultural theoretical idea destined to disrupt the classical worldview. Today, this branch of science is not only the starting point for philosophical reflections, but a concrete force that has changed our daily lives. From communication tools to medical devices, quantum physics has become the invisible pillar of our technological reality.

But how did all this happen? How did we go from the equations of Max Planck and Albert Einstein to lasers and computers?

In 1900, Max Planck, a German physicist, introduced the radical idea that energy is not continuous, but occurs in discrete packets called "quanta." At that time, quantum mechanics was born. Albert Einstein, a few years later, took up the concept to explain the photoelectric effect, a fundamental phenomenon for understanding light and matter. Einstein proved that light itself is made of quanta, or photons.

His ideas not only changed physics but laid the foundation for instruments that are essential today. The laser, for example, is a direct descendant of this insight into the quantum nature of light. In the 1960s, when the first laser was built, it quickly became clear that this invention would revolutionize communications and industry. From surgery to scanning bar codes in supermarkets, the laser is now a ubiquitous presence, but few remember its quantum origin.

Another revolutionary leap was the application of quantum mechanics to solids. In the 1940s physicists discovered that quantum principles could explain the behavior of electrons in materials. This discovery led to the creation of semiconductors and, with them, to one of the most important inventions of the 20th century: the transistor.

The transistor, invented in 1947 by John Bardeen, Walter Brattain and William Shockley at "*Bell Labs*," is the beating heart of modern electronics. Every computer, smartphone and electronic device owes its existence to this little technological marvel.

William Shockley, in particular, was not only a brilliant theoretical physicist but also a controversial innovator. Thanks to his work on semiconductors, Silicon Valley, the world center of technology, was born. In fact, in the 1950s, Shockley founded a small semiconductor company right in the Santa Clara Valley, California. Many of his collaborators later created companies we know well today, such as Intel. From here, Planck's theoretical science became inextricably linked to the industrial world.

Today we are experiencing what many call the "second quantum revolution." If lasers and semiconductors represent the first technological era of quantum physics, quantum computers are its future.

But what makes a quantum computer special? While traditional computers use bits that can only be 0 or 1, quantum computers use qubits, which thanks to the superposition principle can be 0, 1 or both at the same time. This makes them incredibly powerful for certain types of calculations, such as cryptography or simulation of complex molecules.

Again, the ideas behind quantum computers go back to theoretical work on quantum phenomena. Physicist Richard Feynman was one of the first to imagine in the 1980s that it would be possible to harness quantum mechanics to build incredibly new machines. Today, companies such as IBM, Google and Microsoft are making that vision a reality.

This scientific and technological revolution has also had cultural and philosophical repercussions. The idea that the behavior of matter is governed by counterintuitive principles, such as indeterminacy and superposition, fascinates and destabilizes. Many have tried to find in quantum physics a resonance with Eastern philosophies, such as Hinduism or Buddhism, that question the linear understanding of reality.

Interestingly, some of the pioneers of quantum physics, such as Niels Bohr and Erwin Schrödinger, were themselves fascinated by Eastern ideas. Bohr, for example, affixed the Chinese symbol of Taijitu (the famous "yin and yang") to his own noble coat of arms, while Schrödinger was inspired by ancient Vedantic texts in his quest for a unified understanding of nature.

Fundamental principles of quantum physics.

Quantum physics, born in the early 20th century, revolutionized our understanding of the universe. Its fundamental principles challenged the certainties of classical physics and introduced us to a world of uncertainty, interconnectedness and ambiguity. Concepts such as the uncertainty principle, wave-particle duality and the phenomenon of entanglement not only changed science but opened up new possibilities for dialogue with Eastern philosophies, particularly Hinduism.

The uncertainty principle: the uncertainty revolution.

Werner Heisenberg, a German physicist and one of the fathers of quantum mechanics, formulated the uncertainty principle in 1927. This principle states that it is impossible to simultaneously know with absolute precision both the position and the velocity of a subatomic particle. The more precisely we measure one quantity, the less we can know about the other.

Indeterminacy is not a technological limitation, but a fundamental property of nature. In the quantum world, particles do not exist in a definite state until they are observed. It is as if a particle is in a "suspension of possibility." This concept seems to be echoed in ancient Hindu philosophy, especially in Advaita Vedanta, which sees the phenomenal world as a momentary manifestation of a single underlying reality: the Brahman. In this context, what appears solid and definite (the universe) is, in reality, a continuous flow of possibilities.

Wave-particle duality: the ambivalent nature of reality.

Another puzzling idea introduced by quantum physics is the wave-particle duality. Photons, electrons and other subatomic particles behave as waves in some circumstances and as particles in others. The iconic double-slit experiment, first conducted in 1801 by Thomas Young and repeated in the quantum context in the 20th century, demonstrated that a particle of light, when unobserved, generates interference typical of waves. But the moment a scientist attempts to observe which slit it passes through, the particle assumes corpuscular behavior....

This phenomenon opens up radical questions about what "reality" is. Richard Feynman, famous physicist, said:

> *"The behavior of light is impossible to explain in ordinary language. This is the heart of the mystery."*

Duality finds an intriguing parallel in Hinduism. Advaita philosophy holds that the universe is *maya* (illusion), a dance of manifestations that differ according to perception. A common example is that of space and time, which appear distinct but are interconnected. Just as particles are waves and particles at the same time, the physical and spiritual worlds derive from the same essence and coexist in seemingly contradictory ways.

Entanglement: the instantaneous connection.

Einstein called the phenomenon of entanglement "*a spectral action at a distance*." When two particles enter an "entangled" state, their fates become deeply united, regardless of the distance between them. If one of the particles changes state, the other will also do so immediately, as if they were intimately connected by an invisible thread. This phenomenon, demonstrated experimentally in 1982 by physicist Alain Aspect, shakes up the classical concept of separation between objects.

Entanglement has prompted many physicists to question whether the universe is intrinsically connected. A strikingly related concept emerges in Hindu philosophy. In the Bhagavad Gita and Vedic texts, it is stated that everything is interconnected through Brahman, the ultimate reality. The saint Yajnavalkya, in the *Brihadaranyaka Upanishad*, describes the universe as an infinite web where each point reflects the other, similar to the concept of "*Indra's web*" in Buddhism.

Entanglement seems to confirm a millennia-old spiritual insight: the boundary between the self and the universe is an illusion, and all things exist in relation to one another.

Quantum physics offers a journey into a mysterious world where ordinary laws of common sense do not apply. Its fundamental concepts-indeterminacy, duality and entanglement- seem to echo ideas rooted in Hindu philosophy. Although science and spirituality follow very different methodologies, their dialogue invites us to see reality in new ways, connecting humanity's millennia-old past with the frontiers of modern knowledge.

Quantum physics and Hindu philosophy: an encounter between science and spirituality.

The idea that science and spirituality can converse may seem paradoxical. On the one hand, science is based on empirical observations and mathematical calculations. On the other, spirituality arises from subjective experience and introspection. However, with the emergence of quantum physics in the 20th century, these two seemingly distant dimensions have discovered striking similarities. Quantum physics, with its revolutionary principles, has offered a new way of thinking about reality that recalls in some ways the ancient insights of Eastern philosophies, particularly Hinduism and Advaita Vedanta.

When Werner Heisenberg formulated the uncertainty principle in 1927, traditional physics suffered an epistemological crisis. This principle states that it is not possible to know with absolute precision and simultaneously the position and velocity of a subatomic particle. Faced with this uncertainty, the determinism of classical physics gave way to probability, introducing concepts that challenged Western linear thinking.

Now imagine how much this concept may resonate with the sacred texts of Hinduism, such as the Upanishads. In their verses, a similar idea is often explored: ultimate reality, or Brahman, is unfathomable through the senses or rational means alone. Brahman is not an object of empirical certainty, but a truth that can only be intuited through an expansion of consciousness. Heisenberg himself, in the course of some conversations with the poet Rabindranath Tagore, did not fail to point out how naturally Eastern thought seemed able to embrace the paradoxes that quantum physics was revealing.

The dual nature of light and reality.

One of the central concepts of quantum physics is wave-particle dualism. Light behaves now as a wave, now as a particle, depending on the experiment conducted. This behavior, observed since the studies of Max Planck and Albert Einstein, defies the rigid categories of classical science. But is it not in tune with the ancient wisdom of Hindu philosophy?

Hindu thought describes reality as composed of seemingly opposing dualities, which, however, are complementary manifestations of a greater unity. In Advaita Vedanta, this unity is named Brahman, the supreme, indivisible principle. Dualities such as prakriti (matter) and purusha (consciousness) find harmony in one underlying reality. This call to unity beyond divisions is not far from the quantum image of a universe where waves and particles are only manifestations of a deeper, interconnected nature.

Entanglement: the universal interconnection.

Another fundamental principle of quantum physics is the phenomenon of quantum entanglement. Two particles, separated in space, can remain interconnected in such a way that the state of one instantaneously affects the state of the other, regardless of distance. This phenomenon, described in Alain Aspect's experiments in the 1980s, challenges the common conception of spatial separation.

Entanglement finds an interesting echo in Hindu teachings on the concept of universal interconnectedness. According to Vedantic philosophy, all entities in the universe are linked together by an invisible fabric. In this context, the individual does not really exist as a separate entity, but is a reflection of the cosmic totality. Simply put, this understanding is expressed in the famous mantra of the Upanishads, "*Tat Tvam Asi*" ("You are that"). The individual Self (Atman) and the universal Self (Brahman) are one. An idea, this, that also seems to resonate with the concept of interconnectedness rooted in entanglement.

The meeting of East and West.

In the 1920s and 1930s, when quantum physics was emerging, many scientists were struck by the similarities between their findings and Eastern thoughts. Niels Bohr, one of the founders of quantum theory, was attracted to Taoist philosophy and the idea of balancing opposites. Erwin Schrödinger, known for his famous "cat," was also deeply inspired by Advaita Vedanta. Schrödinger declared without reservation that he felt that the Indian view of the spiritual unity of the universe aligned with the emerging scientific understanding of an indivisible universe.

This dialogue between science and spirituality demonstrates how human progress is never confined to a narrow domain. Quantum physics, with its radical insights, has not closed doors but opened others to alternative ways of knowing reality. Hinduism, with its meticulous exploration of the infinite and the human mind, continues to offer conceptual tools to address the fundamental questions that science itself raises.

The dialogue between quantum physics and Hindu philosophy represents an encounter between two different languages, both directed toward the same deep curiosity. What underlies it all? Faced with the mystery of the universe, modern Western science and ancient Eastern wisdom come together, like two travelers who, traveling along different paths, arrive at the same vantage point. This is but the beginning of a conversation, potentially endless, between logic and intuition, between measurement and mystery.

The uncertainty principle: the unpredictability of the quantum world.

In 1927, Werner Heisenberg, one of the fathers of quantum physics, shook the foundations of classical science with a startling revelation: the uncertainty principle. This discovery, elaborated while Heisenberg was working in the nerve center of European theoretical physics, in Copenhagen under Niels Bohr, did not just transform science. It opened a new way of conceiving reality.

The Uncertainty Principle states that it is impossible to know simultaneously, with absolute precision, two fundamental quantities related to each other, such as the position and momentum of a particle. This limitation does not result from a lack of instruments or technology. On the contrary, it is inscribed in the very nature of the universe. The more precisely one attempts to measure the position of an electron, for example, the more nuanced the information about its velocity becomes. This is not a technical problem, but a deep law of physics, which Heisenberg summarized as a universal principle.

Before Heisenberg, physics was dominated by Newtonian determinism. Isaac Newton, in the 17th century, had defined a universe governed by precise and predictable laws. In the classical paradigm, if one knows the initial conditions of a system, one can calculate its future. However, with quantum physics, this dream of total control vanishes.

The quantum world hovers in the realm of probability. We cannot say what a particle will do, only what the probability is that it will do a particular thing. This concept inserts uncertainty as an intrinsic part of nature.

The conceptual revolution introduced by quantum physics finds fascinating resonance in Hindu philosophical texts. According to the *Vedas*, one of humanity's oldest sapiential collections (dating back to at least the second millennium B.C.), the reality we perceive with our senses is only an illusion. The sages of the Vedic tradition called this illusion *Maya*.

However, the Maya is not an illusion in the trivial sense of the word. It is not a deception or something false, but a curtain hiding a deeper and more elusive truth. Similarly, in the context of quantum laws, our attempt to measure and understand the subatomic world comes up against a fundamental impossibility: every measurement changes the system we observe. What remains is a kind of invisible dance, mutable and indefinable, much like the veiled reality described by Hindu philosophy.

The connection between Maya and indeterminacy is not a simple play on symbols. It is a deep parallelism that has fascinated Western physicists and philosophers. Fritjof Capra, noted physicist and author of the famous "*The Tao of Physics*," explored this correspondence. He noted that the contemplative approach of the Indian tradition, which accepts the mystery of reality, harmonizes surprisingly well with quantum physics. Both perspectives embrace uncertainty and suggest that human knowledge has inherent limits.

To illustrate the effects of quantum indeterminacy on the real world, physicists have often used mental experiments. One of the most famous is that of Schrödinger's cat, proposed in 1935 by Erwin Schrödinger, another giant of quantum physics. This experiment stages a cat locked in a box, along with a mechanism that can kill it or leave it alive. The question is: Until we open the box, is the cat dead or alive?

According to the quantum interpretation, until the observation is made, the cat is in a kind of superposition of states: alive and dead at the same time. This paradox illustrates how the quantum universe does not conform to our common sense. Indeterminacy and probability are intertwined in ways that defy our intuition.

Reality hides in the invisible.

Heisenberg's indeterminacy and the Vedic concept of Maya converge on a similar conclusion. The universe is not as it appears at first glance. At the macroscopic level, reality seems solid and deterministic. However, the deeper we get into the microscopic, the more everything dissolves into a maelstrom of probabilities, uncertainties and strange connections.

British physicist Sir James Jeans, in the 1930s, had observed with wonder:

> *"The universe is beginning to look more like a big thought than a big machine."*

Quantum physics thus invites us to reexamine our idea of reality, just as Hindu philosophy invites us to look beneath the surface of appearance. Perhaps, as the Vedic sages would have said, what we ignore is no less real than what we perceive.

From Maya's veil to the uncertainty principle: how reality escapes our control.

Quantum physics and Eastern philosophy meet on surprisingly common ground: the elusive mystery of reality. This 20th-century science, which has profoundly transformed our understanding of the world, seems to resonate with millennia-old ideas found in the sacred texts of Hinduism. In particular, Werner Heisenberg's Uncertainty Principle (1927) and the concept of *"Maya*" from Vedic traditions reveal a conceptual affinity that might seem unthinkable, but which suggests how much human beings in every age have sought to unravel the veil separating ultimate truth from sensory perception.

In 1927, German physicist Werner Heisenberg formulated one of the fundamental principles of quantum physics: the famous uncertainty principle. This principle states that it is not possible to know simultaneously, with absolute exactitude, two properties of a particle, such as its position and its momentum. The more precisely we measure one of these two variables, the less we can know about the other.

Heisenberg discovered that the behavior of matter at the subatomic level is never completely determinable. Electrons, for example, do not move along predictable trajectories like planets around the sun. Instead, they seem to "exist" as a diffuse probability, an indistinct cloud that takes shape only when directly observed. This has been confirmed by fundamental experiments such as those related to the famous "wave-particle dualism," in which light and electrons behave either as waves or as particles, depending on the mode of observation.

This revolutionary vision destabilized the paradigm of classical physics, which was based on a mechanistic universe where everything was determinable. Reality, as science was discovering it, now appeared elusive, unpredictable and, in a sense, illusory. But this idea was not new to everyone. Centuries earlier, in India, the sages of the Upanishads had already described the phenomenal world as something similar: an illusion, a veil hiding the ultimate truth.

Hinduism, with its rich philosophical and spiritual tradition, has developed the concept of "Maya" to indicate the illusory or deceptive nature of perceivable reality. In the Vedas, the oldest sacred texts of Indian culture, and in the Upanishads, the philosophical commentaries that elaborate on them, it is emphasized that the world that appears to our senses is not the ultimate reality, but a partial and illusory representation of Brahman, the absolute and indescribable principle that underlies all that exists.

A significant passage is contained in the Rigveda, where it states:

"Those who know this truth hidden behind the veil of appearance are enlightened."

Here, "veil" alludes to Maya, that which separates our empirical world from essential reality. Maya does not mean only illusion in the sense of deception, but rather the perception of a world that appears real but which, underneath, is composed of something deeper and inherently elusive.

An interesting example is found in the figure of the sage Yajnavalkya, the protagonist of numerous dialogues in the Upanishads. When asked to describe Brahman, Yajnavalkya responded with the famous teaching:

"Neti, neti" ("Not this, not that").

This expression indicates that ultimate reality cannot be defined or categorized, just like the elusive nature of subatomic particles observed by modern physics.

The analogy between the uncertainty principle and Maya should not be interpreted as a direct overlap; it is, rather, a resonance. Both concepts remind us that reality is more complex than our minds or tools can grasp. On the one hand, Heisenberg shows us a microcosmic universe in which the very act of observing alters what we observe.

On the other, Hindu philosophy invites us to recognize the limits of our senses and turn to deeper insight.

Fascinating in this comparison is also how both perspectives recognize the role of the observer. In quantum physics, direct observation of a particle influences its state; in Hindu philosophy, phenomenal reality is shaped by human awareness, but ultimate truth emerges only when sensory appearance is transcended.

A famous intellectual encounter between the Eastern tradition and quantum physics occurred precisely with the pioneers of this discipline. Heisenberg himself, during one of his trips to India in the 1930s, had the opportunity to engage with the Indian philosopher Rabindranath Tagore. On that occasion, Tagore is reported to have explained the Indian conception that the world and consciousness are in an interdependent relationship, an idea that mirrors, at least intuitively, the role of the observer in quantum mechanics.

What emerges from the comparison between quantum physics and Hindu philosophy is an invitation to review our certainties about reality. The universe, both in its subatomic and phenomenal dimensions, seems shrouded in a mystery beyond our control. Twentieth-century physicists and Vedic sages, though through different paths, have conveyed to us the same realization: that which appears solid and unchanging is, in reality, only a reflection, an illusion, or perhaps, as an Indian sage said millennia ago, a dream we dream together.

The wave-particle duality: the elusive nature of matter.

Another pivotal principle of quantum physics is wave-particle duality, theorized in 1801 by English physicist Thomas Young, then mel 1924 by Louis de Broglie, and later verified by experiments such as the modern version of the double slit. This phenomenon shows that a particle, for example an electron, can behave like a wave if no one observes it, while it manifests corpuscular properties (a real "particle") when measured. This paradoxical behavior is at the heart of the famous Schrödinger equation, which describes the quantum state as a superposition of possibilities.

In the Hindu context, the wave-particle duality recalls the dual aspect of Brahman, the absolute, which manifests as both immanent (matter) and transcendent (consciousness or spirit) reality. Just as the particle is both wave and corpuscle, so Brahman can be both the observable substance of the universe and that which transcends it.

In 1801, English physicist Thomas Young demonstrated, with the famous double-slit experiment, that light behaved like a wave. This result seemed to close a long-standing question, debated for centuries by Western science: is light composed of particles or does it propagate like a wave? Young offered a clear answer, but the case was far from closed. With the advent of the 20th century, a revolutionary finding emerged: matter and energy not only behave as waves, but can also manifest as particles.

The first step toward this amazing discovery was taken by Max Planck in 1900, when he introduced the concept of energy quanta, theoretical indivisible "packets" of energy. Then came Albert Einstein, who in 1905 demonstrated that light could behave as particles called photons, explaining the photoelectric effect and opening the door to a new way of looking at the physical world. But it was not until 1924 that Louis de Broglie laid the groundwork for an even deeper conception of reality. De Broglie proposed that not only light, but all matter could exhibit dual behavior: as a particle and as a wave. His theory was confirmed a few years later by experiments such as the one performed by Davisson and Germer in 1927, which revealed the phenomenon of electron diffraction.

The double-slit experiment.

This experiment, in any case, remains the most emblematic example. Imagine a single particle, such as an electron, thrown toward a barrier with two thin slits. If no one intervenes to "observe," the result will be surprising: the electron will form on the sensitive photoscreen placed behind the barrier a banded figure typical of waves, as if it had crossed both slits simultaneously. However, if an apparatus is placed to measure through which slit the electron passes, then the wave figure disappears, giving way to a simple dot, as one would expect from a particle. The behavior of the electron changes depending on the interaction with the observer, that is, depending on whether or not the observer places a passage gauge. But what is the crucial point? Until a measurement is made, the electron is neither particle nor wave: it is both, in a state of quantum superposition.

This paradoxical behavior was formalized by Erwin Schrödinger in 1926, with his famous equation describing the quantum state of particles as a "sum" of possibilities. Until a measurement is made, reality is open, ambiguous. The quantum universe thus challenges our human need for clarity, for sharp divisions, for a world made of visible, measurable, tangible things.

The parallelism with Hinduism.

This concept may seem rooted in modern scientific thought and foreign to ancient philosophy. Yet, to those familiar with Hinduism, the parallel appears surprisingly familiar. Indeed, the wave-particle duality finds its metaphorical echo in the concept of Brahman. In the Vedic tradition, Brahman is the ultimate reality, simultaneously immanent and transcendent. It is what constitutes the material world, but also what radically transcends it, reigning as pure consciousness.

The Vedas, fundamental sacred texts of Hinduism, describe Brahman with cryptic but powerful statements, such as the famous verse:

"Ekam sat, vipra bahudha vadanti."

("Truth is one, wise men call it by different names.").

According to this view, reality itself can appear multiple and dualistic to the human eye while remaining a fundamental unity. Exactly as an electron is both wave and particle, but at the same time neither one nor the other.

The Brihadaranyaka Upanishad, one of the oldest and most fundamental texts in Hinduism, describes the concept of the latent state, in which Brahman is simultaneously everything and nothing, resting in a condition of "*infinite potentiality*." This description seems to recall the quantum state of superposition. Brahman, just like matter in the quantum realm, can manifest in different ways depending on the perspective or knowledge of the observer.

Science and spirituality: a possible bridge?

Niels Bohr, one of the fathers of quantum mechanics, was known for his interest in Eastern philosophies. The Danish physicist even adopted the Taoist symbol of Yin-Yang in his noble coat of arms, seeing in it an image of the complementary dualism that characterizes quantum reality. However, we also find connections in Hinduism, albeit less immediately. The concept of complementarity in quantum physics, according to which seemingly opposite properties such as wave and particle are not contradictory but rather complementary aspects of the same reality, seems to align with the Vedic view of unity in diversity.

The interconnection between science and spirituality becomes even more apparent when we consider the role of the observer in the double-slit experiment. In the realm of Maya, the "veil of illusion" described in Hinduism, material reality is perceived as separate from the universal essence of Brahman. However, this separation is, according to tradition, caused precisely by the limited point of view of the human observer. Similarly, in quantum physics, the observer is not a neutral or passive entity. Its interaction with the system changes the observed reality.

Wave-particle duality is not only one of the fundamental principles of quantum physics. It is also a profound challenge to the way we understand the very nature of reality. Science and spirituality, though arising in seemingly distant contexts and with seemingly distant goals, seem to converge in the discovery of a universal ambiguity, of a multiform reality in which matter, energy and consciousness seem inseparable.

After all, as the great Indian poet and philosopher Rabindranath Tagore wrote,

> *"The world is not made up of things, but of a halo of possibilities."*

A poetic description, perhaps, but not so far removed from Schrödinger's quantum vision. Nor from the philosophical heart of Hinduism.

Wave or particle? Duality and the cosmic interplay of Shiva and Shakti.

Among the most fascinating concepts in quantum physics is wave-particle duality, a phenomenon that has revolutionized our worldview and, surprisingly, finds deep echoes in ancient Indian philosophies. The interplay between the concepts of Shiva and Shakti, immutable and movement, consciousness and creation, seems to echo the same complementarity that dominates the behavior of subatomic matter.

In 1801 Thomas Young conducted a famous experiment known as "the double slit." Young illuminated a screen with a light source and allowed it to pass through two thin slits. The result was surprising: instead of two simple beams of light projected onto the opposite screen, an interference pattern emerged, similar to that produced by waves on water. This showed that light was not only composed of particles, but also possessed a wave nature.

So far everything might have seemed relatively simple. But the real mystery emerged a century later, when scientists tried to repeat the experiment using single electrons. So far everything might have seemed relatively simple. But the real mystery emerged a century later, when scientists tried to repeat the experiment using single electrons. These particles, although sent one at a time, also produced the same interference pattern, suggesting that they behaved like waves. However, by the time they were observed -- with instruments designed specifically to figure out which slit they were crossing -- the electrons acted as particles, canceling out the interference pattern.

This phenomenon, which violated all intuition, is one of the foundations of quantum mechanics. Matter, at the subatomic level, is neither wave nor particle until it is observed. The act of observation seems, somehow, to "collapse" the wave function into a specific reality. But what does all this mean? Scientists have found more questions than answers.

Shiva and Shakti: the cosmic dance of the dynamic and the immutable.

The choice of names is not accidental: in Hindu culture Shiva and Shakti represent the opposition of the two pivotal principles of the universe . The former is spirit, connected to the male figure, while the latter represents the energy of the world, connected instead to the female figure.

In the world of Hindu philosophy, the universe is often described as the result of the interaction between these two supreme principles.. Shiva represents pure consciousness, the motionless and unchanging witness of reality. Shakti, in contrast, is the creative and dynamic energy that shapes the material universe. Separate, neither can exist. Life, time and change result from their interaction. As described in the Shivagamas and Tantras, Shiva without Shakti is pure potential, but devoid of expression. Shakti without Shiva is energy without direction.

This complementarity recalls the wave-particle duality. Matter behaves like a wave when it is not observed: it is pure potential, a field of possibilities. Yet, the moment observation occurs, those possibilities collapse into a concrete reality, like Shiva who, unmoving witness, gives meaning to Shakti's dynamism. Without observation, there is no form; without consciousness, there is no creation.

The wave-particle duality presents a profound question: the role of consciousness in the construction of reality. Prominent physicists, such as Eugene Wigner and John Wheeler, explored this idea. Wigner, in particular, postulated that the consciousness of the observer could be fundamental in determining subatomic behavior. Although this hypothesis remains controversial, it adds another layer of fascinating resonance with Indian philosophy.

In the context of Hinduism, the idea that reality is determined by consciousness is not new. Vedic texts and the Upanishads state that external reality is, in a sense, a projection of the supreme consciousness, Brahman. The latter is both the witness (Shiva) and the active creator (Shakti) of all things. Just as observation in the quantum domain determines whether we will see a wave or a particle, in the same way consciousness, in Hinduism, creates and shapes perceived reality.

Another key concept in Hinduism that is intertwined with quantum physics is that of Lila, the "cosmic game." According to this view, the universe is a playful dance of creation, preservation and destruction, orchestrated by Shiva and Shakti. Nothing is static, nothing is unchanging: everything manifests dynamically through the play of duality.

Similarly, the quantum universe is a dynamic realm of possibilities. Matter does not exist in a fixed state; it fluctuates between different states, just as Shakti fluctuates in constant motion, manifesting reality through the presence of Shiva. The electron acting alternately as wave or particle is a perfect example of this fluidity-a "game" that reveals the interconnected nature of reality.

A bridge between East and West.

Quantum physics, born in the West as a response to the limitations of classical physics, invites us to reevaluate the role of reality and observation in ways that find striking parallels with ancient Eastern visions. At shrines in Varanasi, along the banks of the Ganges, devotees chant mantras in honor of Shiva and Shakti. At the same instant, in a modern laboratory, a scientist watches in wonder as an electron "decides" its state in response to observation.

Perhaps this is not a coincidence. Perhaps it is no coincidence that the West is discovering, through its scientific instruments, truths that the East has been contemplating for millennia through meditation and philosophy. As physicist and Nobel laureate Erwin Schrödinger argued:

"Truth manifests itself in different ways, but in the end it is one."

In other words, the universe is a bridge. The dance of Shiva and Shakti and the behavior of electrons remind us that, beyond cultural differences, there is a profound unity in the search for what is eternal and what is dynamic, what is wave and what is particle.

Entanglement: connection beyond time and space.

Among the most fascinating concepts in quantum physics is "entanglement," or "quantum entanglement." When two particles interact and become "entangled," their quantum state can no longer be described independently. Although separated by immense distances, the particles remain mysteriously connected: the change in the state of one is immediately reflected in the other. Albert Einstein called this phenomenon "*spectral action at a distance*" and considered it one of the greatest enigmas of modern physics.

Albert Einstein, while acknowledging the experimental truth of the phenomenon, appeared skeptical. For Einstein and many of his contemporaries, the idea of instantaneous influence between particles across the boundaries of space and time bumped against the fundamental principles of classical physics. However, since Alain Aspect's fundamental experiments in the 1980s, the reality of entanglement has been confirmed in the laboratory.

Entanglement has a fascination beyond science. It seems to recall ancient philosophical insights into cosmic interconnectedness. Prominent among these is the Hindu thought of Brahman, often described as the immense, invisible web that unites everything. According to the Vedas and Upanishads, the perceived multiplicity of the universe is an illusion (*Maya*), since in reality everything exists as part of a single indivisible entity.

An evocative image of this interconnection is represented by "*Indra's Net*," an ancient Vedic metaphor. Imagine an infinite net, suspended in space, where at each node is a perfectly polished jewel.

Each jewel reflects the whole net, and in each reflection are images of all the other nodes. Each part contains the whole, just as entanglement shows that two particles, although physically separate, exist as part of a single shared reality.

Science meets metaphysics.

Physicist David Bohm, a disciple of the great Robert Oppenheimer and a collaborator of Albert Einstein, proposed a model of the universe that seems to be in deep dialogue with the conceptions of Hinduism. According to Bohm, reality cannot be fully understood through fragmented analysis, but must be seen as a whole. Bohm proposed the idea of an "implicit order," a hidden underlying structure that contains and connects everything. Entanglement, in this framework, is not an isolated oddity but the tangible example of how the world is fundamentally interconnected.

Another scientist, Erwin Schrödinger, known for the very famous cat paradox, was among the first to recognize the philosophical importance of entanglement. Not surprisingly, Schrödinger had a deep knowledge of Vedic philosophy, from which he drew inspiration. He wrote that in quantum mechanics

"the world emerges as a network of interdependent events."

Words that, to an audience familiar with Indian philosophy, seem almost an echo of the ancient sacred texts.

Today entanglement is finding concrete applications in the field of quantum computing, with the development of computers and cryptosystems that exploit this very mysterious property of connectedness. Some of the most advanced experiments on entanglement have been conducted at CERN laboratories in Geneva and in China, where scientists have demonstrated entanglement on scales never before achieved, such as in Earth-orbiting satellites.

Yet, as innovative as these experiments are, they always return to a big underlying question: what is reality? Hinduism and quantum physics, in different ways, invite us to look beyond what appears. Science shows us that separation is, after all, an illusion. Eastern philosophies, for millennia, have been telling us exactly the same thing.

In a world where everything is seemingly divided-people, cultures, particles of matter-entanglement becomes a kind of vertiginous bridge between ancient mysticism and the most advanced science. This is a demonstration that, perhaps, at the heart of the complexity of the cosmos lies a truth as simple as it is profound: nothing is truly separate.

Quantum entanglement and the Vedic concept of cosmic unity.

During the 20th century, quantum physics radically transformed the way we understand the nature of reality. Among its most fascinating and controversial principles, we find the phenomenon of quantum entanglement. Two particles, even if light-years apart, can influence each other instantaneously-a reality so astonishing that it prompted Albert Einstein to call this interaction "spectral."

This concept-which challenges common sense and the classical idea of causality-was later validated by Alain Aspect's experiments in the 1980s. However, the intuition that the universe is an interconnected whole is not new. Many millennia earlier, the Vedic texts of Hinduism described the cosmos as an indivisible network whose beating heart is Brahman.

Quantum entanglement emerges from the foundations of quantum mechanics, a theory developed in the early decades of the 20th century by pioneers such as Werner Heisenberg, Erwin Schrödinger and Niels Bohr. Simply put, entanglement occurs when two particles, after interacting, become connected in such a way that the state of one depends instantaneously on the state of the other, regardless of the distance between them. Typically, two particles born from the same event are said to be "related," like a double quantum jump. If you measured the spin of an entangled electron, you would instantaneously determine the spin of its companion even if it were on the other side of the universe.

Einstein, although the father of the theory of relativity, found this concept difficult to accept. His view of the universe was rooted in local causality, that is, the belief that an event at one point in space cannot directly influence another far away without a physical intermediary. For this reason, he labeled entanglement as "*spooky action at a distance.*" But in 1964, Irish physicist John Bell proposed a concrete experiment to test entanglement. His experiments-verified decades later with increasingly sophisticated instruments-proved Einstein wrong. Entanglement is not an illusion: it is real.

Ancient Vedic philosophers, long before the advent of modern science, developed visions that today seem to anticipate some of the insights that quantum physics is beginning to uncover. At the heart of this philosophy, we find the concept of Brahman. In the Vedic view, Brahman is the ultimate, eternal, infinite and indivisible reality. It is described as the cosmic source from which everything emerges and to which everything returns, but also as the secret web that connects everything in the universe.

The experiments suggested by Bell marked a turning point in modern physics, showing that our universe is not simply governed by visible forces or local interactions. But what can these discoveries tell us about our relationship with the cosmos? Here an evocative dialogue between physics and Eastern philosophy opens up. Vedic texts remind us that the interconnectedness of everything is not just a scientific curiosity. It is a profound insight that spans millennia of human reflection.

For example, the Chandogya Upanishad, one of the fundamental texts of Hinduism, states:

"All that exists is Brahman" (Sarvam khalvidam Brahman).

This statement invites us to recognize the unity of every aspect of reality. Unity not as a metaphor, but as the essential nature of the cosmos. That same unity, reverberated in quantum entanglement, suggests that there is no real separation between what we are (the microcosm of our mind and body) and the vast universe we inhabit (the macrocosm).

It is fascinating to note how the insights of quantum physics, born in the laboratories of Western scientists with advanced instruments, echo the murmurs of Vedic sages who, thousands of years ago, meditated in the forests of ancient India. This does not mean that quantum physics validates Vedic philosophy or vice versa. But it does suggest that both, from different perspectives, seek to answer the same question: what is the deep nature of reality?

Vedic texts often push the reader to confront truths that challenge the duality of the rational mind. Both worlds, the quantum laboratory and the "*philosophical ashram*," invite us to explore the greatest mystery of all: we are all connected.

In these explorations, what emerges is not just a fascination with the technical mystery of physics or the mystical eloquence of the Upanishads. It is the possibility of seeing the human being as a bridge between the infinitely small and the infinitely large. Between entangled particles and the Brahman, a timeless truth is revealed: what is distant is, after all, always close.

The collapse of the wave function and the role of the observer.

Quantum physics has presented us with a startling reality: the subatomic world is not static and determined, but fluid and probabilistic. One of its most fascinating concepts is the "*collapse of the wave function.*" This phenomenon postulates that until a subatomic particle is "observed," it exists in an undefined state, a superposition of possibilities. When the observer intervenes, either with a measuring instrument or with his own eyes, this superposition disappears and the particle assumes a definite state. What, then, is the power of the observer? And, more importantly, what does it mean to "observe"?

The question is not only scientific. For centuries, Indian philosophers and yogis have wondered about the fundamental role of the mind in shaping reality. The parallels with the principles of Hindu meditation, particularly in Raja Yoga and the teachings of Patanjali, prove irresistible. Both, in their own way, suggest that consciousness is the pivotal tool for influencing what we perceive as real.

Erwin Schrödinger, one of the fathers of quantum physics, described the phenomenon of wave function collapse with the famous example of "Schrödinger's cat." Imagine a cat locked in a box, along with a mechanism that can release poison depending on the state of a subatomic particle. The particle can be in two states at once, bringing the cat into a theoretical condition in which it is both alive and dead. The system "decides" which condition becomes real (the live cat or the dead cat) only when an observer opens the box. The more scientists tried to answer this metaphor, the more the question arose: does the human observer, by his act of observation, determine this reality?

This phenomenon is known as the "observer effect." Even in academic circles, philosophers such as Eugene Wigner have speculated that mind itself plays a fundamental role in the evolution of the subatomic world. This premise inevitably leads back to dialogue with ancient spiritual traditions.

Yoga, Patanjali and inner observation.

Raja Yoga, as described in Patanjali's "Yoga Sutras," is based on introspection and observation of the mind itself. Patanjali, in explaining the steps of yoga, writes in a famous aphorism:

"Yogas chitta vritti nirodhah"
(*Yoga is the cessation of fluctuations of the mind*).

Here, the term "*chitta*" means mental consciousness, which is the instrument through which we perceive the world.

For Patanjali, thoughts are like waves on the surface of a lake. When these waves subside, the mind becomes a clear mirror reflecting reality without distortion. This state is achieved through focused meditation and breath control. The subject (the inner self) thus becomes the detached observer, capable of influencing its own perception of the world.

Yogis and sages have long held that the world we see is the result of impressions produced by our minds. This idea is closely related to the concept of "*Maya*" in Hinduism: the illusion of material reality. In other words, what appears solid, concrete and independent is, in fact, the "*collapse*" of our perceptions onto the fluid field of the possible. Consciousness shapes reality through a process of conscious observation, similar to what happens in the collapse of the wave function.

Let us consider another historical example. In 1973, Indian physicist Jagadish Chandra Bose, a pioneer in the study of electromagnetic waves, devoted himself to exploring the coincidences between scientific discoveries and Vedic principles. Bose, inspired by Indian scriptures, saw the universe as an interconnected whole, a kind of "quantum field" that resonated in synchronization with the observer. Many contemporary yogis followed his vision, suggesting that deep meditation not only modulates our consciousness but also interacts, on subtle levels, with the fabric of reality.

Physicist David Bohm, known for his theory of implicit order, also expressed ideas that overlap with Hindu spiritual insights. Bohm argued that everything exists in an implicit state of potentiality until interaction makes it explicit, just as the collapse of the wave function does.

In light of these reflections, the parallel between the observer in quantum physics and the introspective observer in yoga takes on new depths. Both confront a crucial truth: what is observed depends on the observer.

Quantum physics does not claim to explain consciousness, just as Vedic yoga does not purport to do science in the modern sense. Together, however, they offer us a fascinating insight: the universe is not an "outside" separate from our "inside." Meditation practices, such as Raja Yoga, train us in conscious observation, just as quantum physics reminds us that the observer is an integral part of what he or she observes.

Ultimately, both traditions confront us with a fundamental question: are we really just spectators of reality or, by observing it, do we also become its creators?

Quantum probabilism and Karma: uncertainty as cosmic law.

Quantum physics is not just a scientific revolution. It is a conceptual rupture that requires us to rethink the very foundations of reality. At the heart of quantum mechanics is its most puzzling trait: uncertainty.

For centuries, the laws of classical physics seemed solid and unchanging. According to Newton, a cause inevitably produces an effect with mathematical precision. But in the 20th century, physics shattered this certainty. With the advent of quantum mechanics, men like Werner Heisenberg and Niels Bohr demonstrated that, in the subatomic world, rules are dictated by probability and not certainty.

A pivotal concept in quantum physics is Heisenberg's uncertainty principle, formulated in 1927. This principle states that we cannot know with absolute precision both the position and momentum of a particle. At the microscopic level, therefore, the universe seems to play by the rules of chance, creating a kind of dynamic kaleidoscope in which nothing is completely predictable.

This "quantum uncertainty" found one of its most elegant formulations in the Copenhagen interpretation, advocated by Bohr. According to this view, the behavior of subatomic particles is not defined until they are observed. Before measurement, they exist as a superposition of probabilities. Einstein himself was skeptical. Yet decades later, the probabilistic model of quantum mechanics holds, explaining phenomena that classical physics could not even imagine.

Karma: a probabilistic system of reality.

The concept of Karma appears in the sacred texts of Hinduism, such as the Upanishads and the Bhagavad Gita. In the common idea, Karma is a kind of cosmic bank of actions: every good deed generates merit, while every bad deed brings a negative consequence. But Karma is not just a law of "*cause and effect.*" In Hindu thought, Karma does not involve a mechanical and immediate reaction. It acts as a complex web of consequences that depend on multiple factors.

According to the Upanishads, Karma manifests as a collective sum of possibilities. There is never absolute certainty about how or when the effects of a choice will manifest. To a contemporary reader, this conception of Karma is surprisingly reminiscent of quantum probabilism. Just as the state of a particle depends on its interaction with its observer, the consequences of Karma depend on context and a set of variables that elude complete predictability. Even in the Hindu world, then, uncertainty is a driver of becoming.

A powerful image from Hinduism that can illuminate this analogy comes from the symbolism of Shiva Nataraja, the "*Lord of Dance.*" His cosmic dance represents the continuous cycle of creation, destruction and regeneration of the universe. It is a dance of movement and transformation, in which uncertainty is essential to the flow of existence.

Just as Shiva's dance does not follow a linear, predetermined pattern, but responds to a cosmic rhythm of variations and adaptations, so quantum probabilism introduces a dynamic interpretation of reality.

Even physicist Fritjof Capra, in his famous book "*The Tao of Physics*" (1975), pointed out the parallelism between the "fluctuating" movements of subatomic particles and the cyclical nature of Eastern thought. It is no coincidence that Capra used precisely dancing Shiva as a symbol of the marriage of science and spirituality.

Another significant parallel concerns the individual responsibility of the observer. In the quantum domain, the act of observing directly influences the observed system. The famous mental experience of "Schrödinger's cat," proposed in 1935, illustrates how the state of a particle (and thus reality itself) is determined only when someone observes it.

Similarly, Hindu Karma emphasizes that individual actions influence the fabric of existence. Not in a fatalistic or predetermined way, but by a mechanism that leaves room for free will and co-creation of reality. Just as subatomic particles respond to interactions, so Karma invites us to consider each choice as part of a larger, unpredictable system.

While starting from seemingly opposite perspectives, quantum physics and the philosophical thought of Hinduism offer a fascinating view of uncertainty as a creative force. In the West, uncertainty is often experienced as a lack, almost an error to be corrected. Instead, in the Eastern view, it becomes an essential element of the cosmic process.

Quantum physics, with its elegant mathematics and abstractions, and the concept of Karma, with its millennia-old roots in Hindu spirituality, form a striking overlap. Both invite us to consider life as a web of possibilities, rather than a rigid, fixed path.

Quantum decoherence.

Quantum decoherence is a fundamental phenomenon in quantum mechanics that explains how and why a quantum system loses its quantum properties, such as superposition and entanglement, under the influence of the external environment. This process plays a crucial role in the transition from quantum behavior (which governs microscopic systems) to classical behavior (which we observe in the macroscopic systems of everyday life).

In quantum mechanics, a system can exist in a superposition of states, that is, in a linear combination of different possibilities (described mathematically by a wave function). However, when a quantum system interacts with an external environment, its wave function "blends" with that of the environment. This leads to the loss of interference between the system's quantum states, making it impossible to observe or exploit certain quantum effects such as superposition.

This loss of coherence (hence "decoherence") brings out classical behavior: the system begins to appear as if it were in a "defined" state, rather than a superposition of states.

Decoherence occurs when a quantum system interacts with its external environment, which may consist of molecules, thermal vibrations, electromagnetic fields, etc.

The stages of decoherence can be summarized as follows:

Initial quantum superposition.

At first, the quantum system may be in a superposition of states (e.g., an electron in two orbitals at once, or a "live and dead" cat as in Schrödinger's thought experiment).

Interaction with the environment.

When the system interacts with the environment, some information about its quantum state "leaks out." The environment acts as an involuntary "observer."

Loss of consistency.
Because of the interaction with so many degrees of freedom in the environment, the relative phases between quantum states are destroyed. This cancels interference effects, which are the distinguishing feature of quantum mechanics.

Emergence of "classical reality."
The system stops behaving quantum and begins to appear as if it were in a definite classical state.

Decoherence is closely related to the quantum measurement problem. When we observe a quantum system, we force it to "collapse" into a defined state. Decoherence is not exactly the collapse of the wave function, but it explains why, at the macroscopic level, we see the world as being made up of objects that exist in defined states, rather than in superposition.

So, decoherence explains how and why quantum systems appear classical when we observe them directly or when they interact with their environment.

In quantum computation, decoherence is a critical problem. Quantum information can lose coherence very quickly, destroying computation. Creating decoherence-resistant systems is one of the main challenges in quantum technology.

Decoherence still does not fully solve the "wave function collapse" problem, but it clarifies how classical properties emerge from quantum systems.

In summary, decoherence is the process whereby a quantum system loses its quantum behavior due to interaction with the environment, leading to the appearance of a classical world.

Decoherence and Maya: the illusion of concreteness in time and space.

Quantum introduces uncertainty and superposition. However, our everyday life appears stable and concrete to us. This phenomenon, which explains how apparent "concreteness" emerges from an underlying quantum nature, is called decoherence. Remarkably, this idea finds a fascinating parallel in the Hindu concept of Maya, illusion.

Decoherence is a key to understanding how we go from the mysterious to the visible. In the quantum microcosm, particles exist in superposition states: they can be, for example, in two places at once. However, when they interact with the environment (such as light, air molecules or even a measuring instrument), they lose this superposition. They "collapse" into an observable state: a defined object. This process, called precisely decoherence, creates the impression that we live in a classical, stable world, but it is only an impression.

In Hindu philosophy, perceived reality is referred to as *Maya*. The Upanishads, ancient scriptures dating back to the first millennium B.C., describe the physical world as a transitory illusion. According to these texts, what the senses perceive is not the ultimate reality, but a mask concealing the eternal truth of Brahman, or the infinite and immortal substance of the universe. This illusion, just as in decoherence, vanishes when the individual attains full awareness.

Why is decoherence important? Decoherence explains why our world seems orderly despite the underlying chaos. However, decoherence does not destroy superposition; it makes it invisible. This means that the quantum realm never disappears: it remains hidden as a substratum of infinite possibilities. Similarly, in Hindu philosophy, the Maya cycle is never completely destroyed. Illusory reality dissolves and recreates, in an eternal cosmic game of construction and disintegration. This is reminiscent of the concept of cyclic time set forth in the Vedas and Puranas, where the cosmos is born and dies in infinite periods called "*kalpa.*"

According to Hindu scripture, whenever a cycle ends, everything returns to the primordial chaos called prakriti. From this state, the cycle resumes. In modern physics, the universe may follow similar patterns. Some theoretical physicists like Roger Penrose suggest possibilities of cosmic cycles, where the Big Bang represents only one of many beginnings.

A pivotal concept in quantum physics is Schrödinger's famous cat paradox. Let us imagine a cat in a closed box. Until someone opens the box, the cat is alive and dead, that is, it is in a state of superposition. With observation, we get an answer: the cat is either alive or dead. Decoherence explains why in ordinary reality we never observe cats "simultaneously alive and dead." This subjective experience corresponds to what Maya represents in Hindu philosophy: the illusion of stability, where infinite potential states actually exist.

The connection with the Hindu concept of *Lila*, the cosmic game, is equally fascinating. According to Hinduism, the universe is a divine game, an endless dance of creation and destruction moved by supreme energy. Similarly, in the quantum world, particles seem to "play" at the edges of existence, disappearing and reappearing, entering and leaving our measuring instruments.

Decoherence reminds us that nothing is stable forever. Even what we perceive as concrete, such as the boundaries of the body or the solid ground beneath our feet, is born of a complex interaction that, over time, will unravel. This view echoes Hindu cosmology, where the world is temporarily constructed and then dissolves into Brahman. The Svetasvatara Upanishad reads:

> *"As thousands of sparks spring from a flame, so countless worlds arise and disappear from the eternal."*

In the same way, quantum physics suggests that the universe may come into being and collapse in ways that still elude our complete understanding. However, for both physicists and Indian sages, the lesson remains clear: what we see is only a partial representation of reality. Our task remains to explore the hidden meaning beneath the apparent solidity.

Decoherence and Maya show how science and philosophy can converge on universal questions. Whether reality is an illusion, as suggested by the ancient scriptures, or an emergent construction, as indicated by quantum physics, the message is the same. Not everything that appears solid and concrete really is. In the heart of matter, as well as in the depths of the mind, lies mystery. A mystery that invites us to explore, pushing beyond appearances, toward a greater, eternal and elusive truth.

Brahman and the quantum vacuum: the all emerging from nothingness.

Mankind has always sought to answer a fundamental question: where did the world come from? This quest, in the West, has followed two main paths, one philosophical and one scientific. In the East, however, these paths have never really separated. They are paths that have been intertwined for millennia in the works of sages and mystics, especially in the Hindu tradition. And today, thanks to quantum physics, such questions are coming to the fore again in a new light, reminding us how science and metaphysics can be two different languages to illustrate the same reality.

At the heart of quantum physics is a concept as powerful as it is puzzling: the quantum vacuum. At first glance, one might think that the "vacuum" represents nothingness itself. Yet, according to quantum field theory, nothingness is anything but empty. The quantum vacuum is a state of unlimited potential, a kind of invisible and omnipresent energy sea. From this sea arise subatomic particles that appear and disappear in a fluctuating cosmic game. English physicist Paul Dirac described this phenomenon in 1928 with his "*vacuum theory.*" He observed that the quantum vacuum is not really "empty," but a living and fertile matrix, capable of creating particles out of nothing and then making them disappear elsewhere.

This "emptiness," so strange when compared with common intuition, is reminiscent of Hinduism's reflections on Brahman, the universal cosmic principle that pervades everything. In the Upanishads, Brahman is neither existence nor nonexistence, but a reality beyond these categories. It is the principle from which everything emerges, that which remains invisible and indefinable, but which is the matrix of all form and phenomenon.

Ancient Hindu sages described Brahman as the metaphysical womb of all that exists. Ancient Hindu sages described Brahman as the metaphysical womb of all that exists. The evocative image contained in the Upanishads is that of a spider web suspended in a vacuum. From Brahman, as from this invisible web, infinite universes unfold, coming into being, touching existence and dissolving back out again. One of the best-known passages in the Upanishads states:

> *"As sparks spring from fire, so from Brahman worlds are born and everything returns to it."*

The parallels with quantum fluctuations-particles emerging from the vacuum only to disappear and return to its womb-are almost inevitable.

For quantum physicists, the quantum vacuum explains the behavior of particles and, by extension, the structure of the entire universe. For the masters of Hinduism, Brahman explains not only matter, but also man's consciousness and his relationship with the infinite. It is remarkable to see how modern scientific horizons and millennia-old philosophies intersect in describing a reality in which everything and nothing coincide.

Paul Dirac's work, along with that of pioneers such as Werner Heisenberg and Erwin Schrödinger, helped in the 1930s to redefine our understanding of reality. Schrödinger, in particular, was fascinated by Eastern philosophies, which he had encountered during his studies. In a 1925 letter, he wrote:

> *"Some of the most profound concepts of modern physics seem to find an extraordinary echo in Vedanta."*

And not surprisingly. Schrödinger found in the Upanishads the idea that the world we perceive is an illusory manifestation, a game of appearances that conceals a single deep and indivisible reality: Brahman, indeed.

Even in quantum physics, the macroscopic world we see is the result of a complex interweaving of probabilities and quantum waves; what appears solid at the human level is nothing more than an invisible dance of particles and fields. Dirac, studying particles in a vacuum, discovered that there are "antiparticles" with the same mass but opposite charge. A fundamental discovery that earned him the Nobel Prize in Physics in 1933. But aside from the prize, his investigations opened a philosophical door: the vacuum can create pairs of particles that cancel each other out as they manifest, in an eternal cycle of creation and destruction.

India, with its temples and sacred texts, has always seen the cosmos as the result of a delicate balance between order and chaos, manifestation and dissolution. This finds a surprising foothold in quantum physics. Underlying modern scientific thought is the idea of indeterminacy: no physical phenomenon, at the subatomic level, is ever completely predictable. Particles behave probabilistically, not deterministically. This is what Niels Bohr, another of the fathers of quantum physics, called "*the complementarity principle.*" Interestingly, in the Hindu tradition, we find something similar. Brahman expresses himself in the world through a dual reality: creative energy (*shakti*) and pure consciousness. It is a complementarity that allows the cosmos to exist.

Today we live in a world where science and spirituality are often placed in opposition. But perhaps it does not have to be that way. Quantum vacuum and Brahman show us that two ways of observing reality-one empirical, the other intuitive and philosophical-are not necessarily in conflict. Rather, they complement each other. The

Upanishads, with their poetic and mysterious verses, anticipate some questions that still challenge theoretical physicists today. Is it possible that the quantum vacuum is not really empty and that, like Brahman, it carries imprinted memories of infinite potential?

A statue of Shiva Nataraja at CERN.

At one of the most advanced sites of modern science, CERN in Geneva, a detail often catches the eye of visitors. At the entrance to the European Laboratory for Particle Physics, next to the complex accelerators that investigate the deep structure of matter, stands a statue of Shiva Nataraja, the dancing god of Hinduism. This statue represents the continuous cycle of creation, preservation and destruction, a central concept in Indian cosmology. What is surprising is not only the symbolic beauty of the work, but also its connections to concepts in quantum physics that are revolutionizing our understanding of the universe.

Shiva Nataraja is depicted dancing within a circle of flames. His body moves gracefully as his right foot crushes a demon, a symbol of ignorance. One hand holds a flame, the other a small percussion that marks the rhythm of creation. According to Indian philosophy, this dance symbolizes the dynamism of the universe: there is no stillness, only a continuous flow.

Until the 20th century, this view seemed distant from the ideas of classical physics, which conceived of the universe as a kind of predictable and rigid machine governed by immutable laws. With the arrival of quantum physics, however, everything changes. Physicists discover a microcosm in which particles do not behave as stable objects, but emerge and disappear, acting in unpredictable ways. Incredibly, this description seems to reflect the symbolism of Nataraja's dance: a ceaseless, creative motion that embraces chaos and order.

Quantum energy and virtual particles.

One of the central phenomena of quantum mechanics is the concept of vacuum energy. In the quantum vacuum, "nothingness" is not really empty, but a sea of invisible activity. Particles called "virtual" briefly emerge from nothingness and then disappear, in what appears to be an ephemeral dance of creation and destruction. This process, confirmed by experiments such as the Casimir effect, is strikingly analogous to the cosmic cycle described in Hindu mythology.

Scientists at CERN explore this very realm inaccessible to human senses, exploiting particle accelerators such as the Large Hadron Collider (LHC). In 2004, when the statue of Shiva was donated from India to the lab, physicist Fritjof Capra called that statue:

> *"a perfect symbol for the synergy between science and spirituality."*

Fritjof Capra, who has devoted his career to exploring the parallels between Eastern thought and modern physics, argues that Shiva's dance represents quantum energy in an extraordinary way: it is an incessant, dynamic and creative energy.

Another aspect of quantum physics that echoes Indian philosophy is the principle of non-locality. This principle, highlighted by quantum entanglement, describes how two particles can remain connected immediately, even when separated by immense distances. This "action at a distance" defies common sense and the insights of classical physics.

Again, the connection to Eastern thought is fascinating. In Indian philosophy, the concept of universal interconnectedness (the "*all is One*") permeates ancient texts such as the Upanishads. According to these scriptures, there is no real separation between the elements of the universe, only an invisible network that binds everything together. Quantum physics, with its experiments demonstrating quantum entanglement, now makes scientifically concrete a principle that Hinduism has intuited for millennia.

The statue of Shiva Nataraja at CERN is thus not just an ornament or a cultural tribute, but a powerful symbol. It embodies the idea that reality, in both science and spirituality, is moved by complex and interconnected rhythms. As physicists study subatomic particles and probability waves, they rediscover the mystery of the universe as a fluid and unpredictable dance.

In conclusion, Shiva Nataraja's presence at CERN's entrance reminds us that science and culture are not worlds apart. The ultimate goal of both, after all, is to unravel the mystery of the cosmos, to question its origins and to grasp that dynamism which, in ancient Hindu symbolism, was already encapsulated in the steps of a dancing god. As in a dance, scientific discovery and philosophical reflection intertwine, uniting past and future in an endless circle.

"Parallel worlds" in quantum physics and Hindu cosmology.

In a famous paper published in 1957, Hugh Everett presented an idea destined to revolutionize modern physics: the "*Many-Worlds Interpretation.*" According to this theory, every quantum measurement produces a bifurcation of reality, generating a potentially infinite number of parallel universes. Each time a quantum phenomenon is observed, the universe "splits," creating separate realities, each with a different outcome of the observed event. For 20th century science, this view was highly speculative. Yet, on closer inspection, the idea of a multiplicity of universes was not new.

In ancient Hindu cosmology, even thousands of years before Everett, the existence of parallel cosmic dimensions, countless worlds and infinite cycles of time were conceived. This fascinating correspondence between quantum physics and Indian religious texts is not simply a coincidence. Rather, it is a sign of how the human mind, in different cultures and contexts, has tried to grapple with the mysteries of reality, formulating insights that prove more relevant today than ever before.

The multiverse of quantum physics.

Quantum physics has broken the traditional paradigm of classical determinism. In the realm of the infinitely small, that of subatomic particles, the laws governing ordinary reality cease to apply. Electrons behave both as particles and as waves. Photons can exist in multiple states simultaneously, according to a principle known as "*quantum superposition.*"

In the traditional interpretation of quantum physics, called the "*Copenhagen interpretation,*" a particle in a "probabilistic" state collapses into a definite state when its behavior is observed. Everett, however, proposed an alternative explanation: the particle does not choose only one state. All possible states of the particle are realized simultaneously, but in distinct universes. This implies that reality is not unique, but multiple. Each decision, each quantum event leads to a new universe, which continues to exist in parallel with the others.

In Hindu thought, reality is conceived as a complex, layered and infinite system. According to the Puranas, the sacred texts that recount the cosmology and myths of the universe, there are innumerable worlds or "*lokas.*" These worlds can be divided into three main categories: the upper worlds (svarga loka), the middle worlds (martya loka) and the lower worlds (patala loka). Each loka is a distinct plane of existence, inhabited by different beings, with its own physical and temporal laws.

The Vishnu Purana, one of the foundational texts, describes how the universe is cyclical and infinitely manifold. Each universe is born, lives and dies in an eternal process of creation and destruction, known as the cycle of "*kalpa.*" Vishnu, the conservative god, sleeps on the cosmic ocean, and from his breath endless universes emerge, like bubbles forming and dissolving in water. This concept is strikingly reminiscent of the modern image of the quantum multiverse, where universes coexist as multiple bubbles of reality, each governed by its own laws.

The Bhagavata Purana also elaborates on this view. In one of its most famous passages, it is told that Krishna, a divine manifestation, revealed to his devotee Arjuna a cosmic image of infinite realities, each representing a different universe. Krishna states that existence is composed of "*innumerable worlds within worlds*." This expression recalls the modern idea of the multiverse.

Parallels between modern science and ancient cosmology.

Both Everett's interpretation of quantum physics and Hindu cosmology challenge our intuition of a single objective reality. Both systems recognize that the universe may not be unique, but rather one of many.

Interestingly, in the Hindu conception, each *loka* is not only physically separate, but can follow a time completely different from our own. Some worlds have an accelerated time, others a slowed down time. This concept can be likened to the "*proper times*" observed in Einstein's general relativity, another field that has offered insights into the multiverse discussion.

A particularly evocative example can be taken from the ancient city of Dvaraka, Krishna's kingdom on Earth. According to the Mahabharata and the Bhagavata Purana, Dvaraka was an earthly kingdom but with a direct connection to higher dimensions. After Krishna's departure, the city was submerged in the ocean, becoming unreachable and invisible to humans. Today, some archaeologists believe they have located submerged remains off the Indian coast (in Gujarat), but the legend of Dvaraka continues to evoke the possibility of a "parallel" world, an intermediate space between human and divine.

The similarity between Hindu conceptions and the multiverse hypothesis suggests that ancient cosmological insights are not just mere myths, but powerful metaphors to describe the complexity of existence. Joseph Campbell, a celebrated scholar of myths, wrote that

"Religions are codes for teaching mystery."

Perhaps what the sacred texts were intended to teach is no different from what quantum physics discovers today.

In both cases-in Everett's theory of multiple universes or Vishnu's infinite cosmic bubbles-what emerges is a concept of infinity. Reality is not determined; rather, it is a play of possibilities, an endless creative space. As we read in the Bhagavad Gita,

"In the dream of creation, the whole universe appears and disappears like a reflection in water."

Quantum physics and Hindu cosmology, though born in very distant historical and cultural contexts, seem to converge on a fundamental point: reality is more vast, mysterious and multifaceted than we perceive with our senses. Everett's words find an echo in the ancient Hindu scriptures, and remind us that human thought, whether through science or spirituality, has always sought to understand the infinite. The allure of the multiverse compels us to explore these worlds, real or imagined, for, as the poet Rumi stated,

"If one universe closes, a larger one opens."

Quantum consciousness and the Atman: who observes the observer?

There is a moment in the history of human thought when science and philosophy seem to intertwine, like threads in a larger web that attempts to explain the universe and our place within it. On the one hand, quantum physics, with its implications bordering on the incredible, and on the other hand, the ancient sapiential traditions of the East, particularly Hinduism. What is surprising is not so much the fact that such a modern science and such an ancient philosophy can dialogue, but the way in which they do so. The dialogue develops from similar questions that seem to arise from a deep shared insight. One such question, perhaps the most fascinating, is: Who observes the observer?

The measurement problem in quantum physics.

Quantum physics was born in the early 20th century as a response to the limitations of classical physics. Giants like Max Planck and Erwin Schrödinger opened the door to a strange and counterintuitive universe where the rules of common sense crumble. One of the key concepts that have shocked scientists is the measurement problem.

What it means. In the quantum world, particles such as electrons or photons do not exist in definite positions. Instead, they live in a state of superposition, as probability waves. It is only when an "observer" (someone or something that measures) intervenes that these waves collapse into a concrete reality. This principle, known as "*wave function collapse*," is one of the key notions of quantum physics. But what remains unexplained is: who or what causes the collapse?

This is where the problem of consciousness comes in. Many physicists and philosophers have wondered whether the human mind itself, in its act of observing the world, plays a role in the creation of reality. Roger Penrose, a British physicist and mathematician, along with anesthesiologist Stuart Hameroff, has theorized that consciousness might arise from quantum processes within the microtubules of brain cells. If so, the reality we know could be woven from a deep and invisible link between mind and matter.

Atman and "*the observing Self*": an ancient Hindu insight.

While science investigates the role of the observer in physics, ancient Hindu scriptures, particularly the Upaniṣhad, had already addressed a similar question, but through a spiritual lens. In Hinduism, the concept of Atman represents the supreme, unchanging Self, the deepest core of every being. Atman is not simply the individual consciousness, but the universal consciousness that, according to Hindu philosophy, observes and sustains the phenomenal world.

In the Upaniṣhad, a famous dialogue illustrates this idea.

This is the discourse between the young Śvetaketu and his father, Uddālaka. This dialogue, contained in the Chāndogya Upaniṣhad (VI.12.3), introduces the metaphor of the seed of the nyagrodha (fig) tree. Uddālaka asks his son to break a seed and observe what is inside. When Śvetaketu replies that inside the seed he sees "nothing," his father reveals to him:

"In that subtle, invisible essence resides the principle that sustains the great tree. The same principle, 'Tat tvam asi,' is also in you."

"Tat tvam asi" can be translated as "You are that."

This phrase, one of the most famous in Hinduism, implies that the ultimate observer, the deep self, is at one with ultimate reality. There is a silent connection between individual awareness and that cosmic intelligence that governs the universe. An idea, this, that is not so far removed from the quantum question about the role of the observer in "creating" reality.

Not surprisingly, some of the world's greatest physicists have found these parallels fascinating. Erwin Schrödinger, father of the equation of the same name and pioneer of quantum theory, was deeply influenced by Eastern philosophies. Schrödinger was familiar with the Hindu concept of Atman; in fact, he wrote:

"Consciousness is singular, pluralism is an illusion. This is the very ancient Indian thought."

More recently, physicist and science popularizer Fritjof Capra explored the similarities between modern physics and Eastern mystical traditions in his book "*The Tao of Physics*." Capra notes that the principles of quantum physics, with their interconnections and lack of separation between observer and observed, reflect a "universal vision" similar to that of ancient Eastern philosophies.

We can then ask: Is human consciousness just an epiphenomenon of the brain, as a materialistic view suggests, or is it part of a bigger picture? Quantum physics, despite its technological achievements, has not yet resolved this dilemma. Hindu philosophy, on the contrary, accepts that consciousness-the Atman-is the root of all existing.

However, these two approaches have something in common: humility in the face of mystery. A mystery that invites us not only to explore the depths of matter, but also those of mind and spirit. Perhaps, as the ancient Indian philosopher Nagarjuna suggests,

"There is no detachment between the knower and the known. Both express the same intimate reality."

What if the observer is not only a curious mind, but also a spark of the very universe he or she desires to know?

Dialogue between science and philosophy.

Quantum physics, with its revolutionary theories, has redefined the very concept of reality. It is a science that, by its very nature, asks profound questions about the foundations of existence. It is here that begins a fascinating dialogue between modern physics and Eastern philosophies, particularly Hinduism.

However, we need to be clear. The relationship between these two dimensions is not one of equivalence or overlap. Science and philosophy operate on separate planes. Quantum physics builds models based on mathematics and experimental verification. Hinduism, on the other hand, reflects on the nature of reality through meditation and intuition. Yet what these disciplines share is an extraordinary insight: what we perceive as "real" is but the surface of something much deeper.

Before the 20th century, classical physics described a stable, predictable universe governed by rigid laws. Isaac Newton, with his universal gravitation, had laid the foundation for a mechanical, deterministic world: every event had a precise, understandable cause. But everything changed with the advent of quantum theory.

At this point, the connections with Hinduism become extraordinary. The Vedic sages, authors of the Upanishads, had already reflected on the impermanence of the perceivable world. According to Hindu philosophy, what we see and touch is maya, an illusion, a shadow of the true underlying reality. The tangible universe, the macrocosm and microcosm, conceal an ultimate essence, Brahman, or the absolute, eternal and indivisible.

A parallel emerges in quantum physics. Here, too, the observed world hides deeper levels, where material objects dissolve into a lattice of probabilities, relations and interconnectedness. For example, the phenomenon of quantum entanglement has elicited comparisons with the idea, central to Vedic philosophy, of an inextricably interconnected universe,.

However, it is crucial to emphasize that physics is not philosophy. Scientists use mathematical tools and technological verification to explore the physical world. The sages of Hinduism, on the other hand, sought answers through introspection. But the common insight is fascinating: the ordinary world we perceive may be only a superficial manifestation of something deeper and more indivisible.

Great thinkers between science and philosophy.

Some leading figures in the quantum revolution sought meaning in dialogue with Eastern traditions. Robert Oppenheimer, a celebrated nuclear physicist, was deeply influenced by reading the Bhagavad Gita, one of the fundamental works of Hinduism. During the first atomic bomb test on July 16, 1945 in the New Mexico desert, Oppenheimer uttered a sentence from the Gita:

"Now I have become Death, the destroyer of worlds."

This reference was not just a quote, but a reflection on the devastating effects of knowledge applied to the real world.

Another example is Fritjof Capra, Austrian physicist and writer, who in his book "*The Tao of Physics*" explored the parallels between modern scientific thought and Eastern spirituality, suggesting that both share a deeply unified and interconnected worldview. However, Capra emphasized that his intent was not to confuse such different fields, but to draw inspiration from each other's languages.

A common point of reflection between quantum physics and Hinduism is the inherent limitation of human perception. Physicist Niels Bohr, one of the fathers of quantum theory, claimed that the models of physics never represent reality, but simply the way we observe it. Similarly, Hindu philosophy holds that the ordinary individual, trapped in the realm of Maya, is incapable of perceiving the absolute until he or she attains enlightenment.

While physicists study nature with increasingly sophisticated tools, Hindu sages taught that awareness is the main tool for accessing ultimate truth. These are, therefore, two distinct paths, but both remind us how limited our knowledge is and how great the mystery of the universe is.

Quantum physics and Hinduism offer worldviews that, while starting from different assumptions, find a curious affinity in their questions. This is not science confirming religion, nor is it ancient wisdom anticipating modern models. Instead, the dialogue between these disciplines represents an opportunity for scientists and philosophers to find mutual inspiration. Perhaps, as Carl Sagan, the great American astrophysicist, suggested, the investigation of the universe, with a look at the stars or our deepest selves, allows us to understand that

"we are the way through which the cosmos knows itself."

Quantum physics invites us to reconsider our understanding of reality by eliminating the rigid categories of space, time and causality. Hindu philosophy, with its millennia-old wisdom, already reached a similar conclusion through very different paths. Rather than identifying a single and absolute truth, scientists and philosophers invite us to contemplate the inherent mystery of the universe. Perhaps, after all, mystery is what unites us all, in science as in spirituality.

Chapter II. Hinduism, philosophy and spirituality.

Origins and historical development of Hinduism.

Hinduism is one of the world's oldest spiritual and philosophical traditions. With roots dating back more than 4,000 years, Hinduism is more than just a religion. It is an articulated system of beliefs, practices, stories and mythologies that has gradually evolved over millennia. Traces of its origins can be found in the sacred texts, *the Vedas*, true cornerstones of Indian spirituality. These texts not only provide insight into early ritual practices, but also suggest a vision of the cosmos that continues to speak to contemporary human beings.

The Vedas: to the origins of Hindu spirituality.

The Vedas are the key to understanding the historical context of early Hinduism. The Vedas were written in the Sanskrit language about 1,500 years before Christ.

These texts represent one of the oldest literary records of humanity. The term "Veda" means "knowledge" or "knowing." These texts were transmitted orally from master to disciple through complex mnemonic methods, so much so that they are thought to have remained unchanged for centuries before being put in writing.

The Rigveda is the oldest of the four Vedas. It is a collection of hymns dedicated to nature deities such as *Agni* (the god of fire) and *Indra* (the god of rain and war). The hymns testify to the Vedic peoples' strong connection with nature and cosmic forces. An interesting insight is already found in these texts: the universe is perceived as a harmonious whole. This concept, which will find further development in later centuries, establishes a spiritual bridge between the ancient Vedic vision and modern scientific reflections, such as those on quantum physics.

The Upanishads: the transition from ritual to philosophy.

Around the 6th century BCE, a significant transformation redefined the Indian spiritual landscape. While the Vedas emphasized rituals as a means of maintaining cosmic order, the Upanishads, a series of later texts, introduced a more philosophical and contemplative dimension. In these texts, the focus shifts from the outside (rituals, sacrifice) to the inside: self-knowledge becomes the heart of the spiritual quest.

The Upanishads are also known as Vedānta, a term meaning "*end of the Vedas*." Not surprisingly, they represent the culmination of Vedic thought. Among the key concepts of the Upanishads are Brahman and Atman. Brahman is the universal principle, the ultimate reality from which everything derives and to which everything returns. Atman, on the other hand, is the inner self, the divine spark present in each human being. The famous statement "*Tat Tvam Asi*" (You are that), taken from the Chandogya Upanishad, encapsulates this view: the individual and the cosmos are not separate, but deeply united.

Historical transformations and the great essays.

Hinduism is not the product of a single founder, unlike other religious traditions. It is the result of a cultural and spiritual evolution involving many civilizations and people over the centuries. A crucial aspect of this tradition is the inclusive criterion by which it has absorbed outside influences without ever losing its own identity.

Among the great sages who interpreted and promoted Hindu thought we find Yajnavalkya. He was a central figure in the development of the ideas contained in the Upanishads. According to a famous anecdote reported in the Brihadaranyaka Upanishad, Yajnavalkya faced a philosophical dispute in which he expressed the view of Atman as that which survives death and at the same time transcends all duality. This approach was revolutionary, as it shifted the focus from outward practices to deep reflection.

The connection with the present world.

The origins and development of Hindu philosophy are not only about the past; they also offer interesting insights for the present. In an age when quantum physics invites us to rethink reality as an interconnected mosaic of energy and potential, the extraordinary Hindu insight into the link between microcosm and macrocosm seems remarkably contemporary. How far the ideas of Brahman and Atman can remind us of concepts such as entanglement and the fundamental unity of particles is still a matter for investigation and discussion.

Places like Benares (now Varanasi), considered the "*eternal city of the gods*," keep alive the memory of Hinduism's origins. On the banks of the Ganges, a sacred river at the heart of Hindu consciousness, lies the perfect synthesis of antiquity and actuality. In these places, the philosophy of the sacred texts finds a concrete dimension in the daily spirituality of millions of devotees.

In conclusion, to dive into the historical origin of Hinduism is to get in touch with one of the most fascinating keys to understanding reality. From the Vedas to the Upanishads, ancient Indian wisdom still stands today not as a relic of the past, but as a powerful lens for understanding our time and, perhaps, our place in the universe.

Hinduism: a wisdom that spans millennia.

When we talk about Hinduism, we enter an ancient and complex universe. It is the oldest living religion in the world. Its roots go back more than three thousand years, intertwined with the history of the Indus Valley and the Aryan civilizations. This long journey has given rise to a multifaceted tradition that is not limited to religion, but expands into philosophy, art and the daily lives of millions of people.

Hinduism does not have a specific founder or a well-defined moment of birth. Its essence was formed gradually, accumulating layers of meanings and sacred texts through the centuries. This dynamism makes it unique. It is not a monolithic religion, but rather a continuous interweaving of practices, beliefs and worldviews. Its strength lies precisely in its ability to evolve and absorb ideas without losing its core.

The philosophical roots of Hinduism lie in the Vedas, the oldest known texts of the Hindu tradition.

Written in the Vedic period (about 1500-500 B.C.), the Vedas are four collections of hymns and sacred formulas: the Rigveda, Yajurveda, Samaveda and Atharvaveda. The Vedas are considered divinely inspired and form the foundation of the entire Vedic tradition. They contain hymns to the gods, solemn rituals and directions for maintaining harmony between the human and cosmic worlds.

A particularly fascinating example is found in the Rigveda, which includes the famous "*Hymn of Creation*" (*Nasadiya Sukta*). In it, we find a stupendously ambiguous reflection on the origin of the universe:

"Where did this creation come from? Is it perhaps the god on high who knows? Or does he not?"

A profound question that seems to resonate with modern questions in quantum physics.

At the origins of the eternal: the birth of the Vedas and the dawn of Indian philosophy.

Hinduism is called the oldest religion in the world. But the term itself, "Hinduism," is a modern label that embraces a complex web of beliefs, practices, sacred texts and philosophical traditions developed over millennia of history. To understand the roots of this immense spiritual legacy, we must go back in time, between 1500 and 500 B.C., to Vedic India, a mythical and extraordinary era in which we see the Vedas, the foundations of Hinduism and one of the earliest written records of human thought, emerge.

Vedic civilization flourished in the region of the Indian subcontinent, on the banks of the Sarasvati and Ganges rivers. That population was a nomadic, pastoral society bound to the cycles of nature, whose people, the Arya (a term meaning "noble"), left as their legacy a body of sacred texts known as the Vedas, a word derived from the Sanskrit "vid," meaning "knowledge." The Vedas are not simply religious documents: they are poetry, cosmology, ethics, philosophy and a practical guide to life. Through songs and hymns, they sing the relationship of the human microcosm to the macrocosm of the universe.

Among the four Vedas-the Rigveda, Samaveda, Yajurveda, and Atharvaveda-a place of honor goes to the Rigveda, the oldest and most important. Dated between 1500 and 1200 B.C., the Rigveda collects more than a thousand hymns, composed in a rigid language steeped in symbolism, celebrating celestial deities such as *Agni* (the god of fire) and *Indra* (the god of war and rain). These hymns are an opening into the Vedic world, dominated by a deep sense of the sacred that permeates every aspect of life.

"Rta" meaning order, harmony.

The heart of Vedic philosophy lies in the concept of "Rta," a word meaning "order," "truth," or "harmony." Ṛta represents the universal principle that governs both the cosmos and human action. According to this view, the sun rises every day, the seasons change and the waters flow because of the sacred order of Rta. But Rta is not only the law of the universe: it is an ethical dimension. Human beings, to live in harmony, must follow this law through observance of Dharma (moral duty) and sacred rituals.

This sense of order is not imposed by a capricious deity but is woven into the fabric of reality itself. It is a vision that anticipates, in many ways, the principles of the interconnectedness of the whole that characterize some modern interpretations of quantum physics.

"Ṛṣi": the seers inspired by nature.

Who were the authors of the Vedas? They were not mere poets or priests. The Vedas were composed by figures known as Ṛṣi, sages considered "seers" who could perceive and hear (*śruti*) the eternal truths revealed through the cosmos. Their visions were inspired by nature: the infinite horizon of the sky, the crackling of fire, the flow of rivers. In a metaphorical sense, it was thought that these sages did not invent anything, but received the Vedas directly from divine forces.

Many stories in Vedic tradition tell of enlightened sages who lived in deep meditation in forests, places considered sacred and imbued with spiritual energy. One such Ṛṣi, the legendary Vasistha, is connected to the concept of sacred fire (*yajña*) and meditation, and is celebrated for his devotion to Rta.

"Yajña": sacred fire as the link between man and the universe,

The fundamental rite of Vedic civilization is the "yajña," a complex fire sacrifice that holds a definite spiritual and social meaning. Fire, in the Vedic view, is not only a natural element but a mediator between earth and heaven, between human beings and deities. During yajña, people offered milk, clarified butter (*ghi*), grains and other gifts to the flames, reciting Vedic hymns with the knowledge that they were sending a heavenly message to the deities.

Yajña was not an individual act, but a collective one. It brought the entire community together in a common spirit of cooperation, dominated by the belief that sacrifice maintained the harmony of the cosmos through the power of Rta. It was a time when the individual perceived himself as an integrated part of a larger universal scheme.

The hymn of creation: the mystery of the origin of the universe.

One of the most famous passages in the Rigveda is the Hymn of Creation (*Nasadiya Sukta*), an enigmatic composition that questions the origin of the universe with incredible philosophical depth. The hymn reads:

> *"Was there Being or Non-Being? Only He knows, or perhaps not even He."*

This extraordinary question, addressed not to certainty but to mystery, has fascinated philosophers for centuries. The idea that the universe could have arisen from both the whole and nothingness is reminiscent of some key concepts in modern quantum physics, such as the quantum vacuum, which is never truly "empty" but teems with possibilities.

Vedic India has been the cradle of thought that does not separate the sacred from everyday life: nature is divine, the cosmic order is ethical, and man is part of a great whole. The Vedas, and especially the Rigveda, are not only a starting point for Hinduism: they are a universal record of how humanity has reflected on the fundamental questions of existence.

This ancient knowledge, transmitted orally for centuries before being put in writing, continues to inspire. To look to the Vedas is to immerse oneself not only in Indian history but in a cultural and philosophical heritage that has influenced the development of many Eastern spiritual traditions. And even today, amid the mysteries of subatomic particles and questions about the nature of the cosmos, the Vedic verses still seem to whisper, mysterious and modern: "*Only He knows, or perhaps not even He.*"

The Upanishads, deeper spiritual and philosophical dimensions.

Over time, religious reflection shifted from communal rituals to a more introspective and philosophical perspective. Thus was born the Upanishads, the texts composed between 700 and 300 B.C., which mark an extraordinary turning point in Hinduism. The Upanishads gradually abandoned the ritualistic aspect of the Vedas to explore deeper spiritual and philosophical dimensions.

In one of the most famous Upanishads, the Chandogya Upanishad, we find the phrase, "Tat Tvam Asi" (You are That). This expression sums up the idea that the individual soul (Atman) is nothing but a manifestation of the universal reality (Brahman). A principle that, in a symbolic key, seems to anticipate the concepts of interconnectedness and oneness that emerge in modern physics.

Parallel to the Vedas and the Upanishads, two giant epic poems emerged: the Mahabharata and the Ramayana. These texts were composed between the fourth century BCE and the third century CE. Mythology, philosophy and ethical teachings are mixed in these texts. The Mahabharata, with its more than 100,000 verses, is the longest epic poem in human literature. Within it we find the Bhagavad Gita, one of the most revered texts in Hinduism.

The Gita is a conversation between Prince Arjuna and the god Krishna in which spiritual, moral and philosophical issues are discussed. Krishna teaches that duty (*Dharma*) must be performed without attachment to results. A lesson that, to this day, inspires not only devotees but also modern thinkers and leaders, such as Mohandas Gandhi, who called the Bhagavad Gita

"my beacon in times of darkness."

Chronology and places of Hinduism.

Geographically and historically, Hinduism took root in what is now the Indian subcontinent. Sacred places, such as Varanasi (the city on the banks of the Ganges considered the spiritual center of the universe), are still important to those who follow Hinduism. According to an ancient tradition, Varanasi was founded by the god Shiva, one of the main gods of Hindu tradition. Here, people not only pray but also reflect on the impermanence of life by observing cremations on the banks of the Ganges.

The Gupta period (ca. 320-550 CE) also played a crucial role in the development of Hinduism as a complete cultural system. It was during this era that many temples were built, strengthening the link between faith, art and architecture. The Khajuraho temple, built later (9th-10th centuries AD), illustrates with its sculptures a unique understanding of spirituality, where the divine is intertwined with earthly life in a symbolic and refined interplay.

Hinduism is not just a religion, but a mirror reflecting the history and culture of a millennia-old civilization. Its ability to reconcile rituals, philosophy and a spiritual view of the universe has made it a beacon for many scholars, travelers and thinkers. In a world where modernity often clashes with the past, Hinduism shows that it is possible to integrate tradition with innovation.

Rediscovering its origins is not just a journey through history, but an opportunity to reflect on big questions that continue to fascinate humanity: Who are we? What is reality? Can we really understand the laws that govern the universe? These questions, born of the Vedas and deepened by the Upanishads, are the same questions that today fuel a new era of dialogue between Eastern philosophy and science.

Brahmanism and the search for the absolute.

In the evolution of Hinduism, the shift from Vedic ritual religiosity to the philosophical thought of Brahmanism marks one of the most significant turning points in Indian spirituality. This process, which unfolds between the texts of the Brahmanas and the Aranyakas (8th-6th centuries BCE), sees the gradual overcoming of ritualistic materialism in favor of a deeper reflection on the meaning of reality and human existence. The central concept of this turn is Brahman, the absolute and indivisible principle, which is established as the core of Indian metaphysical speculation.

In the Vedic phase, religiosity was anchored in a complex system of rituals (yajña) designed to preserve the cosmic order (Rta) and ensure the prosperity of the clan. Priests (Brahmans) performed elaborate sacrifices, convinced that these practices had power capable of influencing the universe. But over time, this ritual religion began to show limitations. The effort then shifted from form to meaning: the question was no longer how to perform the ritual, but why the ritual could work.

This change is reflected in the Brahmana texts. Here, the priests begin to seek inner connections between man, the universe and rituals. Ritual takes on symbolic significance: it represents a microcosm of reality. However, deeper speculation takes shape in the forest texts (Aranyakas), where the search becomes more personal and mystical.

In these texts, ascetic practices and isolation in the silence of nature become the means to access a higher dimension of knowledge.

The notion of Brahman arose as an answer to the fundamental questions that were emerging in Vedic society: what is ultimate reality? What is the principle that unifies all that exists? In the Brahmanas, Brahman is still linked to ritual, the invisible power that makes it effective. But in the Aranyakas and early Upanishads, Brahman is emancipated from this context, becoming an absolute, infinite and indivisible reality.

The mystic Yājñavalkya, one of the central philosophers of the Brihadaranyaka Upanishad (dated around 700 B.C.), offers an illuminating example. Yājññavalkya discusses with his disciples and his own wife Maitreyi, exploring the meaning of Brahman. In a memorable dialogue, he states:

> *"Everything that is precious is precious not for itself, but for the Atman."*

Here Yājñavalkya introduces a crucial link between Atman (the deep Self of the individual) and Brahman. This link reveals a universal truth: the human soul is not separate from the absolute, but is itself part of it.

The forest as a place of transformation.

A central symbolic element in the rise of Brahmanism is the aranya, the forest. In an increasingly stratified and urbanized India, the forest becomes the place of the inner quest. The forest rituals of the Aranyakas mark a detachment from the material world, social chaos and ritualistic strictures. Solitude fosters introspection, and silence leads to a higher understanding of Brahman.

Asceticism, which already had roots in the Vedic period, takes on a new dimension at this stage. Sages (*rishis*) begin to practice meditation and breath control techniques to achieve higher states of awareness. This intuitive and mystical approach merges with early philosophical reflections, initiating a fruitful dialogue between meditation and speculative logic.

A significant example is that of Venerable Uddālaka Aruni, another master of the Upanishads. In his famous dialogue with his son Śvetaketu, Uddālaka explains the concept of Brahman with a poetic image:

"Just as looking at the seed of the fig tree you cannot see the fig tree, so out of the whole (seemingly empty) emerges the unique and omnipresent being."

This teaching emphasizes the ineffable character of Brahman.

The insights of the Brahmanas and Aranyakas did not remain confined to the distant past, but laid the foundation for much of Indian philosophy and spirituality. The concepts of Brahman and Atman, refined and expanded over time, have continued to influence schools of thought such as Advaita Vedanta, which proclaims the absolute identity between Self and Absolute.

Even today, the notion of Brahman permeates Indian spirituality. The idea that there is a unifying reality, invisible to the eyes but perceptible to the mind and heart, continues to inspire not only mystics and philosophers, but also scientists and thinkers who seek connections between the visible and invisible universe.

Brahmanism represents a crucial moment in the history of Hinduism. It was the shift from ritual action to philosophical and mystical meditation. This transformation not only shaped Indian spirituality, but left a lasting legacy that is still explored today by those who seek to understand the mystery of existence. As Yājñavalkya said:

"He who knows Brahman becomes Brahman himself."

This is a quest that continues, between myth and philosophy.

The Upanishads, the culmination of Vedic thought.

The Upanishads, the centerpiece of Vedic and Hindu philosophy, represent one of the cornerstones of humanity's spiritual thought. Their name, which literally means "to sit near," evokes the image of a disciple standing next to his or her teacher, ready to receive profound and revelatory teachings. Composed between the 8th and 4th centuries B.C.E., at a time of great spiritual transformation in India, these texts mark the decisive shift from a ritualistic religion, based on Vedic sacrifices, to a philosophy of introspection and universal inquiry.

One of the most striking anecdotes from the Upanishads comes from the Chandogya Upanishad, where a master, Uddalaka Aruni, teaches his son Svetaketu the meaning of the identity between Atman and Brahman. Uddalaka invites his son to dissolve salt in a container of water and then drink it. Svetaketu cannot see the salt but clearly perceives its taste. The master then explains:

"Similarly, Brahman pervades all that exists. Although invisible, its presence is evident everywhere."

This simple but profound example has spanned the millennia, guiding generations of thinkers and spiritualists.

The Upanishads are more than just religious texts: they are philosophical treatises reflecting on universal themes. The Upanishads' approach to the cosmos, nature and the human condition makes them incredibly modern. They do not offer dogma but invite reflection and experimentation. "Who am I?" is the central question that runs through their pages, a question that philosophers of every age have asked themselves.

In the Isha Upanishad we find the paradoxical statement that Brahman is "smaller than the smallest, greater than the greatest." This conception impels human beings to transcend their mental limitations to embrace infinite reality. The oscillation between the immanent view (the divine in the world) and the transcendent view (the divine beyond the world) makes these texts particularly rich in insights.

The Upanishads have deeply influenced Indian and world culture and spirituality. They have inspired philosophical systems such as Vedanta, which is based directly on their reflections, and internal dialogues with other religious traditions. When German philosopher Arthur Schopenhauer discovered the Upanishads in the 19th century, he called them "the consolation of my life" and considered them one of the pinnacles of human thought.

In addition to laying the foundations of Hindu spirituality, the Upanishads continue to engage in dialogue with contemporary issues. Their reflections on the nature of the universe and the relationship between the individual and the cosmos have been related to theories of modern physics, such as quantum mechanics. The idea that reality is interconnected and seemingly contradictory-the apparent dualism of wave and particle, for example-recalls strikingly on the concepts set forth in these ancient texts.

In an age such as ours, where many question the meaning of existence, the Upanishads invite us to look within, reminding us that "*You are That*." The insights of these texts, born millennia ago, remain alive and relevant, indicating that authentic wisdom knows no time or boundaries.

From rituals to mantras: how spirituality evolves among the Upanishads.

In the intellectual and spiritual revolution represented by the Upanishads, prayers and rituals give way to meditation and direct knowledge of the self (Atman). This section shows how these "secret scriptures" reformulated the Vedic religious experience, introducing major themes such as Samsara (cycle of rebirth), Karma (law of action) and Moksha (liberation).

Hinduism has its roots in the Vedas, the oldest sacred texts. In this Vedic phase (about 1500-800 BCE), religion was based on Vedic rituals. Through complex sacrifices performed by Brahmins, they sought to maintain cosmic order (Ṛta) and gain divine favor. However, later (about 800 BCE) a spiritual and intellectual revolution profoundly changed the religious experience. The Upanishads, the "secret scriptures," ushered in a new phase: inner quest replaced the centrality of outer rituals.

The Upanishads shifted the focus from prayer and sacrifice to meditation and introspection. Central to this shift is the Atman, the individual self. Through direct knowledge of this inner principle, one discovers a fundamental truth: Atman is not separate from the universe, but identical to Brahman, the absolute.

The great questions about existence are answered in the Upanishads. The concept of Samsara, the endless cycle of birth, death and rebirth, is introduced. This cycle is governed by Karma, the law that every action has a moral consequence, which determines the fate of the soul in future lives. However, the ultimate goal of life is not to remain imprisoned in Samsara, but to achieve Moksha, liberation. This state transcends suffering and delusion, leading to complete union with Brahman.

Before the Upanishads, religion was dominated by complex rituals and sacrifices, which required the supervision of brahmins. However, these texts mark a gradual move away from Vedic ritualism. The importance shifted to mantras, sacred words or sounds that promote concentration during meditation. One of the most famous is the mantra "Om," believed to be the primordial vibration from which the universe was born.

This spiritual transformation democratized religion. Meditation and philosophical thought became accessible not only to experienced priests but also to individuals seeking truth. The knowledge of the

Upanishads, while exclusive to some, was the seed of future practices that inspired schools of thought such as Buddhism and Jainism.

The new approach of the Upanishads influenced not only religion but also culture. The worldview proposed by the Upanishads emphasized unity, interconnectedness and compassion. This profound reflection spread to the arts, literature and thought. Moreover, many of these themes would find echoes in later eras when India entered into dialogue with the Western world through the theosophical movement and modern philosophy.

The Upanishads took Hinduism from a ritualistic religion to a philosophical and spiritual quest. Through meditation, self-knowledge and an ethic centered on Karma, these texts offered a way to understand the divine and the ultimate meaning of existence. Their legacy is not limited to the development of Hinduism, but continues to inspire universal reflections on identity, reality and the unity of being.

Parallel to the religious transformation, ancient India experienced profound cultural and social changes. (6th century B.C.). This period marked an era of change. Urbanization intensified in many regions and new cities sprang up, such as Pataliputra and Varanasi. These centers attracted traders, thinkers and ascetics, fostering a fertile intellectual climate for the evolution of ideas. It was in this context that many of the philosophical and spiritual notions that would define Hinduism were developed.

Cities became crossroads of cultures, stimulating a deep exploration of the meaning of life. People began to ask crucial questions. What is suffering? Is there life after death? How can we break the cycle of birth and rebirth, known as Samsara? In response to these questions, concepts such as Karma, which governs actions and their consequences, and Dharma, which represents moral and universal duty, were developed.

The idea of Moksha, the ultimate liberation from Samsara, became one of the pillars of Hindu philosophy. For many, the ultimate goal in life was to escape the cyclical nature of existence and achieve union with the divine, Brahman. This vision offered answers that went beyond the ritualistic level of the Vedas, approaching more intimate and speculative thinking. These developments are reflected in the Upanishads, philosophical texts that emerged during this era. In them, the human being is no longer just an instrument in rituals, but a seeker of truth.

Among the most iconic places associated with Hinduism, the city of Varanasi, overlooking the sacred Ganges River, stands out. According to Hindu tradition, Varanasì is one of the oldest cities in the world, also known as Kashi, "*the city of light.*" Many believe that dying in Varanasi, immersed in its sacred waters, guarantees liberation from the cycle of rebirth. The city is rich in symbolism. Its banks teem with worshippers performing ritual ablutions, meditating or lighting votive lamps at sunset, creating a spectacle that combines the sacred and the profane.

One of the "*Shakti Pitha,*" the seven most revered holy places, is located here in Varanasi. Ancient poems tell that the god Shiva, in his destructive and regenerative aspect, chose Varanasi as his earthly kingdom. This connection with Shiva, a major deity in Hinduism, reinforced the city's centrality in the religion. The epics of the Mahabharata and Ramayana frequently mention these places, connecting the city to the mythic fabric of India..

The intellectual climate of the period gave shape to new currents of thought that responded to profound needs. One of the defining figures of this era was the sage Yajnavalkya, known for his teachings found in the Brihadaranyaka Upanishads. Yajnavalkya was said to converse with disciples and even his own wife, Maitreyi, on such lofty topics as the immortality of the soul and the nature of the absolute. Maitreyi emerges as a significant figure, highlighting the role of women in philosophical research in ancient India, although the context remained predominantly male-dominated.

Reflection on Samsara also found echoes in other religious currents, such as Jainism and Buddhism, which began to spread during this period. Hinduism, however, maintained a powerful synthesis of ritualism and philosophy, creating an open and flexible tradition.

The Ganges, a sacred river par excellence, represents more than just a waterway. Hinduism conceives of it as a divine gift brought by Shiva to Earth. Countless pilgrims come each year to bathe in its waters, which are believed to have the power to cleanse the soul of sins and allow access to Moksha. The figure of the Ganges is celebrated in texts and ceremonies. It is said that the goddess Ganga, personification of the river, descended from the heavens at the behest of Shiva, who acted upon her descent into its canopies to protect the world.

Varanasi, located on the banks of the river, is the convergence point between the earthly and spiritual dimensions. Every year, the "*ghats*" (the steps overlooking the Ganges) see millions of devotees gather to celebrate religious festivals, such as Diwali or Kumbh Mela.

So, that period (6th century B.C.) was a crucial moment in the history of Hinduism. The growth of cities and the evolution of philosophical thought led to the emergence of a deeper and more complex spirituality. Places like Varanasi and concepts like Karma, Dharma and Moksha allowed the ritual tradition of the Vedas to be combined with a new sense of introspection. This balance between religiosity and personal reflection made Hinduism one of the richest and most enduring philosophical systems in the world. The connection to "sacred territory" and respect for rivers, such as the Ganges, continue to inspire millions, uniting present and past in a timeless tradition.

Spirituality on the way: sacred geography and the original places of Hinduism.

Hinduism, one of the world's oldest spiritual traditions, has its roots in the places where earth, water and sky have met to inspire generations of sages, poets and devotees. Geography is not a mere backdrop in the history of Hinduism, but a living, sacred element steeped in symbolism. The plains of the Ganges and the forests of northern India have been cradles of reflection, prayer and philosophy. In these places, spirituality has been deeply intertwined with nature, resulting in a unique vision of the world and the divine.

The fertile and majestic plains of the Ganges have been the main stage where Vedic traditions have been established. This river, which flows majestically through northern India, is much more than a waterway. To devotees, the Ganges (Ganga) is a mother, a living goddess who purifies and regenerates. According to sacred texts, its waters flow straight down from heaven, and bathing in its currents has the power to wash away sins.

The first spiritual and cultural centers developed on these plains. Sacred cities such as Kashi, now known as Varanasi, became landmarks for knowledge, rituals and meditation. Kashi, called the Eternal City, is considered one of the oldest inhabited places continuously in human history. Here, "*rishis*" (sages) met to study the

Vedas, recite mantras and teach the relationship between Atman (individual essence) and Brahman (universal reality).

Anecdotes enrich this sacred geography. One of the Vedic texts relates that the Sarasvati River, now vanished, joined the Ganges and Yamuna at a sacred point called Prayaga (modern Allahabad). This confluence, or "*triveni sangam,*" is still revered as a symbol of the union of divine forces and the quest for spiritual enlightenment.

Where the bustle of the plains ends, the silence of the forests is born. The dense forests of northern India, populated by towering trees and remote places, were ideal refuges for ascetics and mystics who sought the divine through meditation and introspection. In these isolated spaces far from civilization, the teachings of the Upanishads, the philosophical texts that developed a new way of conceiving the relationship between man and the cosmos, took shape.

An emblematic figure of this contemplative spirituality is Yajnavalkya, a Vedic philosopher who wandered through these forests wondering about the nature of the universe and the self. The forest thus became not only a physical refuge but a metaphor for the inner quest, the journey toward Brahman, the absolute truth.

The Sarasvati River: the lost stream of knowledge.

A key chapter in Hindu sacred geography concerns the Sarasvati River, a waterway as real as it is mythological. Mentioned in the Rigveda as a "mighty, swift and luminous river," the Sarasvati was symbolic of the creative flow of knowledge and purity. This river, which once flowed parallel to the Indus, has gradually disappeared due to climatic and geological changes. Its fate has fueled debates and archaeological research, but its spiritual presence remains intact.

The Sarasvati River has never "died" in the hearts of Hindus. Its myth has made it immortal, an archetype that celebrates the link between nature and culture. This river is often compared to other great river resources of antiquity, such as the Nile in Egypt or the Tigris in Mesopotamia. Just as the Ganges, Indus, or Nile nurtured civilizations, the Sarasvati was also considered a divine womb where not only cities but also religious and social ideas developed.

The connection between rivers and spirituality is not unique to India. The first great human civilizations were born along sacred rivers. Egypt revered the Nile as the source of life and cosmic order. Mesopotamia relied on the fertile banks of the Tigris and Euphrates to nurture its culture and religious beliefs. Likewise, the plains of the Ganges and the now-lost course of the Sarasvati fostered not only material but also metaphysical development.

What makes Hinduism unique is the symbolic intensity attributed to these places. While other civilizations saw rivers as givers of physical life, Indian sages also found spiritual metaphors in them. Water became a symbol not only of physical rebirth, but of karmic purification and passage to a higher reality.

The plains of the Ganges and the forests of northern India are not just landscapes-they are open books on the history of Hinduism. Every river, every tree, and every city has a story to tell. These geographies have shaped generations of thinkers, rituals and beliefs. The Ganges and Sarasvati continue to flow, one with its sacred waters, the other as a memory, but both eternal to the millions who worship them.

Geography, in Hinduism, is never static. It is a journey. An invitation to explore, reflect and discover the divine manifested in the natural world. Ultimately, it is evidence of the deep connection between man, nature and the infinite.

Religion and philosophy

These developments made Hinduism not only a religion but also a highly articulated philosophy. Among its peculiarities is the absence of a single founder and its evolution as an open tradition. A key concept is plurality: God is one, but is manifested in the many. Deities such as Vishnu, Shiva and Devi are different expressions of the same absolute.

In summary, the shift from Vedic rituals to the meditative thought of the Upanishads represents the heart of Hinduism's transformation. Its history is not only a lesson in the past, but also an invitation to explore the depths of spirituality and philosophy. In a world today that seeks a connection between science and spirituality, Hinduism offers examples of insights that seem to anticipate modern concepts, such as the interconnectedness of the universe and the search for what is beyond the apparent.

The heart of Hinduism originates in the Vedas, the oldest texts of this tradition. The Vedas are poems and hymns. They celebrate the forces of nature, such as fire (Agni) and the sun (Surya), invoked through ritual sacrifices. In this period, the relationship with the divine was mediated by priests (the Brahmans), who officiated at complex ceremonies to maintain cosmic harmony.

Over time, however, many began to ask: what is the meaning of these rituals? These questions gave rise to the Upanishads. The Upanishads abandon the emphasis on sacrifices and delve deeper into the mystery of being. They introduce revolutionary concepts, such as Brahman, the absolute principle that underlies the universe, and Atman, the individual essence that coincides with the whole.

While faith and philosophical thought have often been opposed in the West, Hinduism integrates them into a unified search for the meaning of life. The centrality of meditation and introspection is evident. A significant example is the chanting of the mantra "Om," considered the primordial sound of the universe. The Bhagavad Gita, one of the most important texts in the tradition, combines ethical, philosophical and spiritual teachings in an epic form, recounting the dialogue between Prince Arjuna and Krishna, an embodiment of the divine.

Hindu philosophy is distinguished by its openness. There is no single path to the divine, no single dogma. Hinduism recognizes the plurality of spiritual paths, summarized in the statement, "*God is one, but is manifest in the many.*" The main deities-Vishnu, Shiva and Devi-are not separate entities, but different expressions of the same Brahman. Vishnu represents the preserver of the universe, Shiva the transformer, while Devi embodies creative energy, often associated with the figure of the divine Mother.

The transition from Vedic rituals to Upanishad meditation and philosophical thought had profound cultural consequences. This process also reflected the social changes of ancient India, marked by the formation of complex cities and kingdoms. Philosophers and teachers such as Yajnavalkya, a figure from the Vedic era, challenged the conventions of the Brahmans, proposing that true wisdom was not in rituals but in self-knowledge.

Hinduism also mingled with other cultural traditions. Some scholars see a mutual influence between Vedic thought and the cultures of the Indus Valley, a civilization that thrived along the Indus and Sarasvati rivers. In this area, archaeological excavations have revealed seals and statues depicting figures in meditative postures and symbols traceable to deities such as Shiva in his aspect as the lord of meditation (Pashupati).

The teachings of this tradition are not only a legacy of the past, but an invitation to explore the present. The Hindu vision of the interconnectedness of humanity and the universe has found fascinating parallels in modern physics. The concept of an all-pervasive Brahman recalls the scientific idea that the universe is an interconnected system based on energy and vibration. Figures such as physicist Erwin Schrödinger, one of the originators of quantum mechanics, claimed to have been influenced by Hinduism. Schrödinger often cited the idea of the Atman in his writings.

Hinduism, with its openness and plurality, remains a beacon for those seeking answers to the great questions of existence. It is not just a religion that keeps alive the memory of its myths and rituals; it is an open philosophy in continuous dialogue with contemporary thought. In a fragmented world, the Hindu idea of unity-between oneself, others and the cosmos-continues to speak with an ancient, yet extraordinarily relevant voice.

Atman and Brahman: the eternal dialogue on the nature of the self.

One central theme stands out in the Upanishads: the relationship between Atman, the individual soul, and Brahman, the universal reality. This relationship is explored through symbolic dialogues between masters and disciples, creating a timeless reflection on the nature of the self.

The most famous teaching of the Upanishads is summed up in the Sanskrit phrase "*Tat Tvam Asi*," meaning "*You are That.*" This maxim comes from the Chandogya Upanishad, where master Uddalaka Aruni guides his son Svetaketu in a series of enlightening dialogues. Uddalaka explains that the individual's deep self (Atman) is not separate from but identical with Brahman, the ultimate source that permeates the entire universe. "*Tat Tvam Asi*" is not just a theory, but a revelation: each individual is part of a larger, unified reality.

Another famous example is found in the story of Nachiketa in the Katha Upanishad. This young man, in the afterlife, asks profound questions of the god of death, Yama, trying to find out what happens after death. Yama teaches that the Atman is immortal, beyond the body and mind. This understanding eliminates the fear of death and leads to the realization of the true eternal nature of being.

The great masters used imagery and metaphors to expound difficult concepts. In the Brihadaranyaka Upanishad, the sage Yajnavalkya uses the example of salt dissolved in water to explain that, just as salt becomes indistinguishable from water, Atman is inseparable from Brahman. In other words, while the individual perceives himself as separate, he is actually one with the universe.

The dialogues were not simply exchanges of ideas, but true pedagogical journeys. The disciple would begin with questions, often uncertain or confined to a materialist view, and the teacher would respond with patience, gradually leading him to insight into a broader, transcendent vision.

The philosophical scope of the relationship between Atman and Brahman has crossed geographical boundaries. The writings of the Upanishads came to the West in the 19th century, attracted by the interest of thinkers such as Arthur Schopenhauer. The German philosopher described the Upanishads as "the comfort of my life" and found in them a speculative vision that reinforced his metaphysical pessimism. Ralph Waldo Emerson, an American poet and essayist, also admired these texts, being inspired by their emphasis on the interconnectedness of human beings and the universe.

In the East, however, the dialogue on Atman and Brahman profoundly influenced later philosophical schools, such as the Advaita Vedānta of Adi Shankara. Shankara, a ninth-century philosopher, consolidated the identity of Atman and Brahman as the foundation of Vedantic monism. According to him, the perceived world is *Maya* (illusion), while the ultimate reality is Brahman.

The Upanishads teach that self-knowledge is the first step toward freedom. Becoming aware of the identity between Atman and Brahman means overcoming the sense of separation, breaking down the ego and living with a deep connection to universal reality. This message remains incredibly modern, offering answers to existential questions and balancing the personal quest with cosmic interdependence.

Sacred places such as Varanasi or the Himalayan glaciers have been the havens of sages and seekers for centuries, committed to the discovery of the Atman. The charm of these landscapes, combined with the symbolic power of Vedic dialogues, continues to attract pilgrims from all parts of the world. Hinduism offers no pre-packaged answers, but an endless journey toward awareness.

In conclusion, the dialogue between Atman and Brahman in the Upanishads is more than a philosophical exploration. It is an invitation to human beings to recognize their own uniqueness and unity with the whole. And this invitation remains open, still dialoguing with our age.

Chapter III. A parallel between Hinduism and physics.

The Hindu view of the universe.

Hinduism, one of the world's oldest and deepest spiritual traditions, has always proposed a view of the universe as a single, interconnected entity. This conception is found in its founding texts, the Vedas and Upanishads, dating back some 3,000 years. Through poetic and symbolic language, these texts describe the universe as a great "*cosmic unity*" in which everything is intertwined, merged and mutually reflected. For Hinduism, the universe is not simply a collection of material objects. Rather, it is a dynamic ocean of energy, consciousness and connection. This idea, surprisingly, anticipates some insights of modern quantum physics.

The concept of Brahman and Atman: the core of universal interconnection...

At the heart of the Hindu vision we find the concept of Brahman, the supreme cosmic principle. Brahman is described as the primordial source of all that exists. It is eternal, infinite and transcends time and space. According to the Upanishads, everything in the universe originates from Brahman and returns to it in a continuous cycle of creation, preservation and dissolution.

Next to Brahman is Atman, the individual soul. Hindu philosophy teaches that Atman and Brahman are, in their essence, one. This notion is summed up in the famous phrase of the Upanishads, "*Tat Tvam Asi*" (*You are That*). In this phrase, "*That*" refers to Brahman, the ultimate reality. The individual is not separate from the universe, but is an intrinsic and fundamental part of it. The same principle of unity is found in quantum physics, where each particle is linked to all others through laws and phenomena that transcend ordinary perception.

Quantum physics, developed in the 20th century, introduced a different view of the universe from that of classical physics. Quantum phenomena, such as entanglement (quantum entanglement), show how two particles, even if separated by enormous distances, can remain connected instantaneously. This behavior has prompted many scientists to question the deep nature of reality and its hidden interconnectedness.

In Hindu philosophy, an interesting parallel is found in the concept of "*Indra's Net*," an ancient metaphor described in Buddhist and Hindu texts. One imagines the universe as an infinite net, with pearls placed at each intersection. Each pearl reflects all the others, in an infinite interplay of cross-references and interdependencies. The image of this network recalls the idea of the quantum universe as an indivisible web of relationships.

Niels Bohr's principle of complementarity, a pillar of quantum physics, also finds an echo in Hindu thought. Complementarity holds that particles such as electrons or photons exhibit both wave-like and corpuscular properties, depending on the context of observation. This dualism bears a striking resemblance to the dual nature of Brahman, which is both impalpable (*nirguna Brahman*) and manifested *(saguna Brahman*).

Hinduism integrates its cosmic vision not only through philosophy but also in sacred geography and rituals. Places like Varanasi, on the banks of the Ganges, are considered physical reflections of cosmic harmony. Hindu temples, with their unique architectural structure, represent the universe in miniature, where the dome symbolizes the sky and the base represents the earth. These places serve as bridges between the material and the spiritual, embodying the idea of universal interconnectedness.

The Mahabharata, one of the great Hindu epic poems, offers another example. In it, Krishna reveals to Arjuna his cosmic form in the Bhagavad Gita, showing a picture of the universe as a living, pulsating whole, in which the divine is present in every single particle. This view also reflects the central idea of quantum physics: the universe is more than the sum of its parts.

Scientists and philosophers: a dialogue between East and West.

Numerous philosophers and scientists have pointed out the similarities between Hinduism and quantum physics. Fritjof Capra, an Austrian physicist and author of "*The Tao of Physics*," has deeply explored this parallel. In his book, Capra argues that the concepts of interconnectedness and unity found in Eastern philosophical traditions, such as Hinduism, resonate with the findings of modern physics. A other example comes from Erwin Schrödinger, one of the fathers of quantum mechanics, who claimed to have been inspired by Vedantic philosophy. Schrödinger, fascinated by the idea of unity between observer and observed, came to this insight precisely through reading the Upanishads.

Hinduism, with its totalizing view of the universe, is not limited to being a religion or philosophy. It is a thought pattern that invites us to perceive the world as a living organism. In an age when science continues to unravel the subtle connections between particles, the Hindu view of the cosmos seems incredibly relevant today.

Quantum physics and Hinduism, each in its own way, remind us that we are all part of one "cosmic fabric." Whether through the mathematical formulas of a physicist or the sacred chants of a mystic, the message is the same: the universe is not divided, but united in depth. In this way, spiritual tradition and modern science offer us two languages to explore a universal truth that embraces both the visible and the invisible.

Brahman: the indivisible absolute.

At the heart of the Hindu vision is the concept of Brahman, the ultimate and absolute reality. Brahman is not a personal deity, but a cosmic, infinite and all-pervading force. According to the texts of the Upanishads, everything that exists is a manifestation of this unique principle. An emblematic passage in the Chandogya Upanishad reads:

"Everything that exists is essentially Brahman."

This statement emphasizes the idea that every atom in the universe, every individual and every event is part of a larger whole.

In parallel, quantum physics has shown that the subatomic world is not composed of isolated entities, but of interwoven fields of energy. The phenomenon of quantum entanglement, discovered in the 1930s, shows that two particles, even if separated by immense distances, can remain connected instantaneously. This principle closely resembles the Hindu view of universal interconnectedness. If everything is Brahman, then the idea of separation becomes an illusion.

Hinduism, through Brahman, describes the universe as an indivisible unity. According to this view, the distinction between objects, living beings and events is only apparent. Human senses perceive separation, but the Upanishads state that this is *maya*, an illusion. Behind everything is a single essence, transcendent and immanent at the same time. Brahman is not only "outside," in the observable universe, but is also found "inside," as the Atman, the inner Self.

A traditional metaphor to explain this truth is that of clay and pots: pots may have different shapes, but they are all made of the same material. Similarly, each element in the visible universe is a particular form, but the original substance remains unique: Brahman.

Quantum physics, developed in the 20th century, challenged common sense and classical scientific intuitions. Along this quest for the infinitely small, physicists have discovered that the universe is not made up of isolated particles. Everything that exists is made up of complex, interwoven fields of energy. Particles are like "waves" of the same sea, an image deeply reminiscent of the concepts expressed in the Upanishads.

One of the most striking examples is that of quantum entanglement. First proposed by Albert Einstein, Boris Podolsky and Nathan Rosen in 1935 and later confirmed by experiments such as those of Alain Aspect in the 1980s, entanglement shows that two particles can remain connected instantaneously, even if separated by thousands of kilometers. If one particle changes state, its "twin" will do so simultaneously. This phenomenon challenges the classical idea of spatial separation and evokes an underlying interconnected reality, similar to that indicated by Hinduism.

For Indian masters, the sense of separation that dominates the human mind can be overcome through spiritual practices such as *dhyana* (meditation) and *jnana* (knowing). The process enables people to recognize the universal interconnectedness and rediscover their oneness with Brahman.

A famous dialogue between father Aruni and son Svetaketu in the Chandogya Upanishad offers a symbolic teaching: Aruni uses milk from a cow to explain to his son that forms may change, but the basic substance remains the same.

Aruni, with a desire to teach her son a valuable lesson, decides to use a very simple example taken from everyday life. He takes some milk from a jug and starts pouring it into different containers one at a time. First he pours it into a tall, narrow glass, then into a wide, flat bowl, and finally into a thin-necked bottle. After each pour, he asks his son to observe how the shape of the milk changes.

The son, intrigued, notices how the milk takes the shape of the container into which it is poured. At this point, Aruni explains to him:

"You see, milk changes shape because it adapts to the container, but in itself it always remains milk. The essence of milk does not change."

With this simple experiment, Aruni wants to convey a profound principle to his son: in life, many things may change in their outward form, but their essence, what they really are, remains constant. He teaches this not only in reference to objects, but also to life in general, to people and situations. It is a lesson in stability, identity and understanding what is essential beyond appearances.

This teaching applies not only to the visible universe, but also to the invisible.

Quantum physics, even with its Western roots, arrives at conclusions that seem to recall the spiritual insight of Hinduism. Theoretical physicists often report that subatomic reality cannot be observed without being influenced by the process of observation. In other words, the observer and the observed phenomenon are part of a single system. Similarly, the Upanishads state that the human Self (Atman) is not separate from what it observes, but is a fragment of the whole.

Contemporary characters and reflections.

Thinkers such as Erwin Schrödinger, one of the fathers of quantum physics, showed fascination with Eastern philosophies. Schrödinger, who read the Upanishads deeply, often wrote about how the notion of indivisibility in scientific inquiry was echoed in the idea of Brahman. He argued that ancient Indian wisdom could offer tools of understanding for complex phenomena, such as entanglement.

Another scientist inspired by the Hindu vision was physicist David Bohm, who developed the idea of an "*implicit order.*" Bohm suggested that the observable universe is a reflection of an underlying order and indivisible. This perspective is close to the Indian idea of Brahman and Maya, where the material universe is only a surface of a deeper reality.

The Hindu view of the universe, with the concept of Brahman at its core, proposes a revolutionary understanding of reality. Quantum physics, although based on scientific methods, also reveals a universal connection between the elements of existence. These two fields, seemingly so far apart, seem to converge toward the same point: the idea that separation is an illusion and that the universe is one indivisible reality.

In a time when science and spirituality seek common languages, Brahman and quantum physics show that there is no need to choose between intuition and rationality: both can help represent the mystery of existence.

In Hinduism, the universe is perceived as an indivisible unity, an immense web in which each part is intimately connected to the whole. This cosmic vision revolves around the concept of Brahman, the absolute entity that represents the ultimate reality and the fundamental principle of all that exists. Brahman is not just "the divine" in the Western theological sense; Brahman is the All. It is the unique and indefinable essence that underlies and permeates the entire cosmos.

According to the Upanishads, the deepest philosophical texts in the Hindu tradition, Brahman is eternal, infinite and indivisible. It has no form, but everything that has form flows from it. The phrase "*Tat Tvam Asi*" suggests that Atman, the individual self, and Brahman, the supreme reality, are actually one. The individual is not separate from the cosmos but is its full manifestation.

This Hindu perspective surprisingly resonates with some ideas in contemporary quantum physics. For example, the phenomenon known as quantum entanglement demonstrates that two particles, even at enormous distances, remain constantly connected. Any change in the state of one particle instantaneously affects the other, as if both were part of a single interconnected entity. Similarly, for Hinduism, nothing is really separate. In a rather emblematic sentence from the Upanishads, it states:

> *"He who sees the whole in everything and everything in the whole, knows no fear or division."*

For modern quantum physicists, such as David Bohm, this interconnectedness invokes the idea of an "implicit order," an underlying reality in which everything is united at a deep level, even if it appears fragmented by our perception. This concept is not so different from the vision of Brahman proposed by Indian sages thousands of years ago.

Adi Shankara and the philosophy of Advaita Vedānta.

During the 8th century, Adi Shankara, one of India's greatest thinkers, elaborated the philosophy of *Advaita Vedānta.* Advaita means "non-duality" and represents the heart of Hindu thought on cosmic unity. According to Shankara, the separation of individual and universe, of spirit and matter, is an illusion derived from the mind. This illusion is called *maya*. Only through knowledge and awareness can the ultimate truth be perceived: everything is Brahman.

An anecdote relates that Shankara, traveling in India, answered a disciple who asked him about the meaning of reality:

> *"Meditating on the ocean, remember that every wave, even the smallest, is always ocean. To forget this is ignorance."*

This metaphor, as simple as it is profound, invokes the modern idea of a fundamental reality that unifies everything, even what seems separate.

A tangible symbol of this unified vision is found in Varanasi, also known as Kashi, one of Hinduism's holiest cities. Located on the banks of the Ganges River, Varanasi is not only a sacred place but also a microcosm of the universe. According to Hindu tradition, Varanasi exists simultaneously on a physical and a spiritual plane, symbolically representing the point at which Brahman manifests through the becoming of the phenomenal world. The city is often described as the "bridge" between the visible and the invisible, between multiplicity and unity.

Every morning, thousands of devotees bathe in the Ganges in Varanasi, seeking spiritual unity with the universe. This act is not only a ritual but a symbolic recognition of their connection with Brahman, the absolute. It is recounted in the texts that "those who die in Kashi are freed from the illusion of separation" and rediscover identity with the whole.

Hinduism and the holistic universe.

The Hindu view of the universe as an indivisible whole invites reflection on what binds us together, beyond the appearances of differences and divisions. The holistic approach of Indian philosophy encourages thinking that every aspect of the cosmos-from the smallest particle to the most majestic galaxies-has intrinsic significance as part of a universal and infinite web.

Quantum physics, with its recent discoveries, suggests a strikingly similar picture of the world. Despite the enormous temporal and cultural distance between the Hindu tradition and modern science, both push us toward the same insight: the universe is not a fragmented puzzle, but a unison breathing whole.

In this view, the union of physics and philosophy no longer appears to be a utopia, but a necessary dialogue to understand the substance of existence. After all, as an Indian philosopher still unknown to the West stated:

"To those who know Brahman, everything is known."

The Cosmic Dance of Shiva and the dynamics of the universe.

At the heart of Hindu philosophy, Shiva appears as a universal figure, a bridge between spirituality and science. Among his many representations, that of Nataraja, *the Lord of Dance*, is perhaps the most fascinating. His cosmic dance not only possesses deep spiritual significance, but also seems to anticipate some of the dynamics studied today by modern cosmology.

The Nataraja represents Shiva dancing on the body of a demon, Apasmara, a symbol of ignorance. His pose is surrounded by a ring of fire, representing the continuous cycle of creation, preservation and destruction. With one hand she holds the drum (*damaru*), a symbol of primordial sound (*nada*) and creation. Another hand carries the flame, emblem of destruction. The gesture of the right hand (*abhaya mudra*) invites calm, indicating that the cosmic cycle, however tremendous, need not be feared.

According to Hindu philosophy, these cosmic rhythms never end. In the dance of Shiva, the essence of the universe is manifested: a perpetual alternation of birth, death and rebirth. This concept, rooted in sacred texts such as the Shiva Purana, has been echoed in modern understanding of the cosmos, which describes an expanding and contracting universe, comparable to the "breaths" of an infinite entity.

Contemporary cosmology describes the universe as a dynamic reality. The *Big Bang* theory represents the beginning of everything, a primordial explosion reminiscent of Shiva's drum and the first vibrating sound of creation. But the universe does not just expand: according to some theories, such as the *Big Crunch* theory, it may come to a final contraction. This eventual collapse would be reminiscent of the destructive fire in the hand of Nataraja. After this destruction, a new cosmic birth could follow, in a cyclicality that resonates deeply with the Hindu view of time, conceived as an infinite cycle rather than a straight line.

> The famous astrophysicist Carl Sagan emphasized this point. Speaking of the dance of Shiva, he commented that this dance *"represents the continuous flow of matter and energy in the universe, which is incessantly transformed to maintain cosmic harmony."*

Sagan was fascinated by the insight of the Indian texts, in which the dynamics of existence at both macrocosmic and microcosmic levels were already intuited.

Shiva's vision of cosmic dance has found expression not only in philosophy but also in art and architecture. An outstanding example is the Chidambaram temple in Tamil Nadu, one of the most important sacred sites dedicated to Nataraja. Built in the 10th century, this site literally embodies cosmic consciousness.

Inside the temple, the Golden Dance Hall (*Kanaka Sabha*) symbolizes the heart of the universe, a sacred place where matter and spirit meet. Here the eternal movement of reality is celebrated. The architecture itself, with its geometric proportions and details, is inspired by the cosmic concept of order and rhythm. Such cultural impact does not stop with India: today a statue of Nataraja also towers in front of CERN, the *European Center for Nuclear Research*, in Geneva, as a symbol of the continuing dialogue between science and spirituality.

The relevance of the Shiva dance lies in its ability to combine ancient philosophical insights and modern scientific discoveries. For the Indian rishis, the sages who composed the Upanishads and Puranas, matter was neither static nor separate from consciousness. In a strikingly similar way, quantum physics shows that matter is never at rest: subatomic particles vibrate in complex patterns, resonating as in the cosmic dance of Nataraja.

The parallelism between Hinduism and physics is not mere coincidence. It is the result of deep reflection on the cosmos, which unites the minds of ancient spiritual sages and modern scientific researchers. Today, as in the past, the message of Shiva's dance remains powerful: the universe is a living entity, constantly moving, always changing.

This dialogue between tradition and science invites us to look beyond appearances, asking fundamental questions about the origin and destiny of the universe. Like Shiva in his endless dance, we, curious observers of creation, continue to search for the harmony hidden in apparent chaos.

The primordial sound of the universe. The resonance of Om.

In the ancient Hindu vision, the universe is not a simple aggregate of material objects, but a complex texture of energy and vibration, constantly changing. This vision finds its most powerful and universal symbol in Om (or Aum), considered the primordial sound, the first vibration from which everything springs. The Upanishads, the central philosophical texts of the Vedic corpus, describe Om as the eternal murmur that permeates existence, the matrix on which the cosmos manifests. Today, surprisingly, this conception enters into dialogue with quantum physics, which explores the universe precisely as a web of waves, vibrations and frequencies.

According to the Mandukya Upanishad, one of the most profound texts in the Hindu tradition, Om is much more than sound. It is the symbolic and sonorous representation of all reality. In its threefold structure (A-U-M), Om summarizes the three states of human consciousness: waking (*Jagrat*), dreaming (*Svapna*) and deep sleep (*Sushupti*). The last state, or the silence that follows the repetition of Om, represents *Turiya*, pure awareness, that which is beyond time and space.

The vibration of Om is described in the Upanishads as an "eternal sound" (*Shabda-Brahman*), a wave that connects every element in the universe. Interestingly, modern physicists and cosmologists have found parallels with this idea through theories that see the cosmos as a network of vibrations. String theory, for example, holds that every subatomic particle is nothing more than a vibration of a fundamental string. For Hindu thought, this is equivalent to saying that everything in the cosmos, from the macro to the micro, is an expression of the original resonance.

The science of vibration.

Many Hindu practices, such as meditation and mantra chanting, have internalized this vibratory perspective. Indian sages (rishis) believed that chanting Om was not only a spiritual activity, but also a way to become attuned to the universe itself. Today, neuroscientists and physicists are beginning to understand the effect of sound vibrations on the human body. Studies conducted on acoustic resonance have shown that repeated chanting of Om can stabilize the nervous system and influence brain waves, leading to states of relaxation and awareness.

In physics, another interesting parallel is the Big Bang phenomenon, the theory that the universe originated about 13.8 billion years ago. Scientists describe the beginning of the cosmos through an initial expansion phase accompanied by energy fluctuations. Poetically, some theologians and cosmologists have compared this "creative explosion" to the initial sound of Om, which is understood as the vibratory origin of everything.

The importance of Om has never been limited to philosophical speculation. This sound has been an integral part of Indian spiritual practices for thousands of years. It is recited as the introduction and closing of prayers, rituals and yoga. Historical figures such as Patanjali, the great codifier of yoga, have emphasized how Om is a sacred symbol that helps one connect with the absolute (Brahman).

In Indian temples, such as the magnificent complex at Rishikesh on the banks of the Ganges, the collective chanting of Om can often be heard resonating like a continuous wave of energy, capable of uniting the minds of devotees in a common experience of transcendence.

In the field of modern meditation, Om also plays a central role. Celebrities such as spiritual master Paramhansa Yogananda have widely popularized the practice of Om chanting in the West, highlighting its benefits for mental stillness and spiritual growth. Modern mindfulness and yoga sessions often take up this ancient sound to help practitioners achieve inner balance.

The convergence of ancient insights and modern discoveries.

Hindu philosophy invites us to see the universe not as a multiplicity of separate objects, but as a single entity generated by universal vibrations. Quantum physics, for its part, emphasizes that every particle is both matter and wave. The wave-particle duality, one of the best-known concepts in quantum mechanics, thus seems to echo the metaphor of Om: a vibration that gives form to all things, but which in itself is also invisible and immaterial.

This parallel between ancient traditions and contemporary science finds further support in modern physicists and thinkers, such as David Bohm. Bohm theorized that the universe itself is a unified field of energies, in which matter emerges from a kind of vibratory implicit order. This is not far from the idea contained in the ancient Hindu texts, in which Om is the hidden web of reality.

Om is much more than a sound or mantra. It is a symbol that spans millennia and represents the essence of the Hindu view of the universe. Its description as a "*primordial sound*" is echoed today in the most advanced models of quantum physics, showing how the philosophical insights of the Upanishads can still dialogue with contemporary scientific thought. In a sense, Om invites us to contemplate the universe as an infinite symphony of vibrations, in which each element, including us, participates in the great resonance of the whole.

Maya: the illusion of separation.

According to Hinduism, the reality we perceive every day is only a semblance, an appearance intended to deceive our senses. This deception is not accidental: it is Maya, the illusion. Maya, in Hindu philosophy, does not claim that the world is nonexistent or false. On the contrary, it states that the perception of separation between objects, space and time is a veil that hides the underlying unity of reality. Matter appears distinct to us-a tree, a mountain, a river-but everything is an expression of Brahman, the unique and infinite principle that permeates all things.

A clear example of this concept is found in the analogy of the rope and the snake. In the Vedic texts one is invited to imagine a person who, in the dark, spots what he believes to be a snake. He becomes afraid; he feels the danger. Only by turning on a light does he discover that there was no reptile: it was a coiled rope. The snake was an illusion, as is the fragmentation of the world we perceive. When we turn on the light (the light of wisdom or self-realization) we understand the real nature of the universe: a unified web in which everything is interconnected.

This worldview, which has its roots in the Vedas and Upanishads, now finds interesting parallels in modern physics. Quantum mechanics, in particular, reveals that what we perceive as solid, tangible and material is only a Maya-like illusion. Particles, considered the "fundamental building blocks" of reality, are never really stable. Experiments show that they are probability waves: they have no fixed position or definite shape until they are observed.

Take the example of the atom, the foundation of all matter. For centuries, scientists have imagined atoms as small solid objects, resembling tiny marbles. But today we know that the atom is almost entirely empty: the nucleus is surrounded by a "shell" of electrons floating in empty space. What appears as "solid," like a stone or a piece of wood, is, in reality, an incredibly fast dance of energy. This awareness echoes the principles expressed by the rishis, the Hindu sages: perceived reality is not what it seems, and only deeper knowledge can reveal the underlying truth.

Albert Einstein himself, although skeptical of some implications of quantum mechanics, recognized the existence of a universal substratum. In 1935, together with Boris Podolsky and Nathan Rosen, he worked out the famous EPR paradox, which demonstrated how two spatially separated particles could remain instantaneously connected, a property known as *quantum entanglement*. This suggests that physical separation between objects is, in fact, an illusion. Although not directly influenced by Hinduism, quantum theories seem to repropose ancient Hindu concepts in contemporary scientific language.

Hindu texts do not just describe Maya. They also offer tools for overcoming it. Meditation, study of sacred texts and deep introspection enable one to see beyond the illusion. Swami Vivekananda, one of the most influential philosophers in modern Hinduism, described Maya as a dream from which man must awaken. He wrote that

> *"this whole universe is only perception, but within it lies Brahman, the real."*

Vivekananda also brought this vision to the West, bringing Vedic philosophy into dialogue with the scientific materialism of 19th-century Europe.

Similarly, scientists engaged in quantum research seem to be pursuing a form of "enlightenment." It is no coincidence that some physicists have found inspiration in Eastern philosophy. Werner Heisenberg, one of the fathers of quantum mechanics, confessed that after reading Vedic texts he felt "deeply fascinated" by their ability to intuit some truths that science had just begun to discover.

Maya as a universal principle.

Hinduism does not consider Maya an enemy to be fought, but an obstacle to be understood. Maya is a symbolic explanation of how the mind and senses work. It separates, analyzes and fragments the world to make reality comprehensible. However, for Hinduism, the path to enlightenment requires transcending these limitations and recognizing the fundamental unity of existence.

Maya helps us reflect on our modern worldview. When we stare at a screen or look at a city, we see a concrete reality, but we forget that it is composed of billions of atoms in constant motion. The same goes for our self, the part of us that we perceive as distinct from others. Today we know that we are part of one global ecosystem and that the illusion of separation can lead us to destroy the environment and ourselves. Instead, modern physics and the ancient wisdom of Hinduism invite us to see beyond the veil and regain the unity we may have lost.

Vedānta philosophy, largely influenced by the Upanishads, defines Maya as the veil that covers the true nature of reality. According to these scriptures, what we see with our senses is volatile, subject to time and change, a transitory manifestation of something deeper and unchanging. Adi Shankaracharya, an 8th-century philosopher, pointed out that Maya creates the illusion of separation between the individual and Brahman. The empirical world, in his view, is like a dream: real while it lasts, but lacking absolute authenticity.

A famous example comes from the words of Swami Vivekananda, who in his discourses in the West described the material world as a "shadow," a reflection of Brahman. In the same way that surface waves do not represent the entire depth of the ocean, the sensory world conceals the depth of existence.

Quantum mechanics similarly questions the true nature of reality. An iconic experiment, that of the double slit, offers an intriguing parallel. In this experiment, subatomic particles (such as electrons) behave sometimes as waves, sometimes as particles, depending on whether they are observed or not. The very act of observation changes the behavior of the particles, suggesting that the observer and the observed are deeply interconnected.

This phenomenon recalls the Vedantic concept of consciousness as a central factor in the perception of reality. In Hindu philosophy, the world exists in relation to the consciousness that perceives it.

Similarly, in quantum physics, the function of the observer seems to influence the outcome of physical events. There is a deep connection between the phenomenal world and the mind that interprets it.

The question that arises is crucial: is there an objective reality independent of observation, or does everything we see depend on the observing consciousness? The Vedānta answers with the concept of Advaita: there is no dualism, but everything is "*non-dual.*" Brahman is the only truth, while Maya is the mental construction that gives rise to a manifold reality. In the world of modern physics, physicist Erwin Schrödinger, who studied Hindu philosophies in depth, was inspired by these concepts for his reflections on the nature of consciousness.

From a different cultural background, scientist John Wheeler presented a theory called the "*participatory universe.*" Wheeler proposed that the universe does not exist in a definite way until it is observed. This statement is reminiscent of the dynamics of the double slit and seems to pay homage to the principles of the Advaita Vedānta.

The connections between Maya and the observations of quantum physics also extend to the idea that perceived reality is, in a sense, a "collective construction." Traditional teachings describe Maya not only as an individual phenomenon, but as an illusion shared by all who live in the phenomenal world. In modern science, commonly accepted paradigms of reference, such as time and space, could be interpreted as mental constructs. Einstein's relativity itself undermined the idea of absolute time, making the concept of Maya even more relevant.

The parallels between Maya and quantum reality do not suggest that Hinduism is "scientific" in the modern sense. Rather, they show how both camps have a similar awareness: what appears to be real may not be real at all. This awareness invites the exploration of a deeper reality. And this is precisely where science and philosophy meet: in humility before the mystery of the universe.

Although coming from completely different paths, both Hinduism and modern physics seem to converge on an essential truth. The reality we perceive, as the Vedānta says, is a temporary illusion. But illusion can be a vehicle for understanding the infinite.

The Higgs Boson

Imagine a cutting-edge laboratory like CERN, nestled in the Swiss countryside, where some of the world's greatest scientists work to answer a profound and seemingly simple question: what is reality made of? It was out of this ambition that the Higgs boson, better known to the general public as "the God particle," was discovered in 2012. This evocative name, while controversial, has drawn attention not only to modern physics, but also to the ancient mysteries of philosophy. Among these, Hinduism stands out for its striking affinities with some of the most advanced theories in quantum physics.

But what is so special about the Higgs boson? This elementary particle was first theorized in 1964 by physicist Peter Higgs and other researchers. Its existence explains how elementary particles obtain mass, through an invisible field known as the "Higgs field." The experiment that confirmed its existence was conducted at CERN thanks to powerful particle accelerators such as the *Large Hadron Collider* (LHC). To simplify, the Higgs boson is a kind of "cosmic glue," a fundamental element that helps explain how the universe works.

Now let us jump back in time and space to ancient India, where the sacred texts of Hinduism celebrate an equally fascinating concept: the Brahman. Indian sages describe it as the ultimate reality, the universal essence that permeates all things. In the Vedas and Upanishads, there is frequent reference to the fact that everything that exists-from the infinitely small to the immensely large- is a manifestation of a single energy. Doesn't that sound surprisingly close to the idea of the Higgs field?

Physicist and popularizer Fritjof Capra, author of "*The Tao of Physics,"* was one of the first scholars to explore the link between these two worldviews. Capra notes, for example, that quantum physics has disrupted the way we conceive of matter. At the subatomic level, what we perceive as "solid" is actually a vortex of particles that continually appear and disappear. The Vedic sages had reached a similar conclusion, stating that material reality is an "illusion" (Maya), a veil that hides the true divine essence of everything.

A curious anecdote takes us back to CERN, where a joke containing an unexpected philosophical echo is common among scientists. It is said that if someone really wanted to "see" the elementary particles of the universe, it would be like trying to look at the waves of the sea without the presence of water. To understand the metaphor, we have to think that particles, such as electrons or quarks, do not exist on their own, but are manifested by the energy fields that surround them. Similarly, in Hinduism, reality does not exist separate from the Brahman that sustains everything.

Another stimulating parallel is found in theories of time. Quantum physics proposes a model in which past, present and future, at a fundamental level, could be inextricably intertwined. This idea, while complex, recalls the Vedic concept of Kalachakra, "*the wheel of time.*" In this view, time is not linear but cyclical, a cosmic dance that continuously repeats creation, destruction and rebirth.

What, then, drives humanity, whether through modern science or philosophical introspection, to seek answers about the origin of existence? Perhaps it is precisely the desire to understand what unites all levels of reality. In the words of Werner Heisenberg, one of the fathers of quantum mechanics:

> *"The first sips of the natural science glass will make you an atheist, but at the bottom of the glass you will find God waiting for you."*

It is fascinating to think that, from the hypertechnological halls of CERN to the meditations of Indian sages under thousand-year-old banyan trees, we arrive at such similar insights. The only glue seems to be the wonder of discovery: the same principle that makes elementary particles vibrate and inspires the most inquisitive minds, as much today as millennia ago.

Atman and the experience of interconnectedness.

Hinduism is one of the world's deepest and oldest spiritual traditions. At the heart of its philosophy are the concepts of Atman and Brahman. While Brahman represents the absolute, the universal cosmic principle, Atman is the individual principle, the spiritual essence of each human being. However, one of the central insights contained in the Upanishads, the Hindu philosophical texts dating back more than 2,500 years, is that there is no separation between Atman and Brahman. The famous phrase "*Tat Tvam Asi*" states that each individual is inherently part of the universal whole.

This view suggests that all reality is united. There are no fundamental divisions between the individual and the cosmos, between the inner and the outer, between the observer and the observed. The experience of Atman, experienced through deep meditation or philosophical contemplation, leads to the recognition of this interconnectedness.

A significant example comes from the practice of yoga, an age-old discipline that aims to unite body, mind and spirit. Yoga comes from the Sanskrit word "*yuj*," which means precisely union. This union is not only between the different components of the human being, but it is also between the human being and the cosmos.

Modern quantum physics, in a strikingly similar way, suggests that the universe is deeply interconnected.

For example, the quantum non-locality principle demonstrated in Alain Aspect's famous experiments in the 1980s showed how two particles, once correlated, can continue to influence each other even when separated by immense distances. This phenomenon, which defies classical intuitions in physics, is reminiscent of the Hindu view in which all existence is part of a universal network of connections.

In classical physics there was a clear distinction between observer and observed element. Quantum physics, however, suggests that the two are inseparable. Danish physicist Niels Bohr, speaking of the results of quantum theory, declared:

> *"The universe is not made up of separate objects, but of a network of indivisible relationships."*

Similarly, in Hinduism one cannot think of the Atman as separate from the cosmos; each human being is, always, an integral part of the whole. Hinduism emphasizes that interconnectedness is not just an abstract philosophical concept. It is a direct experience that can be lived. In meditative practice, the boundary between self and other dissolves. The perception of being separate from the world vanishes, bringing the meditator to a profound understanding: the Atman and Brahman are one.

If meditation reveals this unified reality to the individual, in the realm of science it is mathematics and experiment that carry the same message. The Hindu vision of the unity between the human being and the cosmos fits surprisingly well with the results of modern science. Both disciplines suggest that separation is ultimately an illusory construct of the human mind.

The cycle of time and the pulsating universe.

In Hinduism, the universe is not motionless or linear. It is a living organism that goes through endless cycles of creation, destruction and rebirth. This conception, described in the Puranas, introduces the idea of the *kalpa*, a cosmic cycle encompassing billions of earthly years. Each kalpa represents the breath of Brahma, the creator: when he exhales, the universe expands; when he inhales, the universe contracts. This strikingly resembles the modern *Big Bang* and *Big Crunch* theory.

Hinduism not only anticipates this idea of a cyclic cosmos, but goes even further. It describes the universe as a living being, whose pulse is the eternal rhythm of the galaxies and whose life pulses with divine energy.

This connection between science and spirituality is not accidental. The idea of the cyclic universe proposed by Hindu texts still holds a unique fascination for contemporary scientists and thinkers. Physicist Robert Oppenheimer, famous for his role in the creation of the atomic bomb, was an emblematic example. Deeply influenced by Indian philosophy, Oppenheimer quoted a passage from the Bhagavad Gita while observing the first nuclear test in 1945:

"Now I have become Death, the destroyer of worlds."

Oppenheimer interpreted the concept of destruction not only as tragedy, but as a natural part of a larger cosmic cycle.

This cyclical idea runs through the entire Hindu philosophy. Not only time, but the universe itself is conceived as a living organism. Its existence vibrates with the rhythm of galaxies, while subatomic particles mark its beat. Each atom contributes to the harmonious dance of what the texts call the Sanatana Dharma, the eternal order.

Brahma's breath and cosmic energy.

In the Puranas, the time duration of a single breath of Brahma is impressive. One day of Brahma is equivalent to about 4.32 billion earth years. When his "day" ends, the universe dissolves into "*pralaya*," the state of cosmic quiescence. Here there is no emptiness, but a dormant energy waiting to awaken. The cosmos, therefore, has no definite beginning or end. It is a continuous cycle, sustained by the principle of universal interconnectedness.

This breath of Brahma finds a striking analogy in modern physics, thanks to the idea of a pulsating universe. Some scientists imagine that the universe can oscillate endlessly, alternating between phases of expansion and collapse, much like the heartbeat of a living being. Indian astrophysicist Jayant Narlikar explored similar theories, proposing an oscillating model that views the cosmos as a cyclic system, in which time itself becomes an infinite variable.

Physics, philosophy and poetry.

Hinduism, however, does not simply describe the cyclical universe in cosmological terms. It envelops it in poetry and spirituality. Each cycle represents a divine work, an emanation of the creative energy called Shakti. In the Bhagavad Gita, Krishna speaks to Arjuna about the eternity of Brahman (the Absolute), describing it as that which exists beyond creation and destruction:

> *"Never born, eternal, unchanging, immortal. Even if the body dies, the spirit survives."*

This eternity resonates with some fundamental questions in physics. Energy, according to the principle of conservation, can neither be created nor destroyed, only transformed. This property of matter and energy echoes the Hindu idea of a cosmos that continuously regenerates itself through cyclic processes.

The link between Hindu philosophy and modern physics, after all, reflects a deep dialogue between East and West. On one side are the ancient *rishis* of India, who conceived of a pulsating universe with startling insights. On the other are contemporary scientists, who, using mathematical models and empirical observations, arrive at similar conclusions.

Albert Einstein, while not directly influenced by Hinduism, sought a unified vision of reality that transcended linear time. His theory of relativity, which connects space and time into one fluid dimension, is reminiscent of the Hindu idea of Maya, the veil that simultaneously unites and separates visible and invisible reality.

The cycle of time and the pulsating universe, then, are not just concepts. They are an invitation to reflect on what binds us to the eternity of the cosmos. They make each moment part of an infinite story whose rhythm is marked by the breath of the universe itself.

The Eternal Return.

Hinduism offers a radically different view of time and space than the Western tradition. In many Western cultures, time is imagined in a linear way: a beginning, a progression, and a final end. Hinduism, on the other hand, conceives of time as cyclical, an eternal return punctuated by cosmic eras, dissolutions and rebirths.

This conception is expressed in sacred texts such as the Puranas and Upanishads, which describe the universe as subject to infinite cycles. There are two pivotal concepts: the Kalpa and the Yuga. The Kalpa represents one "day" in the life of Brahma, the Creator, and lasts 4.32 billion years. At the end of each Kalpa, Pralaya, the cosmic dissolution, takes place. Yuga, on the other hand, divides time into shorter epochs, in a cycle reminiscent of the changing seasons. Each Yuga cycle ends with a time of degeneration before a new rebirth.

According to Hinduism, time and space have no beginning and no end. This concept surprisingly resonates with some modern cosmological theories. Some physicists propose the idea of a cyclic universe, growing and shrinking in an eternal process. The Big Bang theory, for example, hypothesizes that our universe emerged from a cosmic compression, bouncing back to a new expansion. These models come close to Hindu descriptions of cosmic cycles of creation and destruction.

In Hindu mythology, these concepts are not exclusively abstract, but are narrated as vivid events. One famous legend tells of Shiva dancing *the Tandava*, his cosmic dance. This dance represents the destruction of the universe at the end of a cycle, followed by its creation. The symbolism of Tandava directly recalls the cosmic oscillations hypothesized by modern physics.

The cyclical nature of time is deeply engrained in Hindu culture, not only through texts, but also in the places and artistic creations of India. One of the most fascinating examples is the Konarak Sun Temple in Odisha. Built in the 13th century, this temple depicts a chariot drawn by seven horses, symbolizing the motion of the sun and cyclical time. The carved wheels of the chariot are also a detailed calendar, reflecting the passage of the ages.

The presence of the sun, as a symbol of cyclicity and regeneration, permeates every detail of the temple. In this context, the Konarak is not only an artistic work, but a material illustration of the Hindu conception of time as perpetual motion.

In the epic poem Mahabharata, the sage Markandeya recounts a vision of the Pralaya. During the great flood, Markandeya describes how the entire universe disappears, leaving only Vishnu floating on the cosmic ocean in the form of a child on a lotus leaf. This tale is a powerful metaphor for dissolution and inevitable rebirth.

Yajnavalkya, one of the most celebrated philosophers of the Upanishads, explained that "*time is like a cosmic breath.*" The seal of cyclicity is reiterated here: the breath begins, expands, reaches a maximum point and returns.

The Hindu view of eternal return is not a purely cosmological concept. It influences the interpretation of life itself. The idea of Samsara-the cycle of birth, death and rebirth-is a microcosmic reflection of this universal pattern. The ultimate liberation, Moksha, represents the only way out of the endless cycle, a theme that resonates deeply in Eastern philosophy and spirituality.

The similarities between Hindu thought and current cosmological theories raise fascinating questions. Could it be that humanity, through ancient traditions and modern science, is coming to the same understanding of the universe? The cyclical nature of time described in the Vedas may be more than a myth: it may reflect profound insights that predate scientific discoveries by millennia.

Cosmic Intelligence. Hinduism between Order and Entropy.

Hinduism offers an extraordinarily complex view of the universe. Central to this perspective is the concept of Ṛta, the cosmic order. Present in the Vedas, Ṛta represents a universal principle that governs the movement of the stars, the cycle of the seasons and the great dance of life. However, this view does not preclude the existence of chaos. In Hinduism, chaos and order do not oppose each other but complement each other, creating a dynamic reminiscent of modern principles of physics, such as entropy.

In the Vedas, composed about 3,500 years ago, Ṛta is described as the force that sustains the harmony of the cosmos. This invisible law governs the laws of nature and moral rules. In the Rigveda, one of the oldest texts, we read:

> *"Thanks to Ṛta, the sun appears in the sky and the rivers flow to the sea."*

This principle is immutable but dynamic as it allows the flow of existence.

Ṛta is a principle that does not act alone. Vedic deities, such as Varuna, the guardian of cosmic order, and Vishnu, the preserver of balance, play key roles in maintaining the universal system. Vishnu, in particular, is central to Hindu mythology. Through his many incarnations, or avatars-such as Krishna and Rama-Vishnu intervenes at times of crisis to restore harmony.

Order and Chaos: complementary principles

Hindu philosophy interprets chaos not as pure destruction, but as a necessary element for transformation. The myth of Shiva, known as the "transformer," presents this idea with narrative power. Shiva, often depicted in a cosmic dance, the Tandava, destroys what is old to create space for the new. This process represents the complementarity of order and disorder, a dynamic that modern science calls "*entropy*."

In physics, entropy measures the degree of disorder in a system. According to the second law of thermodynamics, entropy tends to increase, pushing systems toward less ordered states. However, life and complex systems develop by processes that create order from chaos, using energy. This scientific idea echoes Hindu thought: chaos and order are not irreconcilable opposites, but components of a cosmic balance.

Another symbolic example of the Hindu cosmic order is Mount Meru, described in the Puranic texts. This mythical mountain represents the "*axis mundi*," the center of the universe. In the Puranas, Mount Meru supports the heavens and connects the earth to the divine. Its central location symbolizes the harmony between the different spheres of existence: the material, the spiritual and the cosmic.

Interesting is the parallel with modern astrophysics. The concept of a "cosmic center" could recall the idea of a central point from which the universe expands, as occurred with the Big Bang. If Mount Meru is motionless, it represents absolute stability amid the dynamism of the universe, an idea echoed in the scientific concept of fundamental laws.

The dialogue between Hinduism and science is not accidental. Both seek to explain the laws that govern the universe, whether through spirituality or empirical observation. Ṛta, in this light, can be seen as a forerunner of the concepts of natural law. Similarly, Vishnu, with his protective role, embodies an approximation to modern visions of a "*fine-tuned universe*" (artfully organized to sustain life).

Hinduism's ability to unite order and chaos, stability and dynamism, thus finds intriguing resonances in scientific thought. These are not forced overlaps, but spaces for reflection that bring a millennia-old philosophy into dialogue with the frontiers of science.

In a universe governed by Ṛta, where even chaos plays a role, human beings are positioned as active participants. Like Mount Meru, we must seek our own balance, balancing order and change, in a cosmos as mysterious as it is intelligent.

The universe as a union of science and spirituality.

The Hindu vision of the cosmos as an interconnected whole continues to inspire not only philosophical thought but also the frontiers of science. The similarity between the teachings of Indian rishis and the discoveries of quantum physics is more than a mere parallel. Both seek to explain the unexplainable: that thin, invisible thread that connects all things.

At the Chidambaram temple in southern India, dedicated to Shiva Nataraja, the Lord of Cosmic Dance, Hindus celebrate this universal interconnectedness. Shiva's dance symbolizes the creation and destruction of the universe, in an eternal cycle reminiscent of the ceaseless flow of energy that quantum physics is still trying to decipher.

Just as the scientist peers into the microscope to understand the mysteries of the universe, the Indian mystic closes his eyes to explore the same depth within himself. Both, in the end, seem to meet in that same truth: that everything is one, and that the universe dances, indivisible, to the eternal rhythm of existence.

The universe according to quantum physics.

For centuries, human beings have been investigating the mystery of the universe. Is it an ordered system, governed by precise laws? Or is it the result of subtle, interconnected energies, where chance and unpredictability play key roles? With the advent of quantum physics in the 20th century, the answers have changed. This discipline has redrawn the boundaries of scientific knowledge, revealing a world where certainty dissolves and mystery reigns supreme. Surprisingly, some of the modern discoveries seem to reflect ancient conceptions of Hinduism, shaped in sacred texts such as the Upanishads.

Quantum vacuum and Brahman: a common origin?

Quantum physics describes reality at the subatomic level as a huge field of possibilities. The so-called "quantum vacuum" is not really empty. On the contrary, it is a sea of potential energies, where particles and antiparticles emerge and disappear in constant motion. This conception recalls the idea of Brahman, the supreme principle of Hinduism. In Vedic texts, Brahman is described as the ultimate reality, an invisible, eternal and indivisible essence that permeates everything. Both, quantum vacuum and Brahman, are not directly observable. Yet, both represent the source of all manifestation.

The most famous Upanishad, the Chandogya Upanishad, states:

> *"The whole universe is rooted in Brahman, everything exists through it, and everything will return to it."*

Quantum physicists, led by pioneers such as Werner Heisenberg, have often cited similarities with this view. Heisenberg himself, after a meeting with Indian poet Rabindranath Tagore in 1929, said he was fascinated by the similarities between Eastern ideas and modern physics.

Another crucial point of contact concerns interconnection. Quantum physics introduced the concept of entanglement quantum entanglement. When two particles are entangled, a change in the state of one is instantaneously reflected in the other, regardless of the distance between them. Albert Einstein called this phenomenon "*a spectral action at a distance*," but today it is considered one of the cornerstones of quantum theory.

In Hinduism, this universal interconnectedness is expressed through the idea that the whole cosmos is an inseparable web. A famous verse in the Brihadaranyaka Upanishad states, "*Tat Tvam Asi*," recalling that each individual (Atman) is an inseparable part of the whole (Brahman). Swami Vivekananda, one of the greatest exponents of modern Indian thought, pointed out that the unity of the universe is a central idea in the Vedic scriptures. In 1893, during the Parliament of Religions in Chicago, Vivekananda spoke of the interconnectedness of all things, anticipating insights that science would verify decades later.

Quantum physics also showed that light and electrons behave either as particles or waves, depending on the experiment. This duality shocked 20th century scientists. How could something be two things at once? Such a paradoxical phenomenon resembles Maya, the Hindu concept of "illusion" that hides the true nature of reality. According to the Upanishads, what we perceive is an illusory representation, created by the mind. Swami Vivekananda explained that the ultimate truth is not what appears, but what is hidden under the veil of appearance.

In the quantum world, observation plays a crucial role. Experiments on Schrödinger's famous cat and Heisenberg's uncertainty principle show that the observer modifies the observed reality. This idea is also echoed in Hindu philosophy, where perception itself is seen as the key to determining what we call "real."

Hinduism embraces a cyclical view of time, divided into eras (*Yuga*) that repeat in an endless dance of creation and destruction. Similarly, quantum physics offers models of the universe in which time may not be a straight line, but rather a malleable and recursive dimension. Recent theories, such as that of the multiverse, suggest the possibility of infinite cosmic cycles, recalling the vision of a universe without beginning or end described in the Puranas and the Bhagavad Gita.

The convergence of quantum physics and Hindu philosophy does not imply that one explains the other. However, both offer complementary views on the mystery of the universe. While physicists use experiments and mathematics, Indian sages explored reality through meditation and introspection. Physicist Fritjof Capra, in his bestseller "*The Tao of Physics*," stated:

> *"Modern physics leads us toward a worldview closer to that of the Upanishads than to Newton's ideas."*

The challenge for humanity today may be to unite these two perspectives. Quantum science deciphers the language of the universe through numbers. Hinduism invites us to experience it as a unity. Perhaps in the dialogue between East and West lies the key to understanding the mysteries that still elude us.

The Newtonian view

he view of the cosmos underwent a revolution in the 20th century. Before then, the perspective of classical physics, as represented by Isaac Newton, dominated. Newton, in the 17th century, described a predictable, ordered universe dominated by mechanical laws. According to this view, everything worked as a perfect machine. The famous example was that of the "*clockwork universe.*" In this metaphor, the universe worked like a cog: each piece moved in accordance with its precise role, allowing deterministic knowledge of past, present, and future.

This idea of order resonated with many Western conceptions of the time and also influenced philosophical and social thought. Scientists believed that once all natural laws were known, nothing would be elusive. Even the future, with sufficient calculation, could be predicted. This mechanical universe gave security. Reality seemed clear, logical, comprehensible.

However, in the early 20th century, this certainty began to waver. Albert Einstein, with the theory of relativity (1905 and 1915), showed that space and time were not absolute. Not only that. In 1927, Werner Heisenberg finally shattered the idea of predictability with the uncertainty principle. Heisenberg proved a shocking fact: we cannot simultaneously know both the position and the velocity of an elementary particle with absolute precision. This seemingly simple principle forever changed the way we look at the universe.

The subatomic universe, that which lies at the base of matter, does not follow Newtonian rigor. It is not mechanical, but probabilistic. Particles do not behave like little billiard balls, but rather as entities that move and vibrate in unpredictable ways. The very act of observing a particle influences its behavior. This introduces a new dimension: the role of the observer in the universe.

In quantum physics, this indeterminacy recalls some ideas from Eastern philosophies, particularly Hinduism. In Vedic texts, the universe is often described as an interwoven infinite web, where everything is connected and ultimate knowledge eludes ultimate human interpretation. Maya (the cosmic illusion) and the cycles of energy in Hinduism reflect precisely the instability and continuous transformation observed in quantum mechanics.

Heisenberg himself had direct contact with Eastern culture. During one of his trips to India in the mid-20th century, he dialogued with the Indian poet and philosopher Rabindranath Tagore, who won the Nobel Prize for Literature in 1913. The conversation focused precisely on the connection between science and Eastern philosophy. Heisenberg, according to a well-known anecdote of his, said he found in Indian philosophical texts a certain "familiarity" with respect to the principles of quantum physics: both suggest the impossibility of absolute knowledge of reality.

Ultimate truth, for Hinduism, is Brahman, an unknowable and transcendent reality. This comes close to the idea that, on the quantum plane, we can never observe the universe in its objective totality.

The German scientist was not the only one to connect such concepts. Erwin Schrödinger, one of the fathers of quantum mechanics, was also deeply interested in Vedic philosophy and the concepts of Brahman and Atman (the individual self). Schrödinger wrote that, according to him, Hinduism understood, long before Western science, that the separation of object and subject is illusory. More recent illustrations of this idea are found in the famous "Schrödinger's cat" experiment, which demonstrates how reality can be "multiple" up to the moment of observation.

Quantum mechanics and Hinduism seem to speak distant languages, but they share one essential point. Both see reality as something mutable, dynamic and elusive. The Newtonian view, so reassuring, is no longer sufficient to describe the complexity of the universe. Modern science has shown that what we believed to be solid and unchanging, such as atoms, is actually a vibrant space of possibilities.

And Hinduism, with its emphasis on cosmic unity and intuitive knowing, offers a profound insight: the universe is a web of ever-changing connections, where knowing means living with mystery.

Perhaps, in this very mystery, we can find a new form of order for the 21st century.

The quantum view.

Quantum physics shows us a world in which randomness governs the foundations of reality. At the smallest scale, that of elementary particles, there are no certainties: there are only probabilities. The famous double-slit experiment, first conducted in the 1920s, revealed a puzzling phenomenon. A particle, an electron, can behave either as a particle (a definite physical element) or as a wave (something indeterminate and diffuse in space). And the choice of either nature seems to depend on the observer. The universe, then, seems to respond to our perception of and interaction with it.

This perspective, which at first glance may seem purely scientific, strongly echoes Hinduism's own view of the universe. According to the Vedic texts and Upanishads, the universe is Maya, an illusion or projection, constantly changing. In Hinduism, there is no sharp boundary between material and spiritual reality: everything is an interweaving of energies. Matter itself is considered dynamic and interconnected with what we may call spirit or consciousness.

An interesting comparison can be made with the Hindu concept of "Brahman," the universal cosmic principle. According to Hinduism, Brahman is the absolute, infinite and indivisible reality. All living beings and phenomena in the universe emerge from this unique reality and are in deep and inseparable relationship with it. This is reminiscent of quantum entanglement, a phenomenon in which two particles, once entangled, remain in a state of mutual connection, regardless of the distance between them. Today entanglement has been demonstrated experimentally and is the basis for the development of quantum technologies, such as quantum computers.

But what does Hinduism say about the role of the individual in the universe? In the Bhagavad Gita, one of the most widely read sacred texts, Krishna states:

"I am the origin of everything. From me everything develops."

Here, the human being is not seen as a separate observer, but as an integral part of the cosmic process. Individual consciousness, or Atman, is an inseparable part of Brahman. Similarly, in quantum physics the observer seems to influence the behavior of particles. The line between the observer and what is observed dissolves.

An anecdote showing the deep connection between these two visions takes us back to the meeting between Niels Bohr, the father of quantum theory, and a delegation of Indian philosophers in 1930. During a discussion regarding the nondeterministic nature of particles, one of the guests suggested that the subatomic reality described by Bohr bore a striking affinity to the cosmic descriptions of Vedic philosophy. Bohr, fascinated, paused to reflect on how such insights could come from different paths: one scientific, the other spiritual, but both in search of ultimate truth.

The "dream" metaphor.

The metaphor of "dream," recurring in Eastern philosophies, also seems to converse with the quantum world. The Upanishad Brihadaranyaka states:

> *"As a spider emits its thread, so Being expands into the universe, and then withdraws from it."*

. Reality is fluid, cruel in appearing stable. A similar view emerges in more modern quantum models, according to which the quantum vacuum from which all matter emerges is itself in constant motion and fluctuation.

These parallels do not claim that Hinduism "knew" about quantum physics. But they do suggest that humanity, through different paths, has sought to understand the same fundamental truths. At the heart of both, there is a revolutionary idea: everything is interconnected. The fate of the smallest electron is linked to the whole. In the Hindu universe, every being, from the smallest insect to the most powerful god, is linked by an invisible thread to the rest of the cosmos.

Perhaps, on closer inspection, the universe is not just a collection of grains or particles. It is a dance, a continuous flow of energies, consciousnesses and possibilities.

The sacred texts of Hinduism describe the world as a dream. Reality is fluid, malleable, constantly changing.

If we move from spirituality to science, we find a surprisingly parallel view. Quantum physics has taught us that the universe, at a fundamental level, is not a collection of concrete objects, but a network of possibilities that emerge and disappear. The "quantum vacuum," the basis of all matter, is not empty at all. It is a pulsating sea of fluctuations that shape particles. It is a reality in constant flux, so much so that some theories suggest that even the universe itself might emerge from these fluctuations, like a dream manifesting out of nothingness.

Similar to the concept of Maya, quantum physics blurs the boundaries between what appears real and what is only potential. Physicist David Bohm, known for his philosophical approach to quantum theory, spoke of the "implicit order" of the universe: a hidden, dream-like level in latency from which everything emerges and in which everything eventually dissolves.

Another central concept in both Hinduism and quantum physics is the role of the observer. In the traditional model of Hinduism, the observer is the Atman, the deep self, which observes the world as one observes a dream. The Atman is not separate from the dream, but is part and partaker of it, yet it can become aware of the illusion and awaken, becoming one with Brahman again.

Similarly, in quantum physics, the observer is a fundamental figure. Wave function collapse experiments have shown that a particle's state (e.g., position or velocity) is not determined until it is observed. In other words, it is as if the universe exists in the state of a dream until the moment an observer intervenes and "chooses" a reality.

Both Hinduism and quantum physics share another extraordinary insight: the interconnectedness of all things. In the Hindu universe, there are no rigid boundaries between one being and another, or between one thing and another. Each entity is intertwined with the rest of the cosmos through an invisible link.

A fascinating parallel is found in the phenomenon of quantum entanglement. When two particles are "entangled" at the quantum level, they form a single system: the fate of one particle is instantly linked to that of the other, regardless of the distance between them.

Through the dream metaphor, we can imagine the universe not as a collection of static objects, but as an ever-evolving dance. In India, this vision is found embodied in the figure of Shiva Nataraja, the Lord of Dance, who represents the cyclical creation and destruction of the universe. Shiva dances on the lingam, symbol of the creative principle, and by his movement causes the birth and dissolution of reality.

Modern physicists also describe the universe as a dance. Some theories suggest that particles are nothing more than vibrations of tiny strings, oscillating and intertwining in ways that create matter and energy. The "music of strings" closely resembles Hindu philosophy, according to which the world is the result of the primordial vibration of the sound Om, from which everything originated.

It is crucial to clarify that the parallels between Hinduism and quantum physics do not mean that ancient Indian sages foresaw modern physics. However, these connections suggest that humanity, at different times and in different ways, has explored the same fundamental questions: what is reality? What is our role in it? What is beyond what we see?

Samkhya: Purusha and Prakriti, being and becoming.

The universe is made up of enigmas. Quantum physics reminds us of this every time we try to grasp the deep nature of reality. Among its most fascinating mysteries, we find wave-particle duality, an intrinsic feature of subatomic particles. Their ability to behave sometimes as particles, solid and definite, and sometimes as waves, fluid and indefinite, defies ordinary logic. On closer inspection, this ambivalence is not so far removed from the thinking of ancient Hindu philosophy, particularly the Samkhya system, which describes the world as a cosmic dance between two complementary principles, Purusha (spirit) and Prakriti (matter). The transition between states is not simple, but it is precisely in this ambiguity that science and spirituality find a surprising meeting point.

To understand quantum duality, we can start with Thomas Young's famous two-slit experiment from 1801. This simple physical proof demonstrates a puzzling fact: a particle of light (a photon) or matter, such as an electron, can behave as a wave or as a particle depending on observation. When an electron beam is sent through two thin slits and not directly observed, the result is an interference pattern typical of waves, with overlapping ridges and valleys. In contrast, if an attempt is made to track the electron's individual path, the resulting pattern draws discrete particles passing through a specific slit. In other words, the very act of observation changes the nature of what is being observed.

This quantum paradox forces us to reflect: is reality objective and already defined, or is it shaped by the interaction between observer and observed object? Quantum science seems to suggest a nuanced answer, reminiscent of the heart of Hindu philosophy.

According to the Hindu philosophy of Samkhya, the universe is the result of the interaction between two cosmic principles. Purusha represents being, the unchanging, the passive, infinite witness. It is the pole of spirit, of pure consciousness, which does not act but observes. Prakriti, in contrast, is the material and dynamic energy, the world of forms, movements and transformations. It is the becoming, the ceaseless activity that constitutes everything we perceive as phenomenal reality.

This relationship between Purusha and Prakriti resembles the implicit dialogue between the wave and the particle. Just as the quantum universe manifests in different states depending on the context, in the Samkhya reality emerges from the interaction between spirit and matter. Matter exists because of the consciousness that observes it, just as consciousness defines itself in confrontation with the visible world. There is no hierarchy: both principles are necessary for the perpetuation of the universe.

Maya and the illusion of reality.

The parallelism extends further when we consider the concept of Maya, found in both Samkhya philosophy and Vedantic texts. Maya, a term commonly translated as "illusion," describes the deceptive and impermanent nature of the phenomenal world. According to Hinduism, what we perceive as stable and final is often apparent. In reality, everything is dynamic, fluid and interconnected. Maya veils to human eyes the true essence of reality, making them believe in a rigid distinction between observer and observed, when in fact everything is a reflection of the whole.

The same is true of the quantum world: until we observe a system, it exists in a state of quantum superposition, a condition in which multiple possibilities coexist. Only with observation does one of these possibilities become actualized. This continuous transition from potentiality to actuality, from possibility to concrete appearance, mirrors the dynamics of Hindu Maya.

Scientist Niels Bohr, father of quantum theory, recognized the value of Eastern traditions of thought. While not religious, the conceptual approach of Vedanta and philosophies such as the Samkhya seemed to echo the way modern physics described the universe. Even Werner Heisenberg, famous for the uncertainty principle, pointed out that the concept of "complementary opposites" in Indian culture anticipated insights that quantum physics would discover only centuries later.

We could say that where physics seeks to understand the universe in its smallest elements, Hindu philosophy goes out of its way to describe the universe in its entirety and in its "cosmic" meaning. However, both disciplines arrive, by different routes, close to the same point: reality is fluid, interconnected and constantly evolving. Wave-particle dualism could thus be interpreted as a modern metaphor for the age-old balance between Purusha and Prakriti.

The encounter between quantum physics and Hinduism is no accident. Both visions invite us to question our certainties and recognize that reality is more complex than it appears. Both the physicist and the ancient Indian sage seek to grasp the truth behind the veil of appearance, reminding us that not everything that exists can be explained in simple terms. Wave or particle? Being or becoming? Spirit or matter? Perhaps, as in the quantum universe and its Hindu metaphors, the answer is both.

Loop quantum gravitation.

Time, as we know and perceive it on a daily basis, is not as linear and uniform as it might seem. Modern physics tells us this, especially through certain theories such as that of "loop quantum gravitation" proposed by Italian physicist Carlo Rovelli. According to this view, time, when analyzed at the microscopic level, might not flow with a constant speed, might fragment, stop or even repeat cyclically. Yet, a very similar idea of "*nonlinear time*" is not that new. Hinduism, one of the world's oldest spiritual traditions, has always conceived of the universe as governed by continuous cycles, where past and future merge into an eternal present.

To understand Hinduism and its relationship to time, we must start with the concept of *kalpa*. A kalpa is a cosmic cycle, an immense duration of time that marks the creation, destruction and regeneration of the universe. According to Hindu sacred texts, such as the Puranas and the Vedas, the cosmos does not follow a linear path, but a cyclical pattern without beginning or end. They speak of eras (*yuga*), which are repeated in a continuous interweaving of birth, decay and rebirth. The idea is poetic and universal: everything from human beings to the stars is subject to this infinite rhythm.

This concept seems to have a striking analogy with some insights from quantum physics. Rovelli's theory, for example, holds that time is not a continuous dimension, but fragmented. At infinitesimal time scales, fundamental structures might behave in unexpected ways, far from our linear perception. It is fascinating to note how an idea derived from mathematical formulas and scientific observation finds an echo in ancient metaphysical speculations.

The Hindu cyclical view is not only cosmology, but also influences daily and spiritual life. Reincarnation (Samsara), for example, is the idea that each soul goes through infinite lives, moving along a cyclic path governed by the laws of Karma. This journey might recall, suggestively, Rovelli's description of time: successive fragments that appear interconnected but not necessarily subject to linear logic.

An interesting parallel is also found in the conception of space-time, which according to Rovelli and other physical theories, does not represent a fixed or universal quantity. Time can speed up or slow down, depending on gravitational fields and conditions in the universe. It is an idea reminiscent of the famous Hindu allegory of the god Shiva, a symbol of destruction and new creation: in his eternal Tandava, the cosmic dance, Shiva creates a rhythm in which space and time are simultaneously born and collapsed.

Perhaps nothing sums up this closeness between modern physics and Hinduism better than Rovelli's phrase:

"Time is a form that reality tells us about."

In Hindu philosophy, time, or more precisely "*Kala*," is often described as an illusion (Maya), a mental construct that human beings use to understand a reality that transcends the mind. In the Bhagavad Gita, one of the fundamental texts of the Hindu tradition, Krishna says:

"I am Time, the destroyer of worlds."

It is a vision that posits time not as a mere ticking of hands, but as a primordial force, both creative and destructive, capable of reshaping the entire universe.

Quantum physics, with its bold and often counterintuitive claims, seems to be setting out on a path similar to that traced millennia ago by Indian sages. Wave-particle dualism, Heisenberg's uncertainty principle and the idea that time is not a universal entity remind us how complex and elusive reality is.

Carlo Rovelli, author of such works as "*Seven Short Lessons in Physics*" and "*The Order of Time*," is a physicist who has been able to cross rigidly academic boundaries and open himself to a broader cultural dialogue. Rovelli himself has stated that he is fascinated by Eastern philosophy. In the eyes of the Italian physicist, time is just one of the patterns that human beings use to interpret the world. This brings him, perhaps unwittingly, closer to what is called the "*Hindu darshana*," or a philosophical vision that invites us to look beyond appearances.

A practical example Rovelli likes to cite is the behavior of black holes. Near a black hole, time slows down to almost a standstill. Normally, this fact is explained through Einstein's general relativity. But at the microscopic level, within the confines of quantum mechanics, time might even "get lost," merging with space. It is an image that leaves room for endless reflections, much like the ancient Hindu concept of "*Lila*," the cosmic game that governs the fate of everything.

The question of what Hinduism has to say to modern science cannot be answered with certainty. But what emerges is the idea that antiquity, with its myths and symbols, can find new meanings in light of contemporary discoveries. Philosophy and physics are not opposites, but two languages for telling a reality that eludes us.

Perhaps, the wisdom is just that: to recognize that time - like everything around us - is much more complex than it seems. And that, in a sense, each step of science seems to be moving closer and closer to a truth that ancient reflections already glimpsed thousands of years ago.

The broken clock: quantum time and the Samsara cycle.

When we look at a clock, the movement of its hands reminds us that time flows in a linear fashion. Past, present and future follow one another as in a straight path. Quantum physics, however, teaches us that this certainty breaks down when faced with the subatomic world. In this dimension, time is no longer a continuous line but takes on strange and unpredictable behavior: it is relative, discontinuous and even reversible. This revolutionary vision surprisingly recalls an ancient and deeply philosophical concept in Hinduism: the cycle of Samsara, the eternal and circular time of reincarnation.

Linear time and circular time.

In Western culture, time has long been imagined as an arrow with a clear starting point and direction. From the Big Bang to the expanding galaxies, the modern scientific narrative seems to confirm this view. However, quantum physics has challenged this conception. Theories such as "*loop quantum gravitation*" suggest that time is not uniform, that it does not flow with the same speed everywhere, and that, at the smallest level, it may even "stop" or "repeat" itself in ways that are hard to imagine.

It is here that Hindu philosophy offers a striking parallel. In Hindu thought, time is not linear but cyclical. The concept of Samsara describes the endless wheel of rebirths, a process with no beginning and no end. Each soul (Atman) goes through a cycle of births, deaths and rebirths until it reaches Moksha, liberation. This eternal cycle contrasts sharply with the Western idea of time as a straight line that goes on without return.

Samsara is not just a metaphor. It is a way to understand reality as a continuous movement in which past, present and future are inseparable. This is confirmed by the Bhagavad Gita, one of the most important sacred texts in Hinduism. In chapter 2, verse 12, we read:

> *"Never was there a time when we did not exist, nor will there be a time when we will not exist."*

This reflection eliminates the conception of time as something that begins and ends. Everything exists in an eternal present.

Similarly, Carlo Rovelli, in his book "*The Order of Time*," explains that time, as we perceive it, is only an illusion produced by our brain. At the fundamental level of reality, that is, at the quantum level, there is no absolute chronological order. Past and future become fuzzy concepts. This brings to mind a statement from Hindu philosophy: time is Maya, an illusion that deceives the senses.

A fascinating example comes from the concept of "superposition" in quantum physics. In this condition, a particle can be in two different states at once until it is observed. This experimentally proven phenomenon suggests that the distinction between past and future is not always clear, especially in the subatomic world. This relativity of time is reflected in the Hindu pantheon, in which deities such as Shiva or Vishnu are associated with the control of time and its cyclical nature.

Another interesting parallel involves black holes. Stephen Hawking has shown that time stops near a black hole. This phenomenon recalls the idea of Kalachakra (the wheel of time) in Hindu cosmology, which introduces the idea that there are points in the universe where time "rolls up" on itself.

Hinduism views the concept of past and future in an intriguing way. Within the cycle of Samsara, what we consider "before" and "after" are only illusions created by the mind. Each life is linked to the previous and subsequent lives in a causal chain, but this chain has no definite beginning. It is a concept reminiscent of the "twin paradox" experiment in Einstein's theory of relativity, in which the perception of time changes depending on movement and gravity. If an astronaut travels through space at a speed close to the speed of light and returns to Earth, he will find that less time has passed for him than for his contemporaries who remained on the planet. Here, too, time turns out to be relative.

Similarly, the cycle of Samsara eliminates any idea of absolute time. What appears to be the "next life" is, in fact, part of an eternal return. This view defies Cartesian logic, but seems to find surprising resonance in modern physics, where concepts previously thought absurd are becoming accepted.

These parallels between Hinduism and quantum physics are not simply fascinating; they offer an opportunity to reflect on how different, seemingly distant disciplines can converge. While Western scientists draw equations to explain the universe, Indian mystics have used the symbolism of the wheel and cosmic dance to capture the same idea. As Niels Bohr states:

> *"Great truths can be expressed either in the language of poetry or in the language of mathematics."*

The idea of the universe as a cycle, without a beginning or an end, might seem foreign to our minds conditioned to think in terms of cause and effect. Yet, both quantum physics and Hindu philosophy invite us to explore a more complex vision, in which time is not a tyrant pushing us toward the end, but a part of an eternal and mysterious dance. A broken clock, in which the beginning and the end blur together.

The illusion of the observer. Quantum physics and Hindu consciousness.

At the heart of humanity's deepest questions (who are we, what is the nature of reality, and what role does our consciousness play) science and philosophy brush against each other, overlap and sometimes seem to answer each other. On one side we find quantum physics, with its surprising but rigorous view of the subatomic world. On the other, Hinduism, a philosophical tradition that has explored the relationship between the individual and the cosmos for millennia. One point of convergence between these two perspectives emerges clearly: the centrality of the observer. What we do, what we choose to observe, seems to modify reality itself. But how far does this connection go?

Quantum physics has revolutionized our understanding of the universe since the early 20th century. One of the most emblematic experiments is wave-particle dualism. The basis of this phenomenon is the so-called "double-slit experiment," a classic of modern physics. Briefly, when an electron is passed through two slits, it behaves like a wave and creates an interference pattern on the screen, overlapping as a water wave would. However, if an observer tries to determine through which slit the electron passes, the behavior changes: the electron behaves like a particle, canceling out the wave pattern.

The mere act of observing thus modifies the result. As if the universe, at its most intimate, waits for the presence of the observer to "decide" what to be. In this regard, theoretical physicist John Archibald Wheeler, one of the great pioneers of this field, proposed the idea of the "*participatory universe.*" According to Wheeler, the universe cannot be described as something separate from human beings. Reality is manifested, or, rather, materialized, only through interaction with the observer.

Wheeler extended this idea with the famous "*choice delay*" experiment, which shows that the observer's choice can retroactively affect the behavior of quantum objects, even after they have already moved through time and space. Reality, then, does not appear to be a rigid and fixed structure, but something fluid, shaped by interaction with consciousness.

These revolutionary views of modern physics find striking correspondences in Hindu philosophy, especially in Advaita Vedanta, one of the most fascinating and profound schools of Indian thought.

According to this doctrine, elaborated by the philosopher Shankara in the seventh century, the world we perceive is nothing but a projection of Maya, an illusion that conceals the only authentic reality: Brahman, the absolute, infinite and indivisible principle. At the same time, the individual (the Atman) is actually identical with this Brahman. There is no division between the microcosm of individual consciousness and the macrocosm of absolute reality.

For Shankara, understanding this truth means awakening from the great illusion, realizing that the universe and everything around us is nothing but a manifestation of our consciousness. It is a thought that comes close to Wheeler's view: the universe exists in relation to those who observe it.

Hinduism, with its millennia-old texts such as the Upanishads, offers poetic language to describe this relationship. In the Chandogya Upanishad, it is stated, "*Tat tvam asi*" (You are That). These three words sum up the whole of Advaita Vedanta: individual consciousness is no different from the universal principle that pervades the whole universe.

Connecting the two worlds, it is interesting to note that in quantum physics the observer is not a mere spectator but an active participant. Similarly, Advaita Vedanta does not see the individual as a separate entity from the rest of the universe, but rather as the very creator of his or her perception of reality. The famous phrase of mathematician and logician Alfred North Whitehead is illuminating in this context:

> *"A universe that can be fully explained without consciousness is a universe that, by definition, does not exist."*

Wheeler's choice delay experiment, where the observer's decision seems to "retroactively" influence quantum behavior, takes this insight to an uncanny level. Words such as those in the Bhagavad Gita, one of the main texts of Hinduism, almost seem to echo the same concept:

> *"The universe is in me, and I am in the universe."*

The Hindu notion of Maya, which describes the material world as an illusion, finds parallels with the paradoxical nature of the quantum world. Just as Maya conceals the true nature of Brahman, quantum phenomena seem to hide a deeper reality that eludes our direct understanding.

For Shankara, Maya is not something negative: it is the veil that allows us to experience phenomenal reality and challenges us to recognize its limitations. This perspective resonates with Wheeler's thinking: the universe we observe is the result of our active participation, an illusion that is continuously shaped by our act of observing and choosing.

If there is one lesson that science and philosophy can learn from each other, it is perhaps a sense of humility in the face of what we do not fully understand. Quantum physics, with its mathematical language, is exploring what Hindu sages have intuited for millennia: reality is not fixed, but an interweaving of awareness and participation.

John Wheeler, with his "participatory universe," and Shankara, with his philosophy of non-duality, remind us that the observer (and his awareness) are at the center of all that exists. We are not separate from the universe; we are the universe awakening to itself through our eyes. And, perhaps, in that fine line between knowing and being, lies the ultimate mystery of what we call reality.

Chapter IV. The nature of mind.

The mind in Hinduism.

Hinduism, with its thousands of years of history and philosophical richness, offers an extraordinarily complex and fascinating view of the mind. The mind, in the Indian view, is not simply a rational faculty or a set of thoughts. It is at the heart of an inner struggle that binds human beings to the cycle of birth, death and rebirth (Samsara), but at the same time can unlock the path to ultimate liberation (Moksha).

According to Hinduism, the mind is closely related to the concept of Maya. Maya is not an illusion in the trivial sense, but rather a filter that conceals the true nature of the world. The mind, having come into contact with the senses, creates a representation of reality that appears concrete but is actually deceptive. In sacred texts such as the Upanishads, it is stated that the mind allows one to experience the phenomenal world but diverts the human being from the direct experience of Brahman.

An emblematic passage comes from the *Kena Upanishad*, where it says:

> *"That which cannot be thought of by the mind, but through which the mind itself thinks: this is Brahman."*

Here, the mind appears not as an infallible means, but as a limit to the understanding of the absolute, a filter that must be passed through.

The mind also plays a key role in conditioning the human being to the cycle of Samsara, the continuous flow of birth, death and rebirth. The individual self's emotions, desires and identifications with the material world chain the soul (Atman) to this wheel. The Bhagavad Gita explains this conflict poetically and dramatically. Arjuna, the protagonist warrior, a symbol of the individual deceived by the senses, receives the teachings of Krishna, the guiding deity who represents divine wisdom. Krishna stated:

> *"The undisciplined mind is your worst enemy, while the controlled mind is your best ally."*

This verse emphasizes the importance of training the mind to overcome attachment to the transitory world.

One of the main practices offered by Hinduism to deal with this dynamic is meditation (*dhyana*). According to Raja Yoga, described in Patanjali's Yoga Sutras, the mind is like a lake. When its waters are rough, it is not possible to see the bottom. The true self can be perceived only by calming the oscillations of thoughts, called *chitta vritti*.

The meditative process teaches how to concentrate, slow down mental activity and ultimately transcend it. Indian mystics, such as Ramana Maharshi, stressed the importance of self-inquiry to overcome the mind and access truth. Maharshi asked his disciples to continuously question, "*Who am I?*" This question, Ramana explained, brings the mind to a halt, revealing the Atman hidden beyond ego and thought.

Another key point in Hindu philosophy is the role of the mind in attaining Moksha, the state of ultimate liberation. In Moksha, according to Vedic texts, the mind ceases to be an obstacle and becomes a tool for unveiling the supreme reality. Traditions such as Advaita Vedanta, founded by Shankara in the 8th century, emphasize that the mind must be "loosed" into the direct experience of Brahman, the infinite. Shankara likened the mind to an eclipse: it can obscure the light of the sun (Brahman), but it can never destroy it.

In Varanasi, a holy city par excellence, the concept of transcending the mind is also manifested in ritual. Here, at sunset, hundreds of lamps are lit along the banks of the Ganges River. This spectacle, which concludes with the chanting of the mantra "Om," reminds us that the mind, though subject to the darkness of illusion, can kindle the light of truth.

Interestingly, the description of the mind as a filter that creates the appearance of reality has intriguing analogies with quantum physics. The observer, according to quantum theory, influences the behavior of subatomic particles. Even in modern physics, "reality" seems to be deeply dependent on the point of observation. This parallelism, while not to be understood as a scientific identification, fascinates scholars and thinkers, so much so that it has been the focus of debate for decades. Physicists such as Werner Heisenberg, who visited India and studied Vedic philosophy, noted philosophical similarities between the concept of universal interconnectedness and quantum theory.

The mind, in Hinduism, is a powerful tool but also an obstacle. Like the glass of a lens that can be clear or opaque, the mind governs the perception of reality. Overcoming its limitations means moving beyond Samsara, discovering the eternal and infinite Brahman. In an age when quantum physics challenges the certainties of science, dialogue with these ancient philosophical insights not only enriches human understanding but invites reflection on the wonders hidden behind the veil of thought.

The illusion that imprisons man in samsara.

Mind, in Hindu philosophy, is a construct deeply intertwined with the concept of Maya, the cosmic illusion that imprisons human beings in a distorted perception of reality. This illusion is not only passive but active: it acts through the thoughts, emotions and fleeting identifications that bind people to the phenomenal world.

The ancient sages of Hinduism, such as Patanjali in the Yoga Sutras, wrote that the mind is like a surface of water. Every thought, desire or fear is a ripple, a wave that disturbs its stillness. Patanjali called these disturbances "*citta-vritti*," or mental modifications. When the mind is agitated, these waves prevent one from seeing clearly one's true nature: the Self, or Atman, concealed beneath the illusions of Maya.

Hinduism directly links the centrality of the mind to the perpetuation of Samsara, the cycle of birth, death and rebirth. According to this view, the identification of the individual with the mind reinforces the sense of separation from the absolute (Brahman). Every thought that arises feeds Karma, the chains that bind the soul to a continuous return to the phenomenal world. Liberation (Moksha), therefore, is not achieved by escaping from the world, but by passing through the veil of Maya and dissolving identification with the conditioned mind.

Adi Shankara and Ramana Maharshi: lessons on the mind.

A central teacher of this view was Adi Shankara (8th century CE), the philosopher who consolidated the Advaita Vedanta system. Shankara preached the fundamental unity of Atman (the Self) and Brahman (the Absolute). Through stories and metaphors, he explained that the mind is like a snake perceived instead of a rope: an illusion. Once the rope is recognized, the snake vanishes.

Ramana Maharshi (1879-1950), one of India's greatest modern sages, expanded this view with the "who am I?" method. Maharshi invited his disciples to question the origin of their thoughts. He showed that thoughts dissolve the moment one recognizes the Self as the unchanging witness, beyond the mind. Maharshi lived on Mount Arunachala, a sacred place in Tamil Nadu, which he described as a living symbol of the Self.

What connects this to quantum physics? Scientist Fritjof Capra, in his famous book "*The Tao of Physics*," pointed out how some insights from Hinduism recall concepts from the quantum world. For example, Heisenberg's uncertainty principle states that we cannot simultaneously know with precision some fundamental properties of a particle (such as position and velocity). This uncertainty represents, in a sense, an "illusion" in our ability to understand reality directly and definitively.

The concept of non-locality, demonstrated by Alain Aspect's famous experiment in the 1980s, also resonates with the Hindu view that all things are interconnected and dependent on a single underlying reality. The universe, at the quantum level, behaves as a unitary system in which the parts are inseparable. Similarly, in Hinduism, Maya creates an illusion of separation where, in truth, everything is one.

The dialogue between Hinduism and quantum physics is not an attempt to overlap religion and science, but to explore areas of convergence in insights about the universe and mind. In the universe, according to quantum physics, reality eludes our total understanding. Similarly, Hinduism teaches that what we see is only a partial reflection of truth, a game of Maya. Only deep work on the mind-as demonstrated by the great masters-can lead us beyond illusion to genuine freedom.

The mind and Samsara: the great entrapper.

In Hinduism, the mind is seen as a powerful but ambivalent tool. On the one hand, the mind enables the person to discern phenomenal reality. On the other hand, it is the mind that creates the illusion of the material world, a condition called Maya. This illusion is what keeps the individual trapped in Samsara, the eternal repetition of rebirths.

According to Vedantic texts, the human mind, with its desires, attachments and fears, constructs an egoic identity (*ahankara*). This ego, believing itself to be distinct from the rest of reality, feeds the sense of separation. For example, Adi Shankara described ego and mind as a dust-covered mirror: the ultimate reality (Brahman) is always present, but the mind clouded by ignorance (*avidya*) prevents its full perception.

One of the most famous metaphors used to explain the illusory role of the mind comes from the Vedas: the image of a rope mistaken for a snake. In darkness, a rope can be mistaken for a snake, generating fear and anxiety. In reality, the snake does not exist. Likewise, Hinduism says, it is the mind that perceives a dualistic and illusory reality, while the ultimate essence is one.

Ego, mind and eternity: the journey from ahankara to the Self (Atman).

Quantum physics and Eastern philosophies, while belonging to seemingly distant worlds, share a striking similarity: both question the usual sense of reality. While physicists wrestle with the paradox of superposition and the interdependence of observers, Hindu philosophy investigates mind as a veil obscuring the true nature of being. An intriguing point of intersection is found in the role of the ego.

According to Hinduism, the mind is enslaved by an entrenched delusion, embodied by ahankara. This term, which in Sanskrit means "*I-maker*," denotes that part of the mind that is identified with the ego. To the modern scientist, this could be likened to the reductionist concept of "*individual consciousness.*" Ahankara, however, is not the authentic source of identity. For the ancient Hindu sages, it is a mental construct, transient and conditional, that separates the human being from his Atman: the eternal, immutable and universal Self.

Quantum physics, with its enigmatic and counterintuitive language, has shown a subtle world in which what is observed cannot exist without an observer. Erwin Schrödinger, one of the fathers of quantum theory, stated with great admiration for Eastern wisdom that

> *"Hindu wisdom has never separated mind and matter as we do in Western thought."*

This reference is no accident: in the Bhagavad Gita, a pivotal text in Hindu philosophy, Krishna teaches Arjuna that the mind is both the enemy and ally of the human being, depending on how it is trained and enlightened.

The mind, in modern physics, could thus be seen as a bridge. On the one hand, it connects observable material events; on the other, it comes to question the observer himself, bringing us close to the mystery of consciousness. This is where we enter a subtle dialogue with Hinduism, in which the whole experiential world is filtered by mental activity, but enlightenment is achieved by going beyond the mind itself.

The heart of Hindu teaching on this theme lies in overcoming the ego, or ahankara. The Bhagavad Gita describes the ego as the main architect of human suffering. The more the individual attaches himself to the egoic self, the further he moves away from the eternal Self and inner peace. Adi Shankara, in his famous commentary on Advaita Vedanta, describes ahankara as "*the shadow in the darkness*," which has no reality of its own and vanishes only in the light of direct knowledge of the Self.

Indian master Ramana Maharshi, a pivotal figure of the 20th century, had a profound approach to approaching this journey to the Self. He taught the method of "*atma-vichara*," or "self-inquiry." Ramana invited his disciples to ask themselves a seemingly simple but actually revolutionary question, "Who am I?"

This meditative exploration gradually dissolves identification with the ego and leads back to the Self, often described by masters as "*endless silence.*" An anecdote related to Ramana relates that when asked by a student how to achieve enlightenment, he replied:

> *"Recognize the one who asks the question. That self that seeks to know is nothing but the veil."*

Another key reference is the master Nisargadatta Maharaj, who in his teachings insisted on awareness of the starting point: the universal feeling of being, or "*I am*." According to Nisargadatta, dwelling in this state of consciousness without identification, one comes to intuit the unlimited nature of the Self, gradually dissolving the illusion of ahankara.

Hinduism differs from many Western traditions in how it perceives time and mind. In this philosophy, mind is not a property of the brain, but rather a transient phenomenon related to the cycle of Samsara, the continuous succession of birth and death. At the same time, Atman is eternal and inexhaustible. This principle is celebrated in the writings of the Upanishads, which state:

> *"Atman is neither born nor dies, nor can it be reduced to fragments."*

To better understand the meaning of this inner journey, one may recall a crucial statement by the mystic poet Kabir:

> *"I am the Self within you. Where else will you look?"*

The Hindu experience invites not only philosophical speculation, but a meditative and practical path in which the mind is not repressed, but used as an instrument toward the infinite.

The modern debate on consciousness.

The parallels between modern theories of consciousness and those proposed by Hindu philosophy appear increasingly relevant. In contemporary neuroscientific research, consciousness is often spoken of as an emergent phenomenon, a concept akin to that of "Maya" or cosmic illusion described in Hinduism. At the same time, physicists and philosophers, such as David Bohm and Amit Goswami, have proposed cosmic models that echo ancient Vedic wisdom, associating the quantum universe with a kind of "unified connectedness."

If quantum physics teaches us that reality depends on the observer, Hinduism adds that the observer himself must dissolve to contemplate the eternal. The mind, with its limitations and potentialities, becomes the fundamental field of this inquiry. The ahankara is not an enemy to be annihilated, but a door to be passed through. Beyond, there remains the Self, still and free, the eternal witness of all transformations.

This path does not belong exclusively to one faith or science, but stands as a universal possibility, a challenge for anyone who wants to go beyond the visible and discover the hidden core of reality, there where matter and spirit, mind and universe merge into one Essence.

The mind as a troubled mirror.

Comparisons between Eastern philosophies and quantum physics find fertile ground when it comes to the mind and its nature. For centuries, Hinduism has offered sophisticated philosophical models of the workings of the mind. Among these, Patanjali's Yoga Sutras occupy a prominent place. The famous verse "*Citta-vritti-nirodha*" ("*The cessation of the fluctuations of the mind*") summarizes an entire philosophy of the mind as a mirror troubled by incessant waves.

According to Patanjali, the fundamental problem is "*vritti*," mental agitations. These fluctuations, likened to ripples on a pond, are caused by desires, attachments and ignorance (*avidya*). He states that to experience ultimate truth, the mirror of the mind must be made still. Only then can absolute reality be reflected in it, free from distortion.

This idea, more than two thousand years old, finds a surprising echo in the language of quantum physics. At the subatomic level, reality does not follow the deterministic laws of classical physics. The quantum model describes a fluctuating universe, where matter is both wave and particle. In this scenario, the role of the observer-what in philosophy is referred to as the mind-is crucial. The very act of observing changes the behavior of particles, as shown by iconic experiments such as the double slit.

It is not difficult to discern a parallel. If for quantum physicists it is the observing mind that conditions reality, for the Indian tradition it is again the mind that must be quieted in order to unveil deep reality: the connection between Atman, the individual self, and Brahman, the cosmic absolute.

The mind as a mirror and the practice of stillness.

The path to quieting the mirror of the mind is known as *dhyana*, or meditation. This practice is rooted in the Vedas and developed along yogic and Vedantic traditions.

Ramana Maharshi, one of the greatest sages of the 20th century, was born in 1879 in Tiruchuli, South India. Ramana spent much of his life at the foot of the sacred mountain of Arunachala, which he considered a manifestation of Shiva.

Ramana offered spiritual seekers a simple but radical practice: self-inquiry (*atma-vichara*). He asked them to place at the center of meditation a fundamental question, "*Who am I?*" This question brought attention to the beholder of all else, to that underlying awareness that is the true self, or Atman. Ramana taught that the mind stops agitating when it ceases to identify with the ego (*ahankara*). Only then does the individual recognize that Atman and Brahman are one.

Arunachala Mountain became a powerful symbol of the inner quest. Even today, thousands of pilgrims go there each year to undertake *the pradakshina*, the ritual walk around the mountain, hoping to gain a glimmer of the peace that Ramana embodied.

Resonances with quantum physics.

Quantum physics, while having a scientific rather than a spiritual approach, suggests that reality is not composed of solid objects, but rather of a field of probability in which consciousness plays a determining role. Werner Heisenberg himself, one of the fathers of the uncertainty principle, was fascinated by the similarities between Eastern thought and his scientific findings. Heisenberg noted how ancient Indian philosophies suggested a fluid reality in which unity underlies apparent diversity.

We can imagine the mind, in the Indian view, as a kind of "*personal quantum field.*" When the field is agitated by ego impulses, the worldview becomes fragmented. But, through meditation, this field returns to its original condition, an indistinct unity.

An interesting anecdote that can be connected to these themes concerns Nisargadatta Maharaj, a spiritual master contemporary to Ramana Maharshi. A disciple asked him one day how to achieve absolute peace. Nisargadatta replied with these startling words:

> *"Try to understand that everything you see and experience is in the mind, not in reality. When the mind is silent, truth is revealed."*

This wisdom is more relevant than ever. In an age when science itself abandons the certainties of objectivity to explore the paradoxes of quantum reality, Eastern philosophies offer a context for reflecting not only on the nature of the universe, but on our perception of it. Perhaps what quantum physics and Hinduism are trying to tell us is that the universe, like the mind, reflects who we are. After all, to know unity, we must first know ourselves.

Non-dualism and the liberation of the mind.

The philosophy of Advaita Vedanta, one of the pillars of Hinduism, is based on the revolutionary idea of non-dualism. This view is condensed into the famous Sanskrit phrase "*You are That*" According to this doctrine, there is no separation between our inner self (Atman) and universal reality (Brahman). Everything is One, undivided and interconnected.

If quantum physics investigates the interrelatedness of all particles in the universe, Advaita Vedanta takes it a step further: not only is everything connected, but this connection has never been fractured. According to this view, the division between observer and observed, subject and object, is only illusory. And this is where the mind comes in.

In Advaita Vedanta, the mind is seen as a two-faced tool: on the one hand, it is the main obstacle to grasping universal unity. On the other, it is the means by which illusion can be transcended. This obstacle is represented by *Ahankara*, the ego, which fragments perception and creates the sense of separation between self and the world.

The ego, fueled by thoughts, desires and identifications, is what binds human beings to the cycle of Samsara, the constant wandering between suffering and delusion. However, through spiritual practice and meditation, the mind can be used to free itself from its own chains. The key is to come to see it for what it is: a shadow consisting of its own contents, not the true self.

Swami Vivekananda, one of the most influential voices in modern Hinduism, used a simple and powerful image to explain this concept. He said:

> *"When the mind becomes calm like a clear lake, then the Mount of Truth is clearly reflected on its surface."*

This Mount represents Brahman, the absolute. The restless mind, on the other hand, is like a mirror moved by the wind, distorting what it reflects, creating the illusion that we are separate from the whole.

To achieve this inner calm, the teachings propose practices such as meditation, scripture study (*shastra*) and self-observation. One of the most profound tools is *Atma-vichara*, or self-inquiry, promoted by mystic Ramana Maharshi.

Ramana Maharshi (1879-1950) left a unique spiritual legacy. A silent master, Ramana pointed to a direct method of overcoming the tyranny of the mind. This approach, called atma-vichara, was based on investigating the self through an only seemingly simple question, "*Who am I?*"

According to Ramana, every thought or emotion starts from the idea of a personal "I." Investigating the origin of this "self," one discovers that it is nothing more than an illusion created by the mind. The true self, the pure Atman, remains unchanged beyond the fluctuations of the psyche.

If Advaita shines in its spiritual aspects, modern researchers find interesting parallels with quantum physics. The mind, according to Advaita scriptures, creates a veil that hides the unique and indivisible reality, just as the observer modifies the observed according to Heisenberg's uncertainty principle. The findings of contemporary physics, which cast doubt on the rigid distinction between matter and energy, seem to echo the philosophical principles that Indian sages expressed centuries ago.

A fascinating concept is what physicist Erwin Schrödinger, deeply influenced by Eastern philosophies, clearly expressed: consciousness is not individual, but universal. Rational Western thought has tried to understand consciousness as a product of mind; Advaita responds, however, that mind itself is a product of consciousness.

From self-knowledge to the experience of the infinite.

For millennia, Hinduism has wondered about the true nature of mind and self, addressing questions that today even Western science, particularly quantum physics, is beginning to explore. At the heart of this spiritual quest is the practice of atma-vichara, or self-inquiry, which promises to lead from mental fragmentation to the experience of the infinite. This perspective, which spans the ancient texts and mystical experiences of India, reveals surprising connections to contemporary thinking about consciousness and the nature of reality.

One of the most famous examples of atma-vichara is the life of Ramana Maharshi, one of the greatest spiritual masters of the 20th century. Born in 1879 in Tiruchuli, a small village in southern India, Ramana had a sudden awakening at the age of sixteen. As he lay stretched out, believing he was about to die, he turned his attention to the fundamental question, *Who am I?* He meditated on the fact that it was neither body nor mind, but something deeper and unchanging: the Supreme Self, called Atman in Vedic texts.

With this insight, Ramana abandoned his ordinary life and settled at Mount Arunachala, a holy place for Hinduism. There, he experienced a condition of total mental silence, which was not a sterile emptiness, however, but an infinite openness, a direct connection with what many would call the Divine. For the master, the mind, once stripped of its egoic projections, became a channel for pure awareness.

According to the Hindu tradition, the human mind is essentially a neutral instrument. However, through identification with the ego (ahankara), it generates an illusory separation between the self and the world. Nisargadatta Maharaj, another great master of the 20th century, explained:

> *"The mind creates the prisoner by imagining the prison. Abandon identification and you will find that you are free all along."*

This statement is echoed in "*atma-vichara*," where the question "*Who am I*?" becomes a sword to cut through the incessant flow of mental identifications. It is not a matter of denying the mind, but of transcending it, of recognizing that, at its deepest level, it is a mirror of the self, not the self itself.

Interestingly, some of these insights find parallels in modern quantum physics, which challenges conventional ideas about reality and mind. Physicists such as Eugene Wigner and Wolfgang Pauli have raised questions about the role of the observer in the creation of the quantum world. The idea that thought or awareness influences material reality is reminiscent of the Hindu principle that the mind, once purified, can perceive the infinite that is always present.

The very concept of reality as an interconnected, nonlocal network explored in quantum theory appears aligned with the nondualistic traditions of Hinduism. Advaita Vedanta, perhaps the most important school of Hindu philosophy, holds that there is no real separation between subject and object, between mind and universe. As master Adi Shankaracharya said:

> *"Brahman is all that is. Grasp it, and you will find that the individual Self does not exist as a separate entity."*

On an experiential level, the practice of *atma-vichara* has transformative power. It leads to inner stillness, reduces anxiety and inner divisions, and allows us to explore our highest spiritual potential. By meditating on who we really are, according to Ramana Maharshi, we stop being slaves to mental fluctuations. In this way, the mind is not eliminated, but reintegrated into the Self as a tool for experiencing the totality of existence.

A modern practitioner might describe this process as "resetting" the mind, eliminating unnecessary programs that separate us from the living present. It is a path that requires courage and patience, but promises the greatest gift: freedom.

At a time when neuroscientists and philosophers debate the origin of consciousness, atma-vichara makes an essential contribution. Rather than looking for consciousness in neural circuits or the brain, Hinduism sees it as the very substratum of reality. It is a view that challenges the reductionist underpinnings of Western science, but also attracts increasing attention among researchers.

Thinkers such as David Chalmers, who has called consciousness "*the hard problem*," recognize that science does not yet have tools to explain how subjective experience emerges from neuronal activity.

Perhaps, as the Hindu sages suggest, the answer lies in consciousness itself, which is not a product of the mind but its origin.

The practice of atma-vichara is both ancient and revolutionary. It challenges what we think we know about ourselves, opening the door to a dimension of existence free from mental and egoistic limitations. Through figures such as Ramana Maharshi and Nisargadatta Maharaj, we see that transcending the mind is not an escape from reality, but the means to rediscover its true essence.

In a universe that modern science describes as interconnected and pulsing with possibilities, perhaps there is no substantial difference between the silence of the realized Self and the quantum field. Both invite us to recognize that behind the appearance of the manifold lies a single infinite reality.

Inner experimentation and mind control.

Hinduism offers practical tools for understanding and transforming the mind. Meditation (*dhyana*), for example, is considered one of the main paths to bring the mind to a state of stillness, reducing the agitation created by distracting desires and thoughts. The texts of the Bhagavad Gita, one of the most studied works in Hinduism, offer key teaching in this regard. In Chapter VI, Krishna explains mind control to Arjuna in words of extreme clarity:

> *"The mind can be your best friend or your worst enemy.*
> *Dominate the mind, and you will be free from the bondage of samsara."*

Another example is provided by the system of Patanjali, who in his Yoga Sutras describes the mind as an instrument that, if not purified, creates chaos. The goal of yoga, according to Patanjali, is to bring the mind to a state of citta-vritti-nirodha: the silence of mental fluctuations.

The human mind is one of the deepest mysteries of the universe, an enigma that fascinates both modern science and ancient spiritual traditions. Hinduism, one of humanity's oldest philosophies, offers tools for exploring the mind that prove remarkably akin to the musings of quantum physics. The two approaches, while coming from different backgrounds, share a common idea: reality emerges from the way the mind perceives and organizes experience.

Hinduism views the mind as a powerful tool, but also as a potential obstacle. Through meditation, the mind is disciplined. This does not mean suppressing it, but observing it without identifying with thoughts. When the process is brought to maturity, a state of deep calm is accessed, which in quantum physics could be likened to the unified field: a still reality from which the energy fluctuations that form the material world emerge.

Another pillar of Hinduism is Patanjali's "*Yoga Sutras*," a collection of aphorisms dating back to the second century BCE. In this text, Patanjali lays out the steps to bring the mind to a state of balance. His goal is the attainment of "*citta-vritti-nirodha*," or the "*silence of mental fluctuations*." Patanjali notes that the mind, if unchecked, functions like a broken mirror: it refracts thoughts and emotions in a chaotic manner, giving the illusion of a fragmented reality.

From a practical point of view, Patanjali proposes eight stages of yoga (ashtanga yoga), which include physical postures (asana), breath control (pranayama), and mental concentration (dharana). Silence within is gained gradually by training the mind to purify itself, just like a river that slowly becomes clear when the agitation of its waters ceases.

Quantum physics has discovered that, at the subatomic level, reality is influenced by the observer. Experiments such as "wave-particle dualism" show that a particle can behave as matter or as a wave, depending on how it is observed. Similarly, Hinduism teaches that the universe is shaped by the human mind.

A central figure in recent history who has explored the relationship between mind and reality is Niels Bohr, one of the fathers of quantum mechanics. Bohr was fascinated by Eastern spirituality and often consulted sacred Indian texts. The *yin-yang* symbol, which he kept in Bohr's personal coat of arms, represents complementarity, a central concept in quantum phenomena. This same complementarity is evident in Hinduism, where the interplay of opposites (prakriti and purusha) represents the balance between matter and spirit.

What makes Hinduism unique is its invitation to inner experimentation. Insights are not accepted as dogma, but are considered attainable results through practice and self-inquiry. The example of the sages of ancient India is illuminating: they lived in hermitages surrounded by nature and devoted their lives to observing the mind. Places such as *ashrams* along the Ganges River, even today, are centers of soul-searching, where practitioners from all over the world experience the millennia-old teachings.

An interesting example concerns Erwin Schrödinger, scientist and creator of the quantum wave equation. Schrödinger was deeply inspired by the Upanishads, Indian philosophical texts that investigate the nature of being. He found fascinating the concept of Brahman, the ultimate and indivisible reality, which resonates strikingly with the idea of a quantum field uniting all that exists.

Hinduism and quantum physics seem to converge on a crucial point: mind is not only an element of reality, but also its creator. While modern physics uses technological tools to observe the subatomic world, Hinduism invites us to turn these same tools inward, exploring the "microuniverse" of consciousness.

The two disciplines are not enemies, but allies. Ultimately, they offer an answer to the question humanity has always asked: who are we, and what role does the mind play in the creation of the universe? Perhaps, as Hinduism suggests, understanding the mind means understanding the whole.

The mind as a mirror of unity.

The idea that the mind is a mirror, capable of reflecting the underlying unity of what we perceive as reality, is central to the philosophy of Advaita Vedanta. This current of Hinduism, rooted in millennia and the sacred texts of the Upanishads, invites us to look beyond appearances. The human mind, while seemingly complex and creative, is only an illusion that hides the fundamental truth. The real Self, or Atman, is identical to the universal principle, Brahman.

In a sense, Advaita Vedanta comes close to modern theories of quantum physics, which challenge ordinary perceptions of the world. According to quantum physicists, what we call "objective reality" is a fluid expression driven by the interplay of energy, observation and awareness. As in Eastern thought, quantum physics also indicates that there is a hidden unity behind the multiplicity of the world.

Nisargadatta Maharaj, one of India's greatest spiritual masters of the 20th century, summarized the nature of mind in a simple but powerful image:

> *"The mind creates the world and names it. But you are beyond the mind. We are the clear sky, and the mind is but a passing cloud."*

For Nisargadatta, the mind is nothing more than a tool by which human beings interpret and construct their perception of reality.

However, this tool is limited because it obscures the unlimited awareness that is the true nature of the individual.

This image is reminiscent of the quantum concept of the "underlying field." In the language of physics, we often speak of a unified field, an invisible reality that connects every particle and every event. It is a unity that, in its essence, is not altered by visible fluctuations or observable phenomena. Like clouds in the sky, however, the surface of things distracts the observer from the deep dimension of reality.

One of the most pronounced differences between the Western view of the mind and that of Hinduism lies in the focus of work that addresses it. In the West, the mind is often seen as an entity to be developed, to be nurtured with ever-changing knowledge and skills. Psychology, neuroscience, and pedagogy have focused for centuries on the evolution and improvement of rational thinking.

Advaita Vedanta calls for a radical reversal: not to improve anything, but to go beyond it. For Vedantic thought, it is not about accumulating knowledge or fighting negative thoughts. On the contrary, spiritual observation is about recognizing that the mind, in its constant activity, distracts from fundamental reality. Meditation, or dhyana, is a key tool for this process. Not to change the mind, but to understand its illusion.

The Upanishads, the philosophical texts that form the heart of Hinduism, reinforce this view. The Mandukya Upanishad, in particular, offers a famous metaphor: the phenomenal world is like a rope mistaken for a snake. In darkness, ignorance (or avidya) creates an illusion. Only with light - awareness - can one see that there is no snake.

Another significant example comes from Adi Shankaracharya, the great Advaita Vedanta philosopher who lived around the 8th century. Shankara argued that the world is like a dream. Everything that appears real ceases to exist when one recognizes the illusory nature of the dream itself. This concept is not far from current thinking in quantum physics on the role of the observer.

The intersection of Advaita Vedanta and quantum physics has fascinated many contemporary thinkers. Fritjof Capra, author of "*The Tao of Physics*," was among the first to show the points of contact between Eastern mystical traditions and the insights of modern science. Capra recognized that both Indian sages and contemporary physicists come to the same conclusion: the world we perceive is a reflection, not the ultimate reality.

Scientist David Bohm, famous for his theory of implicate order, also echoed similar ideas. Bohm described the universe as an indivisible whole, where what appears separate is only a fragmentation imposed by the mind. For Bohm, the "whole" is hidden behind every manifestation of reality. This description is strikingly close to the conception of Advaita Vedanta, where everything is Brahman, unique and indivisible.

Hinduism proposes practices to discover this unity beyond mental flow. Meditation is central, but the study of sacred texts and dialogue with a spiritual master (guru) can also help. The famous 15th-century poet-mystic Sant Kabir described the mind as a monkey jumping from branch to branch: unstable, disordered. But Kabir, in his poetic works, showed that love of the divine and concentration on the Self lead to stillness and recognition of truth.

The mind, according to Hinduism, is a mirror capable of reflecting reality. For this to happen, its distortions must be eliminated. The challenge, however, is not to transcend the mind through struggle or improvement. It is the simple recognition that behind its clouds lies a clear and unchanging sky. This teaching, which spans millennia of Eastern wisdom, is now also being rediscovered by contemporary science. Philosophy and physics point to one great truth: unity has always been there. It is just up to us to stop and observe it.

Beyond the mind: a timeless perspective.

In the vast philosophical landscape of Hinduism, the mind is seen as a bridge, but also as a veil. It is a bridge because it allows us to explore the universe of perceptions and experiences. But it is a veil because, at its borders, lies a deeper dimension, beyond the illusions of ordinary thought. It is a perspective that today proves remarkably close to the insights of quantum physics.

Quantum physics, the same physics that revolutionized our understanding of the universe with figures like Niels Bohr and Werner Heisenberg, teaches us that we cannot observe reality without changing it. The observer is not a neutral appendage; he influences the experiment itself. This principle finds something like a spiritual echo in Hinduism. The question here is not simply scientific. It is existential: who is the observer? Who is looking behind the mind?

For ancient Indian thinkers, the observer is both an enigma and the heart of existence. Such age-old texts as the Upanishads describe the mind as a tool, but not as the essence of the human being. One of the most famous metaphors we find in the Katha Upanishad compares the human being to a chariot: the senses are the horses, the mind replaces the reins, the intellect is the driver, while the "observer," the Atman, is the silent passenger. The crucial point, according to the Hindu sages, is to stop identifying with the driver or the mind, and discover the passenger.

This journey is an immense challenge, because the mind is not only a bridge to knowledge; it is also a deceiver. Hinduism defines this deception by the term Māyā, which means "illusion." Māyā is not a condemnation: it is the very condition of the phenomenal world, the "cosmic game" (līlā). The human being, a prisoner of this illusion, tends to become confused, to believe that the mind and its thoughts are the ultimate reality. But this, say the masters, is a superimposition that keeps us from the deeper truth.

A perfect example of this effort to transcend the mind is the work of Ramana Maharshi, a South Indian saint and philosopher who lived between 1879 and 1950. Maharshi, who spent much of his life in the meditative silence of Arunachala, a sacred mountain, invited his disciples to ask themselves a radical question, "Who am I?" This is not a rhetorical question, but a direct practice to unmask the mind and its illusions. According to Maharshi, the authentic self, or Atman, needs no thoughts to exist: it is eternal, unchanging, and not bound by time or space.

Quantum physics, a modern parallel of this insight, leads us to explore a universe in which time and space seem to be dominated by probabilities and nondeterministic events. An experiment such as the double-slit experiment shows that a particle can behave either as a wave or as a corpuscle depending on its interaction with the observer. This contradictory interplay of multiple realities and possibilities may find a fascinating resonance with the Hindu idea of a truth beyond our cognitive tools.

The Bhagavad Gītā, one of the most widely read sacred texts in India, explores this dynamic with extraordinary clarity. Krishna, in his dialogue with Prince Arjuna, speaks of the mind as a tool to be mastered, not denied. Krishna offers a radical vision: to overcome suffering, it is necessary to free oneself from attachment and identification with thought. The disciplined mind is not an enemy, but becomes an ally on the path to spiritual freedom.

> *"He who has conquered the mind is calm in the face of heat and cold, pleasure and pain, victory and defeat" (Bhagavad Gītā, 6:7).*

Looking at Hinduism, we find that this vision is not only philosophical but also practical. Traditional techniques such as yoga and meditation have a specific purpose: to quiet the noise of the mind in order to glimpse what is deeper. The silence of meditation appears, then, as the inner equivalent of a "quantum singularity," a point at which the categories of time and space dissolve.

Hinduism proposes to us a timeless reality, which is not limited to the conceptual structures of thought. The observed universe is only a projection of the mind, but behind the mind lies the eternal. The mind is a fundamental tool for our daily life, but it is not who we are. Overcoming it, in Hindu terms, is the attainment of Moksha: liberation from all constraints.

Beyond the mind, there is neither past nor future. There is the eternal present, a reality that cannot be described but only experienced. And that, with an intonation that links us to the depths of quantum physics, may be the only truth.

Consciousness and quantum physics.

Quantum physics, described by many as the most mysterious frontier of modern science, has challenged established views of reality. Its theories redefine concepts such as matter, energy and even consciousness. In parallel, Hinduism, one of the world's oldest philosophical traditions, has been addressing similar questions for millennia, especially regarding the nature of mind and the role of consciousness in the universe. These two seemingly distant worlds surprisingly find points of contact.

The nature of mind in the Upanishads.

Hinduism considers the mind a powerful tool, but also an obstacle to understanding the ultimate reality, Brahman. This reality, according to the Upanishads, is the universal and unchanging principle underlying all that exists. However, the mind often hides this truth with its limited perceptions and incessant workings, which Hindus call "*manas*." To discover the true nature of existence, Hinduism calls for silencing the mental noise through techniques such as meditation and dhyana (contemplation).

A famous passage in the Mundaka Upanishad uses a significant metaphor:

> *"The mind is like an arrow that must be shot accurately toward the target, that is, Brahman."*

This idea is somewhat reminiscent of the quantum concept of the observer, according to which an observer's perceptions can influence the observed reality.

An interesting parallel between Hinduism and quantum physics emerges in the notion of Maya, the "veil of illusion." According to tradition, Maya is that which distorts human perception, making what is actually unitary seem separate. Quantum physics, through the double-slit experiment, highlights how a particle can behave as a particle or a wave depending on the observation. Here, reality seems to take on different forms, as if "shaped" by the observer's perspective.

In the 20th century, physicists such as Erwin Schrödinger found fascinating analogies between Maya and the behavior of subatomic particles. Schrödinger, who was deeply influenced by reading the Upanishads, wrote that "*the separation of subject and object is only an illusion.*" This view reflects the heart of non-dualist Hindu thought, known as Advaita Vedanta, that all reality is interconnected.

Another central aspect is the importance of consciousness. Hinduism, with the concept of Atman (the soul or self), regards consciousness as the manifestation of the union between the human microcosm and the universal macrocosm. In quantum physics, consciousness is a controversial but inescapable topic. Some scientists, such as Eugene Wigner in the 1960s, hypothesized that conscious observation was necessary to make the collapse of the wave function "real," that is, to determine the actual state of a particle.

From this perspective, the role of human consciousness resembles that attributed by Hinduism to the awakened mind. Indian philosophy holds that when the mind recognizes the illusory nature of the ego, it opens up to an understanding of unified reality. Quantum physics, while not embracing a metaphysical explanation, suggests that consciousness might play a non-marginal role in the fundamental structure of the universe.

In building bridges between these two worlds, personalities such as Fritjof Capra have played a key role. In his 1975 book Capra emphasized the similarities between ancient Eastern traditions and the recent discoveries of quantum physics. Capra visited India several times and found in the teachings of Vedanta a rich source of inspiration for understanding the mystery of matter and mind.

Symbolic places also tie these reflections together. The research center founded by J. Robert Oppenheimer at Los Alamos, where the atomic bomb was developed, became a symbol of how scientific insights can also intersect ethical and philosophical dilemmas. Hinduism and quantum physics address mind and consciousness with different but complementary approaches. One proposes tools to transcend illusion and discover universal unity. The other arrives at similar conclusions, but through mathematical and physical lenses. For those seeking answers to the riddle of reality, this dialogue between ancient philosophies and modern science remains one of the most fascinating investigations of our time.

Does consciousness create the cosmos? From wave function to Advaita Vedanta.

The nature of the relationship between consciousness and the universe is one of the most fascinating questions in human thought. This dialogue takes place in seemingly distant contexts: modern quantum physics and the ancient insights of Advaita Vedanta, a current of non-dualistic thought in Hinduism. Both fields, in different ways, interrogate the role of the observer in the creation of reality.

In quantum physics, the wave function describes a probabilistic state, a realm of pure possibilities. It is with observation that one of the potential states manifests, becoming a concrete reality. This concept, central to the Copenhagen interpretation, took bolder forms with Eugene Wigner's interpretation and later in physicist John Wheeler's "*Participatory Universe.*"

Wigner, a physicist and philosopher of science, hypothesized that human consciousness was an active agent in the collapse of the wave function. According to him, without the conscious mind, the universe would remain in an indeterminate state. This suggestion introduced a radical idea: the universe could depend on the intervention of consciousness to exist in definite form. John Wheeler expanded this perspective, proposing that the universe was not simply observed, but co-created by observers in a continuous cycle of interaction. His idea of a "participatory universe" implies that the act of measuring or observing does not passively register reality, but constitutes it.

These quantum insights, though born out of physical-mathematical contexts, find a surprising echo in Advaita Vedanta.

Advaita Vedanta, the nondualist school founded by Indian philosopher Adi Shankara in the 8th century, teaches that phenomenal reality is Maya, an illusion. Maya does not mean that the world does not exist, but that what we perceive as independent reality is ultimately devoid of true essence. The only absolute truth, according to Advaita, is Brahman, the supreme, indivisible, featureless reality. The individual self (Atman) is also Brahman, but deluded by Maya, it experiences existence as separate from the totality.

This view finds a curious parallel in quantum physics. If reality does not exist in itself but is manifested only by the intervention of the observer, then it can be considered an illusion until the moment of its "creation" through measurement. In the context of Advaita, the one who observes (the observer) and that which is observed are not really separate. Both share the same essence.

This encounter between quantum physics and Advaita is more than an academic confrontation. It can be seen as an attempt to understand mind, the cosmos and their relationship as one. Wheeler suggested that the universe itself was one big feedback loop, in which each observer contributes to the construction of reality. Adi Shankara, centuries earlier, pointed to a similar principle, saying:

"The world is like a dream, real for those in the dream, unreal for those awakened."

A famous anecdote from Indian culture illustrates this idea well. It is told that the ancient sage Janaka, after an intense dream in which he lived as a poor man, asked his spiritual master:

"Am I the king who dreams of being poor or am I a poor man who dreams of being king?"

The master replied:

"Neither of them. You are the consciousness that observes both."

This story reflects the heart of Advaita: the observer is the fundamental reality, not what is being observed.

The concept that consciousness is central to reality finds modern reflections not only in physics, but also in the philosophy of science and the arts. Nobel laureate in physics Roger Penrose pointed out how some aspects of consciousness could be related to the quantum structure of the brain, although his Orch-OR theory is still debated. Meanwhile, writers such as Fritjof Capra popularized the connection between modern science and Eastern wisdom, rekindling interest in the spiritual dimension of scientific thought.

In the cultural sphere, this dialogue has influenced film and literature. Films such as The Matrix (1999) reintroduce the idea that the perceived world is an illusion (Maya), and that a "higher reality" is accessible by entering a deeper level of awareness.

The idea that consciousness creates the cosmos remains central to the debate. Quantum physics suggests that emergent reality is not something "out there," but a construction dependent on the observer. Advaita Vedanta comes to the conclusion that the observer and the universe are one. Both seem to suggest that the separation between the observer and the observed is an illusion, and that consciousness is the point of origin of all reality.

As Wheeler wrote,

"The universe is an act of participation."

Similarly, Adi Shankara proclaimed:

"Everything is one. The self is the whole."

Two seemingly distant perspectives, but converging in one fundamental question: what if the essence of the universe is not separable from the essence of those who contemplate it?

Is the Universe aware?

A fascinating idea has begun to take shape in physics labs and university classrooms: the universe could be conscious. On the one hand, quantum mechanics raises profound questions about the links between matter and mind; on the other, Hinduism's texts, such as the Vedas and the Upanishads, offer a vision of total unity called Brahman, described as the ultimate and universal reality. Can there be a dialogue between these two perspectives?

In 1935, Erwin Schrödinger proposed a paradoxical idea to illustrate the oddities of quantum mechanics. Imagine a cat locked in a box, in which its fate is tied to the outcome of a quantum event: a particle may or may not decay, and the cat's life or death depends on it. In quantum terms, until someone opens the box, the cat is "alive and dead" at the same time, that is, it is in a "*superposition of states*."

This paradox underscores a crucial point in quantum theory: the observer plays a fundamental role. Only at the moment we observe does the quantum state "collapse" into one of the defined possibilities. But who or what is the observer? Is it simply a mechanical experiment, or does it take a conscious mind to complete the process?

This question is intertwined with ancient Hindu reflections. In Vedantic philosophy, Brahman is regarded as the "supreme witness" (*Sakshi*). He is the conscious presence that observes and sustains every manifestation of the cosmos. Just as observation in quantum physics transforms possibilities into reality, Brahman represents that awareness that allows the universe itself to exist.

Some scientists have wondered whether consciousness is an emergent phenomenon of the human brain or something more fundamental. Physicist John Wheeler, one of the fathers of quantum theory, advanced the idea that the universe requires conscious observers to truly exist. He called this view a "participatory universe" ("participatory universe"). In other words, the universe itself could be constructed by the interaction of matter, energy and consciousness.

A similar idea is found in sacred Indian texts such as the Upanishads, in which Brahman is not separate from the universe but is the very fabric of it. A famous passage in the Chandogya Upanishad states, "*Tat Tvam Asi*" ("*You are that*"), expressing the unity between the individual and the cosmos. Here again, consciousness is not confined to the human being, but is the very substance of the universe.

A famous experiment, known as the *double-slit experiment*, further demonstrates the fascinating connections between quantum physics and consciousness. In the case where a particle, such as a photon, passes through a barrier with two slits, its behavior depends on observation: if it is observed, it behaves like a particle; if it is not observed, it exhibits the characteristics of a wave.

This phenomenon raises a fundamental question: does the act of observing alter reality? According to some interpretations, it seems so. The mind of the researcher could collapse the system to a defined state. If so, awareness would not just be a property of reality, but an active force that transforms it.

In Hindu philosophy, all this finds striking parallels. Brahman is described both as infinite and static and as dynamic in its manifestation in the world. Even Maya, the veil of illusion, dissolves only at the moment of interaction with self-awareness, a process very reminiscent of the idea of the observer making quantum reality tangible.

Both physics and ancient texts converge on a crucial point: the separation of mind and world is an illusion. This insight was already evident in centuries past when Schrödinger himself, a scholar of Indian philosophy, wrote:

> *"This life you live is not only a part of the whole universe, but in a sense it is the whole universe."*

In exceptional cultural moments, modern physics seems to echo ancient insights. The Geneva conference in 1975, for example, discussed these ideas formally for the first time, thanks to physicists such as Fritjof Capra who sought convergences between science and spirituality. Meanwhile, in India, millennia-old Vedic schools continued to teach that the universe and consciousness are expressions of the same eternal principle, Brahman.

Questions about consciousness and the ultimate nature of reality remain unanswered, but the perspectives offered by quantum physics seem to make millennia-old insights from Hindu texts relevant. We are all part of a universe that is perhaps not only alive but also conscious.

If Schrödinger's cat teaches us that reality depends on observation, and the Vedas show us that everything is Brahman, we might indeed be active participants in a cosmos whose fabric is awareness itself. The question remains to what extent this awareness is individual or universal. Science and spirituality, once again, seem to dance together in the great mystery of reality.

What is reality? The quantum question.

Quantum physics has revolutionized the way we think about reality. Born in the early 20th century, this discipline has demolished the certainties of classical physics, unveiling a subatomic world that defies everyday logic. Phenomena such as superposition of states and quantum entanglement force us to revise the very concept of reality. But what does it really mean to "exist"? And what is the role of the observer in defining what we perceive as real?

One of the central questions in quantum physics revolves around the behavior of particles at the microscopic level. Experiments show that a particle, before it is observed, does not have a well-defined location or state. It is in a condition of "superposition," where it appears to be in several places or states at once. But when an observer comes into play, this ambiguity disappears. The particle seems to choose only one of its possible states.

This phenomenon is described as the "collapse of the wave function." But who or what causes this collapse? For many physicists, the observer himself is the key. Eugene Wigner, one of the leading figures in quantum physics and a 1963 Nobel laureate, pushed this idea to a radical conclusion: according to him, human consciousness may be the determining factor. It is the act of observation--by a conscious subject--that transforms the potentiality of the particle into concrete reality.

Wigner, with a mixture of audacity and insight, placed the human mind at the center of the interaction between the subatomic world and perceived reality. For many, this is a fascinating vision, but also a disturbing one: it implies that without conscious observers, the universe could exist only as a sea of indeterminate possibilities.

To illustrate this ambivalence between reality and observation, we can recall the famous Schrödinger's cat paradox, formulated in 1935 by Austrian physicist Erwin Schrödinger. Imagine a cat locked in a box with a vial of poison that can be released as a result of the decay of a radioactive particle. Until we open the box and observe, the cat is neither alive nor dead, but in a superposed state. It is only human observation that defines, with certainty, the animal's fate.

This thought experiment not only deepens the mystery of quantum superposition, but also opens deep philosophical doors: is there a reality independent of our observation? Or do we, with our consciousness, shape it?

The questions raised by these experiments find a surprising echo in some currents of Hinduism, particularly in Advaita Vedanta, a philosophical school that promotes a nondual view of reality. According to Advaita, ultimate reality - called Brahman - is not separate from consciousness. In fact, consciousness and reality are one. This thought is reminiscent of the quantum concept of the role of the observer: as in physics, reality in these spiritual doctrines depends on the mind's point of view.

Physicist John Wheeler, Wigner's successor, proposed an equally provocative idea: the principle of a participatory universe. According to Wheeler, the universe is not an unchanging entity that simply exists. It is, rather, an entity "under construction," continually defined by observations and interactions. The key question, for Wheeler, then becomes: can there be a universe without observers?

This is where physics and philosophy meet. In many Hindu texts, the universe is described as a "cosmic game" (Lila) enacted by primordial consciousness. The observer is not a mere spectator, but an essential participant in the creation of reality. If this idea is true, then the universe itself could be considered a manifestation of consciousness.

The implications of quantum physics are not only scientific. We also touch the existential plane. If consciousness does indeed play a fundamental role, this could change the way we view life. Reality is no longer something fixed and independent, but a changing field defined by the interaction between mind and matter.

Eastern cultures, particularly Hinduism, have always considered the mind to be central to understanding the nature of the universe. Today, thanks to quantum physics, these insights may find a dialogue with modern science. It is an invitation to reflect on ancient questions:

who are we really? And how much does what we call "reality" depend on our point of view?

Consciousness and reality in Hinduism.

Questions about the nature of reality and consciousness are not foreign to the great spiritual traditions. In Hinduism, these questions emerge as central themes, particularly in Vedantic thought and especially in Advaita Vedanta. This philosophy asserts that everything that exists is an expression of universal consciousness, called Chit or Purusha. According to this school, consciousness is not a by-product of the brain. Instead, it is the fundamental, unchanging principle that permeates the universe and manifests itself in the form of everything we perceive as physical reality.

Vedic texts and the Upanishads emphasize the distinction between what is real and what is illusory. Material reality, known as Prakriti, is often described as maya, an illusion or cosmic game projected by consciousness. In other words, the observable universe has no independent existence. It is totally intertwined with universal consciousness. This concept is expounded in the famous verse of the Upanishads:

> *"Brahman is the real, the world is unreal, the individual is no different from Brahman."*

These ideas find surprising resonance with some fundamental questions in quantum physics. In particular, the effect of the observer raises the question of the relationship between mind and matter. In modern physics, the simple act of observation can influence the behavior of reality at the subatomic level. The famous double-slit experiment is an emblematic example. Scientists have found that a photon or electron behaves like a wave or a particle depending on whether it is observed or not. The observer, therefore, seems to play an active role in determining phenomenal reality.

Physicist Eugene Wigner, the 1963 Nobel Laureate, was fascinated by this paradox. Wigner, also known for his interest in Indian traditions, questioned the role of consciousness. He proposed that the human mind was not just a passive spectator of the physical universe, but a key element in defining it. Wigner declared:

> *"I cannot formulate the laws of physics without referring to consciousness."*

Wigner himself admitted that he found inspiration in Eastern philosophical systems, including those of ancient India. He recognized that Hinduism, with its emphasis on the unity between observer and observed reality, had advanced ideas akin to those that modern physics was beginning to explore.

According to Advaita Vedanta, there are not two separate realities: one of the observing subject (consciousness) and another of the observed object (the physical world). There is only one indivisible reality, called Brahman. The separation of subject and object, in the Vedantic view, is a product of ignorance (Avidya). Meditation and spiritual study aim to dissolve this illusion, enabling the individual to realize that his or her consciousness is identical with Brahman.

Even in quantum physics, the idea that the subject cannot be extracted from the equation of reality is finding more and more ground. The concept of "quantum entanglement" suggests that particles can remain connected beyond space and time. This universal interconnectedness, which defies the laws of common sense, seems to recall the Upanishads' statements about the underlying unity of existence.

Physicist David Bohm, another important thinker of the 20th century, proposed the concept of "implicit order" as a model for the universe. He suggested that there is a hidden, more fundamental reality from which the phenomenal world emerges. This view is reminiscent of the Vedantic distinction between Brahman (ultimate reality) and Maya (apparent reality).

Interest in the philosophical insights of ancient India is not limited to theoretical physicists. The likes of Carl Jung, the Swiss psychoanalyst, explored the symbolism of Indian traditions in relation to the mind. Fritjof Capra, author of "The Tao of Physics," devoted entire chapters to parallels between Eastern traditions and modern physics. His work showed how the holistic perception of reality, consciousness and the universe is common to both fields.

This convergence of ancient philosophy and modern science invites deep reflection. Not only about the laws of physics, but also about the meaning of life itself. As Wigner himself noted, understanding consciousness is not only the task of science. It is a challenge that requires the collaboration of philosophy, religion and even art.

Ultimately, Hinduism reminds us that the reality we observe may only be a reflection of something greater. An image projected from the infinite mirror of universal consciousness.

David Bohm's approach: between physics and mysticism.

Another key figure in this dialogue is physicist David Bohm, known for his work on implicit order and his interest in the holistic dimensions of reality. Bohm argued that the universe was not a series of separate objects, but rather a huge connected totality, a kind of "implicit order" from which the perceivable world, or "explicit order," emerges.

Bohm was deeply influenced by Hinduism and Eastern thought in general. His idea of an "indivisible totality" is reminiscent of the Vedantic view of Advaita, which literally means "non-duality." For Vedantic masters, the physical world and consciousness are two aspects of the same fundamental principle, which is called Brahman. Bohm dialogued intensely on these issues with figures such as Jiddu Krishnamurti, an Indian philosopher who questioned the separation of observer and observed. These kinds of dialogues offered the physicist new insights into his model of physics.

David Bohm, one of the most revolutionary theoretical physicists of the 20th century, united science and philosophy in a way that few others did. Born in 1917 in Pennsylvania and active at a time of extraordinary developments in quantum physics, Bohm did not stop at equations. He went further, exploring the deeper meaning of reality. His key idea was that of a connected universe, an indivisible unity in which what we see is but a small part of the whole.

According to Bohm, reality is divided into two levels: the "implicit order" and the "explicit order." The former represents a hidden, deeply connected dimension in which everything unites into a totality without separations. It is, so to speak, the invisible architecture of the universe, a deeper plane where the distinctions between matter and consciousness, between observer and observed, lose meaning. The "explicit order," on the other hand, is the world perceived by the senses: the manifest reality, made up of seemingly separate objects, emerging from the implicit order just as a figure emerges from a background.

An illuminating example offered by Bohm is the metaphor of the hologram. A hologram is a three-dimensional image, but each part of it contains the information of the whole image. Similarly, according to Bohm, each fragment of the universe contains the essence of the totality, a concept that finds remarkable parallels with Hinduism and, in particular, with the philosophy of Advaita Vedanta.

Advaita Vedanta, one of the most influential currents in Hinduism, is based on the idea of a nondual reality (Advaita literally means "not two"). According to this view, the phenomenal world-that perceived by the senses-is illusory (Maya), while the fundamental principle, called Brahman, is the only reality. Brahman represents both matter and consciousness, united in one indivisible essence.

This philosophy provided Bohm with a profound vocabulary for reflecting on the nature of reality. The idea of a hidden unity in the universe, similar to the "all is one" concept of Advaita, found fertile ground in his dialogues with Jiddu Krishnamurti. The Indian philosopher, known for his iconoclastic approach to meditation and mindfulness, rejected the separation of observer and observed, insisting that only by understanding such unity can human beings come into contact with truth.

Bohm and Krishnamurti dialogued extensively between the 1960s and 1980s, exploring these issues in both public and private meetings. Krishnamurti, in particular, emphasized the role of the conditioned mind, a theme that resonates deeply with the implications of Bohmian physics regarding perceived reality. In their dialogues, often recorded in locations such as Brockwood Park in England, Bohm displayed not only an intellectual curiosity but also a spiritual inclination, rare among scientists of his time.

Bohm's ideas highlight how modern science can approach Eastern mysticism. He saw quantum physics as a tool not only for describing natural phenomena but also for investigating the limits of human thought. He argued that separation-so deeply ingrained in Western culture-was an illusion. For Bohm, as for the Vedantic masters, any duality, whether matter/spirit or time/eternity, is a form of misunderstanding reality.

A concrete example of Bohm's holistic approach is found in the phenomenon of quantum entanglement. This phenomenon, in which two particles remain connected at even extreme distances, suggests that the entire universe can be considered an integrated system. Bohm interpreted this not as a mechanical curiosity, but as a fundamental clue to a hidden order.

David Bohm invites us, through his science and his dialogue with Eastern philosophies, to rethink the idea of reality. At the end of his life he said:

> *"Science does not exist separate from humanity: it is an expression of our existence."*

This phrase condenses his message, a call to see physics, mind and spirituality as parts of a single human adventure. In this, Bohm seems almost like a modern *rishi*, a sage crossing the frontier between the visible and the invisible, between what we know and what we intuit. Like Brahman in Advaita, Bohm's implicit order represents an indivisible reality, the hidden heart of the universe.

Mind as illusion or tool?

At the heart of Indian philosophy, the Upanishads trace a fascinating view of reality. These texts, composed between 800 and 300 B.C., describe the mind as a dual force. On the one hand, the mind is an illusion, trapped in the game of maya, the "veil" that obscures the true nature of the universe. On the other, it is also a valuable tool: a bridge that, if harnessed wisely, can lead beyond illusion to liberation, Moksha.

Similarly, quantum physics presents us with equally profound questions. In this field, reality is never absolute. Subatomic particles exist in a state of superposition: they are simultaneously in multiple places or states, but stop being so when an observer measures them. This phenomenon, known as wave function collapse, has prompted some physicists to question the role of the human mind in the process of observation.

The concept of Maya, central to Hinduism, recalls the elusive nature of reality as described by quantum physics. In the best-known philosophical work of the Upanishads, the "*Brihadaranyaka Upanishad*," we read that the visible universe is like a web, a construction that conceals the ultimate truth, Brahman. At the subatomic level, the world behaves similarly. Matter, at its most fundamental level, appears more as energy or probability, rather than as something solid and definite.

In both contexts, therefore, the observer is crucial: his presence is not neutral, but influences reality. This point of contact between physics and Hinduism has fascinated many modern thinkers.

One of the contemporary scholars who has brought these perspectives together is quantum physicist Amit Goswami. Born in India in 1936 and educated in the West, Goswami developed the concept of monistic idealism. This theory holds that consciousness, not matter, is the basis of the universe. Goswami believes that observations from quantum physics confirm this philosophical position: the observer, as consciousness, plays a direct role in creating reality as we perceive it.

In an interview, Goswami told how reading the Upanishads had inspired him to explore the link between this ancient knowledge and new scientific discoveries. For him, the mind is both a tool that can free us from the illusion of matter and the force that risks confusing us if not well directed. He said:

> *"Consciousness, not matter, is the ultimate foundation of the universe, and it is through the mind that we can perceive it."*

The theme of Maya and the mind as an instrument also resonates in Indian culture through symbols and places. In the Kanchipuram temple in southern India, a famous legend tells of a king who, after meditating for a long time, discovered that the entire visible universe was contained within his mind.

According to tradition, the king in question may have been Rajasimha Pallava, ruler of the Pallava dynasty, which ruled Tamil Nadu between the 7th and 8th centuries. Rajasimha was known not only for his military exploits but also for deep spiritual devotion. Legend has it that as he retired inside the temple for a long meditation, he experienced a state of superhuman awareness. At that moment, the king realized that the external reality--made up of stars, mountains, planets and even the temple itself--was nothing more than a reflection of his mind.

This insight did not merely mark the king's life. It was celebrated by the sages of the time as a direct experience of what, in Hindu texts, is called the nature of *Brahman*: the ultimate reality that permeates the entire cosmos.

This concept comes close to what physicists call the "holographic universe": the idea that the whole of reality can be a three-dimensional projection from a more fundamental state, just like a hologram.

Knowing the Self in the quantum multiverse. Hinduism and Eugene Wigner.

Modern science and ancient spiritual traditions seem to be distant paths, separated by centuries of history and methodologies. However, there is an extraordinary dialogue that has begun to emerge between the philosophy of Advaita Vedanta, a pillar of Hinduism, and quantum physics. The concept of "*unity of the Self*" (Atman= Brahman) and Eugene Wigner's interpretation of the role of consciousness in the collapse of the wave function stand out in the center of this meeting.

Advaita Vedanta, one of the most prestigious schools of Hinduism, affirms the fundamental identity between Atman (the individual Self) and Brahman (the universal absolute). According to Vedantic masters such as Adi Shankaracharya, who lived in the 8th century, the idea of separation between objects, beings and minds is an illusion, it is Maya. Only pure consciousness exists, eternal and infinite.

This philosophical conception, developed in a spiritual context, focuses on personal realization, inviting individuals to "*know the Self*" in order to overcome the illusions of the world. Meditation becomes the bridge to recognize this unity, removing ego barriers.

Centuries later, in the heart of the 20th century, physicist and Nobel laureate Eugene Wigner proposed a revolutionary idea. In the context of quantum mechanics, he suggested that human consciousness was essential for the "*collapse of the wave function*." This, simplifying, is the critical point at which a subatomic particle manifests as a definite reality only when it is observed. For example, as an electron manifesting at a specific position

In the celebrated Copenhagen interpretation of quantum physics, the role of the observer was fundamental, but Wigner pushed the concept further. He suggested that the presence of a simple measuring instrument was not enough. A conscious observer, capable of perceiving and processing information, is needed.

Wigner's insight found a surprising echo in the Vedantic concept of universal consciousness. Advaita teaches that consciousness is the only fundamental reality; similarly, Wigner attributed to consciousness the power to influence matter itself. Both, while operating in very different contexts, seek to answer the same question: what is the nature of the observer?

To better understand this connection, one can imagine the quantum multiverse as the Vedantic Maya: an infinity of potential possibilities, which only through consciousness are manifested into a perceivable reality.

Eugene Wigner was not only a brilliant theoretical physicist, but also a thinker deeply influenced by philosophy. Born in Budapest in 1902, he developed a keen interest in metaphysics, inspired by Europe's growing interest in Eastern traditions. In the 1920s and 1930s, the dialogue between physicists and philosophers was animated by debates about the essence of reality. Niels Bohr, a pioneer of quantum physics and contemporary of Wigner, visited India in 1930 to engage with Rabindranath Tagore on the relationship between science and the spiritual.

Tagore, a poet and philosopher, had close ties with Vedantic thought, and his poetic view of reality seemed to find deep agreement with emerging quantum physics. This cultural exchange influenced not only Wigner but also other giants of science such as Erwin Schrödinger and Werner Heisenberg. Schrödinger, in particular, cited the concept of Atman in some of his reflections.

Knowing the Self in the multiverse of possibilities.

This convergence of Advaita Vedanta and quantum physics offers a fascinating and provocative insight. Knowing the Self, for the Vedantic tradition, means transcending the illusions of the mind and recognizing one's own nature as infinite consciousness, identical to universal reality. In the language of quantum physics, it could be described as observing oneself in the context of a multiverse of possibilities, where consciousness is the creative act that shapes reality.

Wigner himself reflected on the fact that physics, without taking consciousness into account, is likely to remain incomplete. He said:

> *"It is not possible to formulate the laws of quantum mechanics in a fully consistent way without referring to consciousness*

This statement, while controversial, underscores the importance of the observing subject as a fundamental variable, not only in quantum equations but also in the big questions about the nature of existence.

The journey between Hinduism and quantum physics does not seek to merge two worldviews, but to recognize where they mirror each other. Advaita Vedanta offers a spiritual path to discover the unity of the Self, while Wigner's insights open the door to an understanding of consciousness as the engine of quantum reality.

In a world where science and faith often seem to be pitted against each other, this dialogue reminds us that both can be valuable tools for exploring the mystery of being, inviting us to know not only the universe but also the Self that observes it.

David Bohm's approach to reality and links with Indian philosophy.

David Bohm, one of the most brilliant theoretical physicists of the 20th century, is remembered not only for his contributions to quantum physics but also for his revolutionary thinking about the nature of reality. Bohm, born in 1917 in Pennsylvania, was a countercultural thinker, pushing beyond scientific formulas and seeking a language that could describe the universe as a deeply interconnected entity. His theories, known as implicit and explicit order, speak not only to physics, but also to the ancient philosophical traditions of India, particularly the holistic principles expressed in the Upanishads.

At the heart of Bohm's scientific philosophy is the idea that reality is not fragmented, but one continuous flow. Bohm describes the universe through two main levels: the explicit order (the surface world we perceive) and the implicit order (a hidden, underlying reality that sustains what appears on the surface). The explicit order represents what we see, touch and measure: the material world, definable by scientific laws. The implicit order, on the other hand, is the realm of potential, a deeper, omnipresent dimension from which everything emerges.

This distinction is strikingly reminiscent of the Hindu view of the universe described in the Upanishads. In the Indian sacred texts, phenomenal reality (Maya) is seen as an illusory representation, sustained by a hidden and unified substratum, the Brahman. Just like Bohm's implicit order, Brahman is the ultimate and indivisible reality where all things are united in a common matrix.

The mind as observer.

For Bohm, the role of the mind and the observer is central to the understanding of reality. Even in quantum physics, the act of observing seems to influence the behavior of subatomic particles. This conundrum, related to the double-slit experiment, is one of the deepest mysteries of the quantum world. Bohm proposes that the mind and the physical world are not separate, but are part of the same flow of energy and information.

This notion of mind as an integral part of reality seems to resonate with Indian philosophy. In Advaita Vedanta traditions, human consciousness (Atman) is considered a direct reflection of Brahman. The separation of observer and observed is therefore an illusion. This perspective, both in Bohm's physics and in Hinduism, suggests that the universe is an interconnected whole, in which mind plays a crucial role in making manifest what exists.

In the 1970s and 1980s, Bohm engaged in a series of dialogues with Indian philosopher and spiritual teacher Jiddu Krishnamurti. These profound and enlightening meetings were held largely at Krishnamurti's educational center in Brockwood Park, England. Bohm and Krishnamurti addressed fundamental issues, such as the nature of thought, the ego and the possibility of a mind free from fragmentation.

Krishnamurti, known for his nondogmatic ideas, argued that the division between subject and object was the result of conditioned mental activity. Bohm, as a scientist, caught a bridge between the philosopher's insights and the study of quantum reality. In their dialogues, the idea emerges that the mental fragmentation, so typical of modern human beings, stems from the inability to see reality in its fundamental unity.

A small significant episode happened in 1980, during a meeting in Ojai, California. Krishnamurti used the metaphor of a river to describe the mind in a state of silence: the river flows without interruption, reflecting the wholeness of reality. Bohm saw in this image a parallel with his idea of "holistic flow," where the universe is not made up of separate fragments, but of a continuous movement of energy and meaning. Bohm called this "*holomovement*."

Bohm's work and his exchanges with Krishnamurti created a bridge between Western science and Eastern wisdom. Bohm himself visited India in the 1960s and was fascinated by the depth of the local spiritual traditions. One of his major insights was that the West, with its analytical and fragmented approach, could benefit immensely from the holistic thinking of the East.

Unfortunately, Bohm did not enjoy the full academic recognition he would have deserved. His ideas were too innovative for many of his contemporaries. Nevertheless, his thinking continues to inspire new generations of researchers who seek to unite science, philosophy and spirituality.

From Schrödinger to Krishna: Hinduism and introspections of quantum physics.

Erwin Schrödinger, Austrian physicist and pioneer of quantum mechanics, is known for the famous "cat paradox." Less well known, however, is his deep connection with Indian philosophy. Ideas contained in the Upanishads and the Bhagavad Gita deeply influenced his thinking, prompting him to reflect on consciousness, the unity of reality, and the role of the observer in the universe.

Schrödinger was born in Vienna in 1887. From childhood he showed an early aptitude for mathematics and physics. During his formative years, however, his interest broadened beyond scientific rigor. A lover of the arts and philosophy, he found any attempt to explain the nature of reality fascinating. The turning point came in the 1920s, when Schrödinger, searching for answers that science could not provide, came across translations of ancient Indian scriptures.

One of his favorite texts was Vedanta, a philosophical collection based on the Upanishads. Schrödinger was struck by the idea that ultimate reality is a single, indivisible entity: the Brahman. This view brought him closer to the idea that apparent separations (between subject and object, observer and observed) are illusory. As he wrote in his memoirs:

> *"Awareness is not an individual property. It is a universal phenomenon that cannot be fragmented."*

Schrödinger was fond of quoting the Bhagavad Gita, one of the fundamental sacred texts of Hinduism. In the Gita, Krishna explains to Prince Arjuna the mysteries of life and death, revealing to him that every being is part of one eternal reality. This concept resonated with the quantum view that Schrödinger was developing. Quantum physics showed that subatomic particles do not exist in separate states until they are observed, suggesting that reality is intrinsically interconnected. In a 1925 letter, Schrödinger wrote:

> *"I am fascinated by how modern physics seems to be getting closer and closer to the worldview put forward by the ancient Indian sages."*

His statement was not accidental. He saw parallels between the indeterminacy of quantum particles and the Hindu conception of Maya, the veil of illusion that hides the true nature of reality.

Consciousness as the foundation of reality.

One of the most revolutionary aspects of Schrödinger's thought was his interest in consciousness. In contrast to many of his contemporaries, who regarded consciousness as a by-product of brain activity, Schrödinger thought that consciousness was fundamental to understanding the universe. This idea found deep roots in Indian thought.

In Vedanta, consciousness, or Atman, is described as the essence of all living beings. Schrödinger interpreted this view scientifically, developing a unified conception of mind and the universe. He often wrote that the boundary between the perceiving subject and perceived reality is arbitrary. For him, the external world exists only insofar as it is experienced through consciousness.

In later lectures and essays, Schrödinger often took up his studies of Hinduism, arguing that science could not ignore the role of the mind in the formulation of natural laws. He claimed, in an almost poetic tone, that the universe is a "*cosmic game*," a flow of energy in which all distinctions are temporary.

During his time in Arosa, a Swiss village, Schrödinger devoted himself to writing many of his more philosophical works. In a small chalet surrounded by alpine nature, he read English translations of the Upanishads and deepened his knowledge of Eastern philosophy. It was in that setting of stillness and introspection that some of his most revolutionary ideas were born.

Another key moment in Schrödinger's life was his contact with other intellectuals influenced by Hinduism. In Switzerland he met Rabindranath Tagore, the Nobel Prize-winning Indian poet and philosopher, with whom he discussed similarities between Eastern views of life and new scientific discoveries. Tagore urged him to bring science ever closer to the human heart, advice that Schrödinger held dear for the rest of his life.

Schrödinger was not the only scientist of his time to be inspired by the East, but he was probably the one who drew the boldest conclusions. For him, quantum physics was not just a numerical description of nature, but a lens for exploring the human condition.

His reflections, such as the famous cat mind experiment, raised profound questions about the meaning of reality and the role of observation.

The influence of the Upanishads is evident in many of his writings. Beginning with the idea that time and space are mere constructions of the mind and ending with the belief that there is an ultimate and indivisible reality, Schrödinger was one of the first scientists to build a bridge between Eastern spiritual insights and the rigid logic of Western science.

Erwin Schrödinger represents an extraordinary example of how science can dialogue with spirituality. His insights, fueled by the ancient philosophies of India, helped him formulate some of the most important theories in modern physics. In his famous book "*What is Life?*", Schrödinger wrote:

> *"The great mystery is not that the universe is made of matter, but that it is aware of itself through us."*

Today, his thought continues to inspire scientists and philosophers, reminding us that to truly understand the cosmos, we must first look within.

Consciousness, quantum physics and reincarnation.

The human mind, a profound and eternal mystery, continues to be the focus of reflections from both science and ancient Eastern traditions. On the one hand, quantum physics unveils an unpredictable universe where the observer plays a crucial role. On the other, Hinduism, through the concept of reincarnation, explores the continuity of consciousness beyond physical life. Can these two seemingly distant views find a point of contact? Some scholars, including quantum physicist Amit Goswami, believe so.

In recent decades, quantum physics has shown how reality, at the subatomic level, is not as objective as once believed. Experiments on photons and electrons, such as the famous double-slit experiment, have revealed that the behavior of matter changes according to observation. In other words, the presence of the observer determines the manifestation of reality. Amit Goswami, a leading figure in quantum physics and author of books such as "*The Self-Aware Universe*," has suggested that consciousness is not a product of the brain, but a fundamental reality, the "*fabric*" on which the universe stretches. Goswami states:

"The world does not exist independently of observation. Consciousness is the ground of all being."

This approach goes against the classical materialist view but finds striking similarities with Indian thought. In Hinduism, in fact, consciousness is seen as primary to matter. The individual self, or Atman, is intrinsically connected to Brahman, the universal, eternal and all-pervading consciousness.

In Hindu philosophy, reincarnation is a central component. According to the doctrine of Karma, every action performed by an individual leaves a mental imprint that determines future experiences. It is not the body that is reborn, but a subtle continuity related to consciousness, which transfers from one life to the next. This cycle of birth, death and rebirth is known as *Samsara.*

Here a possible parallel with quantum physics comes into play. Modern experiments suggest that reality is not static but a continuous process of collapse of possibilities, influenced by observation. Similarly, one could see Karma as the continuous collapse of potentialities caused by the acts and thoughts of the individual. Consciousness, then, would be the guiding thread, capable of shaping not only the present but also the future.

One of the most fascinating experiments illustrating the fundamental role of consciousness is Schrödinger's "cat paradox." In the famous thought experiment, the cat is simultaneously alive and dead until an observer determines the outcome. This "potential dualism" resembles the Hindu concept of Maya: sensory reality is only an illusion, influenced by the intervention of consciousness.

In parallel, ancient texts such as the Bhagavad Gita explain that the soul (*jiva*) never dies; on the contrary, it moves from one body to another like a man changing clothes. This process, governed by Karma, creates a cyclical reality where past, present and future are intertwined. In the quantum context, even time is not linear but relative, making the cyclical idea of reincarnation more in tune with contemporary scientific findings than the Western deterministic view.

Another parallel between reincarnation and modern science emerges from the studies of Ian Stevenson, a psychiatrist who documented thousands of cases of children who claimed to remember previous lives. Through detailed testimony and verifiable evidence, Stevenson raised questions about the nature of memory and consciousness. Could these memories be quantum records on a deeper level, beyond the physical brain? Goswami himself speculates that consciousness grounds a unified field where information can be preserved, similar to the Indian concept of "*Akashic records.*"

Spiritual traditions and modern science seem to converge on one crucial point: reality is inherently influenced by consciousness. Whether it is the Hindu Samsara cycle or the collapse of the quantum wave function, both perspectives recognize a central role for the mind, not as a passive spectator but as a co-creator of experience. As Aurobindo, Indian philosopher and mystic, put it:

"The mind is the bridge between spirit and matter."

With the discoveries of modern science, we are perhaps discovering that this bridge is not as narrow as previously thought, but a broad passage that connects past and present, East and West, scientific subjects and spiritual realities.

The question remains open: does consciousness create reality or does it observe it? Perhaps, as in Hinduism and quantum physics, there is no one answer, but an eternal cycle where every observer is also a creator.

The web of reality: Hindu philosophy, vacuum physics and cultural dialogues.

Reality is a web, a fabric that connects everything. This view, elaborated for centuries by Hindu philosophy, finds striking parallels in quantum physics today. On the one hand, the Akasha, the Vedic concept of cosmic space, represents the universal matrix that unites everything. On the other, the quantum vacuum of modern science is not empty at all, but a vibrating structure that generates and connects subatomic particles, revealing itself as "a full vacuum." At first glance, these seem like two worlds apart. However, a bridge may exist between these millennia-old traditions and contemporary physics.

In Hindu philosophy, akasha is the fifth primordial element, in addition to earth, air, fire and water. It is not empty space, but a subtle and dynamic reality that permeates and connects everything that exists. Ancient texts, such as the Vedas, describe akasha as the basis of cosmic manifestation, the meeting point between the visible and invisible worlds. Swami Vivekananda, the Indian mystic who brought Hinduism to the West in the late 19th century, likened the akasha to an eternal background, within which the phenomenal world is formed. For Vivekananda, the universe would not exist without an invisible unifying field, an underlying web.

Quantum physics, between the 20th and 21st centuries, has revolutionized our understanding of the vacuum. In the quantum model, what we call "vacuum" is actually a field filled with energy fluctuations. These fluctuations generate virtual particles and fundamental forces. According to theoretical physicist David Bohm, this substrate is akin to an "implicate order," a hidden level of reality that gives rise to everything we see. It is no accident that Deepak Chopra has often used the concept of quantum vacuum to explain the interconnectedness of human consciousness with the cosmos. Chopra, an Indian-born contemporary physicist and thinker, interprets the vacuum as a point of contact between modern physics and ancient spiritual insights related to the idea of an indivisible universe.

In the 20th century, thinkers such as Vivekananda and physicists such as Nikola Tesla openly suggested a cross-disciplinary dialogue between science and spirituality. Vivekananda met Tesla in 1893, during the Chicago World's Fair. Both shared an interest in subtle energy as the basis of all things. Tesla sought an "*unlimited cosmic energy*," while Vivekananda spoke of universal energy as an expression of akasha. Both, in different ways, sensed that reality is not reducible to separate entities. Rather, it is an interconnected network.

Today, modern science is confirming some millennia-old insights. Studies of quantum entanglement, for example, show that particles light-years apart can be simultaneously connected, as if acting within a single matrix. This discovery, which for Einstein was "*frightening*" and hardly acceptable, resonates with the Indian idea of cosmic unity. Alain Aspect's 1982 experiment, which confirmed the phenomenon of entanglement, ignited new discussions about the hidden complexity of the vacuum.

Deepak Chopra continues to emphasize the link between the Hindu non-dualistic view and contemporary quantum principles. In his books and lectures, Chopra proposes that the universe is a unified field of potential. This is an interpretation that echoes Advaita Vedanta, one of the most influential philosophical schools of Hinduism. According to Advaita, only one indivisible reality exists: the "Brahman," the supreme consciousness. Everything that appears separate is, in truth, an illusion.

A fascinating example of this idea emerges from the practice of meditation, which has also been studied by Western researchers. During deep meditation, the mind seems to synchronize with a state of minimal activity, similar to quantum vacuum fluctuations. Some neuroscience is beginning to suggest that conscious attention behaves like the collapse of a wave function, a mechanism underlying quantum physics. The mind, in other words, would not be a stranger to the web of the cosmos, but an integral part of it.

The convergence of physics and philosophy is not just academic. It prompts us to rethink the way we perceive the world and ourselves. The idea that the whole universe is connected at a deep level is not just spiritual. It is becoming a scientific and cultural reality. Places like the *Chopra Center* in California or *the Indian Institute of Science in Bangalore* are advancing this multidisciplinary dialogue, exploring how these ideas can influence our relationship with nature, technology and mental health.

Just as the canals of the Indus irrigated cultures for centuries, so the flow of knowledge between East and West is charting a course for a new global dialogue. Thanks to the insights of minds like Swami Vivekananda and the discoveries of modern physics on the quantum vacuum, a simple but powerful message is emerging: everything is interconnected.

Meeting points?

Hinduism and quantum physics might seem worlds apart, separated by the invisible boundary between rationality and spirituality. However, upon closer examination, surprising points of encounter emerge. Both place at the center of their discourse two of humanity's deepest questions: what is reality? and what is the nature of consciousness?

Hinduism, particularly through the Advaita Vedanta school, states that the universe is an indivisible unity. Everything is Brahman, the absolute, the eternal substratum from which everything emerges. The perception of separation between the individual and the universe is considered an illusion, called maya. In a strikingly similar way, some concepts in quantum physics suggest a deeply connected universe, a hidden totality underlying visible reality.

A fundamental point of contact is found in the role of the observer. In classical physics, the universe exists independently of the observer. But quantum mechanics, through experiments such as the famous double slit, has shown that the act of observing changes the behavior of particles. An electron, for example, behaves as a particle or a wave depending on the presence of the observer.

Hungarian physicist Eugene Wigner, in the 1960s, explored the idea that human consciousness might be an integral part of the collapse of the wave function, that mysterious process that transforms a superposition of quantum states into a definite reality. This view seems to resonate with the concepts of Advaita Vedanta, according to which the mind itself participates in creating the experience of reality.

As the Indian sage Adi Shankaracharya stated in the 8th century AD,

"what we see as separate is merely a projection of our mind."

Quantum physics, almost a thousand years later, seems to be approaching the same insight.

Another point of convergence concerns the idea of an intrinsically connected universe. Physicist David Bohm developed the theory of implicit order in the 1980s. According to Bohm, the visible universe is only one manifestation of a deeper, hidden, indivisible reality. It is like looking at the surface of an ocean without noticing the deep currents that animate it.

This theory is reminiscent of the conception of Brahman in Hinduism, the immutable and omnipresent essence that unifies every element of existence. The example of the "*sutratma*," the "thread of the soul," is emblematic. According to texts such as the Upanishads, this invisible thread runs through every being, connecting it to the divine source. Bohm did not use the language of spirituality, but spoke of the same interconnectedness, of a universe that is greater than the sum of its parts.

Consciousness as the key to the cosmos.

Consciousness is a theme shared by both quantum physics and Hinduism. For science, the mind is still a mystery. Physicists try to understand whether consciousness simply emerges from the brain or whether it has a more fundamental nature in the universe. Some scholars, such as Roger Penrose and Stuart Hameroff, have speculated that consciousness may be linked to quantum processes within the microtubules of brain cells. This theory, known as "*Orchestrated Objective Reduction*," suggests that consciousness might be embedded in the very structure of reality.

Meanwhile, Indian sages describe consciousness as the unchanging core of the self. It is the Atman, the universal soul, which is not separate from Brahman. Through practices such as meditation, one can experience this absolute unity. Some Indian mystics, such as Ramana Maharshi, spoke of states of consciousness in which all separation between subject and object disappears, leaving a pure experience of being.

Science does not yet have tools to prove these experiences, but the implications of quantum physics, with its emphasis on the role of consciousness, suggest that the boundary between mind and matter may be less rigid than previously believed.

This dialogue between Hinduism and quantum physics remains open and fertile with possibilities. Neither perspective has all the answers, but together they create a unique space for reflection.

Hinduism invites us to go beyond appearances and recognize the unity hidden behind diversity. Quantum physics, albeit in other terms, pushes us in the same direction. Perhaps, as Heisenberg suggested, when we explore the depths of nature:

"...we do not find matter, only waves of probability."

And in the end, as the ancient Indian sages used to say, those who look at the universe without preconceptions discover that the observer, the observed, and the act of observing itself are one and the same. A lesson that, today more than ever, unites East and West in a common search for the ultimate meaning of reality.

Chapter V. Time and impermanence.

Time in Hindu texts.

Time, in the Hindu view, is not a journey with a definite beginning and end. It is not a straight line advancing toward a single destination, as is often the case in the Western tradition, deeply influenced by Judeo-Christian philosophy. In Hinduism, time is an infinite circle, a perpetual flow that repeats itself through cyclic phases. This concept, known as "*Kalachakra*" (translated as "*the wheel of time*"), represents the very essence of impermanence and continuous change.

Vedic scriptures and the Puranas, among the oldest records of Indian culture, describe time as divided into cosmic cycles called *yugas*. Each yuga is a phase in a greater cosmic cycle called a kalpa. A kalpa, in turn, is part of a day of Brahma, the creator according to Hindu mythology. To understand these cosmic proportions, a single Brahma day is said to last 4.32 billion human years. Brahma night, of equal duration, represents the destruction of the universe before the cycle begins again.

This view is not only philosophical but also deeply symbolic. The universe, according to Hinduism, goes through a constant alternation of creation, preservation and destruction, represented by the three main deities of the Trimurti: Brahma (the creator), Vishnu (the preserver) and Shiva (the destroyer). Time itself is also considered a divine manifestation, often identified with Kala, a form of the god Shiva. Kala is not only time, but also its destructive power: a force that reduces everything to impermanence.

In the Kalachakra concept, everything repeats, but not in a static way. Each cycle brings with it transformations, like seasons that always return, but never identical. The past is not lost, but intertwined with the present and the future, in a movement that eludes linearity. In major Vedic texts, such as the Rigveda, we find hints of this conception. For example, one passage in the chant reads:

> *"The wheel of time turns, forming the days, passing through the night and bringing back the dawn."*

The epic poem Mahabharata elaborates on the cyclical nature of time by recounting the Dharma-yuga and the kali-yuga, epochs in which virtue and righteousness gradually decline, only to begin anew. According to Hindu cosmology, we are currently living in Kali Yuga, the darkest of cycles, characterized by conflict, ignorance and loss of tradition. At the end of this cycle, the universe will be destroyed to be recreated anew.

Impermanence as a universal principle.

The idea of impermanence in Hindu time is closely related to the concept of Maya, illusion. The reality perceived by living beings is transient, illusory and destined to change. The passage through time cycles serves, in fact, as a metaphor for spiritual evolution, death and rebirth, and the path to breaking free from the cycle of reincarnations (Samsara).

This concept is also found in the Bhagavad Gita, where Krishna, a manifestation of Vishnu, instructs the warrior Arjuna. Krishna states:

"Just as the soul goes from an infant body to an adult body and then to an old body, so it takes on a new body after death. The wise man is not troubled by these changes."

The similarity with cyclical time is obvious: death is not an end, but only a transitory phase toward renewal.

Differences with Western weather.

The Western conception of time, influenced by Greco-Roman and Judeo-Christian traditions, is based on a finite timeline. Human history has a definite beginning, as narrated in Genesis, and an end point, the Last Judgment. This linear model provides a narrative structure of salvation and progress. In contrast, the Hindu view does not contemplate a definite conclusion. The cycle of the universe will continue without an ultimate end, in an eternal and unceasing movement.

In philosophical terms, the difference lies in the perception of reality. In the Western view, history has a progressive purpose and a meaning aimed at the future. In Hinduism, however, the emphasis is on harmony with the cycle itself, rather than on the control of time. The observation of the passage of time, in Eastern culture, is often accompanied by an acceptance of its impermanence.

Today, quantum physics is surprisingly intertwined with these ancient concepts. Cyclic patterns hypothesized in cosmology, such as the theory of the oscillatory universe, recall the idea of Kalachakra. The notion of nonlinear time, which emerges from studies of spacetime and quantum paradoxes, also offers a conceptual bridge between Hindu philosophy and contemporary science.

This interconnection between philosophy and science shows that the idea of cyclical time is not just a religious construct, but a perspective that invites humans to reflect on their position in the universe. In Hinduism, as in modern science, time is impermanent but at the same time eternal. It is an incessant flux that demands not only to be understood, but also to be accepted.

Time without end: the concept of eternity in Vedic texts.

The concept of eternal time, known as "*ananta*," occupies a fundamental place in Vedic philosophy. In the Vedas and Upanishads, time is described not only as a linear measure, but as a cyclical and illusory reality, part of the larger web of maya, the appearance that conceals the true nature of existence. Understanding time is seen as a crucial step on the spiritual path that leads beyond relativity toward Brahman, the absolute.

In the Vedic view, time does not continue rectilinearly, but manifests itself in cosmic cycles. These cycles are divided into yugas (cosmic epochs), which are repeated in an eternal order. This cyclical process is closely related to the idea of creation, maintenance and dissolution of the universe, often represented by the cosmic triads of Brahma (creator), Vishnu (preserver) and Shiva (destroyer).

The concept of ananta indicates that the universe, while eternal in its continuous regeneration, remains bound to the domain of relative time. For example, in the cycles of time (kalpa) described in the Puranas, we observe an idea of infinite universes arising and dying out, confirming the transitory nature of phenomenal reality.

According to Vedic texts and particularly the Upanishads, time is an integral part of Maya, the illusory relative reality. In this context, Maya includes space, change and duality, elements that bind the individual to the perception of movement and transformation. The Mundaka Upanishad states:

> *"Two birds, inseparable companions, sit on the same tree; one eats the sweet fruit, while the other watches impassively."*

Here, time could be interpreted as the fruit of the cosmic tree, which the individual being tastes, while the higher Self (Brahman) remains beyond temporal experience and dualistic illusion.

Vedic philosophy strongly exclaims that time, while eternal and cyclical, belongs to the relative dimension. Through spiritual practice-such as meditation, yoga and self-knowledge (Atman)-the individual transcends the limitations of time and attains Brahman, the absolute and eternal reality. At this level, cyclicality dissolves into perfect unity without beginning or end.

The Mundaka Upanishad describes this state when it states that Brahman is the place where all the currents of creation merge:

> *"He who knows that inconceivable, eternal, causeless, colorless, invisible and indescribable achieves supreme peace."*

The concept of eternal time in the Vedic texts is not limited to a cosmological view. It is also a guide for the spiritual practitioner, who seeks to understand the link between material time and spiritual eternity. Access to Brahman requires recognizing that time, in its cyclicality and relativity, is an obstacle that must be overcome to understand the true nature of reality.

In summary, Vedic thought invites reflection on eternity not as a mere temporal extension, but as the state of complete transcendence, where all duality, including that of time and eternity, vanishes in the experience of the absolute.

Brahma's clock: creation, destruction and cosmic rebirth.

Imagine a clock that marks not hours or minutes, but cosmic rhythms, scanned on time scales that defy our understanding. This is how Hinduism describes the cycles of the universe: through the image of "Brahma's day and night." According to Vedic and Puranic texts, the entire cosmos lives, dies and is reborn in an eternal dance within intervals that last billions of years. A mythical tale, to be sure, but somehow curiously close to contemporary scientific hypotheses exploring the expansion and contraction of the universe.

According to the Hindu scriptures, the deity Brahma-the Creator-is not eternal, but subject to cycles of life. For this divine figure, each "day" corresponds to 4.32 billion years; the same applies to each "night." During Brahma's day, the universe is active: life, matter, energy manifest. But with the coming of night, everything dissolves into a cosmic rest called "*pralaya*."

This immense rhythm is divided into further units of time such as kalpas and yugas, terms that reflect a precise articulation of the cosmos. Each yuga, for example, spans hundreds of thousands of years and represents a gradual spiritual and moral decline of the world until a complete reset.

The figures are frightening in their immensity. An entire Brahma day (24 hours, divided between day and night) lasts a whopping 8.64 billion Earth years. After 100 "Brahma years," the entire universe ceases to exist, and even Brahma himself dissolves to make way for a new cycle. This idea of infinite creation and dissolution is a fundamental key in the Vedic texts and mirrors the principle of impermanence, the basis of the entire Hindu philosophy.

The cosmic cycles described in the Puranas and Vedas have often been considered pure symbolism. Yet with the development of modern physics, some scientists have begun to glimpse surprising parallels. The Big Bang theory, for example, recounts the expansion of the universe from an initial point. But many post-Big Bang hypotheses postulate a future "*Big Crunch*," in which the universe might return to its starting point and then, perhaps, expand again.

Hinduism seems to anticipate this idea. The universe has neither a linear beginning nor a definite end. It is a continuous roll, similar to the cycle of a giant breath. In some texts such as the Śrīmad Bhāgavatam, the expansion and contraction of the universe are described as parts of the inhalation and exhalation of Mahavishnu, a divine principle even greater than Brahma.

People such as Carl Sagan, renowned astrophysicist and popularizer of science, have expressed their astonishment at these ancient theories. Sagan, in his book "*Dragons of Eden*," reflected on how Eastern traditions had a temporal vision immensely closer to the vastness of the cosmos than Western traditions.

> *"One day of Brahma is almost as long as the estimated age of the Earth," he said. "How could they have conceived something so enormous?"*

Thus Sagan wondered.

Hinduism, through the Brahma myth of day and night, delivers an underlying message that accompanies its entire philosophy: change is inevitable. Nothing is unchanged in time, not even the stars or galaxies. This impermanence is not cause for anguish but for acceptance. Destruction is always a prelude to new creation.

An evocative anecdote tells of Brahma himself, who one day wondered how insignificant his role in the cosmos was. Legend has it that Vishnu showed him a boundless immensity, full of Brahma of all kinds: some just born, some in full swing, many already dead. There is no single Creator in the universe, but countless. Each cycle is about not one, but infinite universes being born and dissolving.

What meaning does this have for us, human beings, who measure our lives in decades? Hinduism invites us to see life as a fragment of a much larger cycle. Our daily anxieties, our fears of the end, fit into a universal pattern that goes beyond the individual.

This idea has also been echoed in contemporary literature. Arthur C. Clarke, in the short story "*The Star*," describes an astronaut observing a nebula remnant of a mega stellar explosion. Here, as in Brahma cycles, destruction generates new beauty, new future.

Brahma's clock doesn't just measure time. It influences our perception of the universe as something fluid, never fixed. It is a perspective light years removed from the linear, finalistic view typical of Western culture. We, so small in the grand scheme of the cosmos, can learn to look at time not as a prison, but as a dance, in which each moment is necessary, ephemeral and - like everything in the universe - unrepeatable.

The four Yugas: the cycles of humanity.

The great cosmic cycles are not uniform. Within each kalpa there are shorter phases that mark the history and destiny of humanity. These are the four eras called yugas, each with well-defined characteristics.

Here is a diagram of the duration of the four yugas in human years:

Satya Yuga: 1,728,000 years.

Treta Yuga: 1,296,000 years.

Dvapara Yuga: 864,000 years.

Kali Yuga: 432,000 years,

Adding these up, a Maha Yuga covers 4,320,000 years.

In each age, the world goes through a gradual moral and spiritual decline that is reflected in the very duration of the eras: each successive one is shorter than the previous one, like an hourglass gradually losing sand.

The era called "Satya Yuga."

The first yuga, Satya Yuga, is considered the golden age and lasts for 1,728,000 y.. It is the time of absolute truth (satya means "truth" in Sanskrit) and righteousness. According to the texts, at this stage humanity lived in perfect harmony with cosmic laws, the Dharma. There was no need for organized religions or laws, because everyone lived spontaneously in accordance with righteousness.

In Satya Yuga, humans were said to be long-lived, honest and fully realized spiritually. The scriptures describe Satya Yuga as a time when the gods were visible and well-disposed toward human beings. One of the myths associated with this era is the creation of the world itself, when Vishnu, the Preserver, establishes the universal laws, and Brahma, the Creator, shapes the cosmos.

One of the most fascinating aspects of Satya Yuga is the perfect connection between human time and cosmic time. Every action directly reflected the course of the universe. The origin story in the Rigveda, the oldest text in Hinduism, celebrates this very unity. But Satya Yuga does not last forever. Impermanence, a cardinal principle of Hindu philosophy, begins to manifest itself with the arrival of the second age.

The era called "Treta Yuga"

The "Silver Age" marks the beginning of decadence. With a duration of 1,296,000 years, this era introduces the first divisions and conflicts. The mythical epic of the Ramayana, with the god-king Rama as its protagonist, is set in this phase.

Treta Yuga is the age when moral decadence begins. Here, Dharma, which in Satya Yuga held up the world with four pillars, is supported on three pillars. Virtue is still present, but it has lost some of its strength.

In Treta Yuga, time takes on a new dimension: a distance emerges between humans and the divine. Humans must strive to maintain moral order and face the first real manifestations of greed, selfishness and conflict. The legend of Rama, with his dedication to Dharma despite personal sacrifices-such as forced exile in the forest-demonstrates the gradual deterioration of universal harmony.

The era called "Dvapara Yuga"

The "Bronze Age" marks a steeper descent. It lasts 864,000 years. Human beings become less virtuous and more selfish. Hindu mythology places the famous epic poem Mahabharata, with the great war of Kurukshetra, in this era. It is here that Krishna, incarnation of Vishnu, reveals the Bhagavad Gita, the philosophical text that is still a pillar of Hindu spirituality today.

Dvapara Yuga represents the twilight of virtue. With Dvapara Yuga, humanity further loses balance and purity. The Dharma now stands on only two pillars, symbolizing the moral fragility of this era.

One of the most significant passages in Dvapara Yuga is the dialogue between Yudhishthira, the elder brother of the Pandavas, and the Yaksha, a divine creature who tests him. When Yudhishthira answers the Yaksha's questions, the sense of morality that humanity is already losing emerges clearly. One of the central questions is:

"What is the most wonderful thing in the world?"

Yudhishthira replies:

"That every day men see death occurring around them, but they continue to act as if they were not mortal."

This reflection expresses the fragility of the human condition. In Dvapara Yuga, time becomes a painful and inescapable force, constantly reminding man of his own impermanence and the ephemeral nature of his endeavors.

The era called "Kali Yuga."

The "*Iron Age*" is our 432,000-year era dominated by chaos, corruption and loss of values. According to the texts, the Dharma remains on one leg, unstable and precarious. This is the time when illusion (Maya) dominates, pushing humanity toward materiality and selfishness. For some scholars of Indian tradition, we are about 5,000 years from the beginning of the Kali Yuga, (a period that began around 3102 B.C.). In that year, according to legend, Krishna left Earth.

Kali Yuga is evidently the era in which we find ourselves today. This is the age of complete moral and spiritual decadence, in which the Dharma stands on a single pillar. The name Kali has nothing to do with the black goddess Kali, but comes from a term for discord and conflict.

Kali Yuga is evidently the era in which we find ourselves today. This is the age of complete moral and spiritual decadence, in which the Dharma stands on a single pillar. The name Kali has nothing to do with the black goddess Kali, but comes from a term for discord and conflict.

Hindu texts, particularly the Vishnu Purana, describe Kali Yuga as a time of great suffering. Human beings live short lives dominated by greed, lies and selfishness. The Bhagavata Purana predicts that toward the end of this era, moral values will completely collapse and the world will be covered by "*aDharma*" (the opposite of Dharma).

Many signs of Kali Yuga are recognizable in the contemporary world: loss of spiritual connection, political corruption, economic greed, and disregard for the environment. Not surprisingly, this conception of time in Hindu philosophy is sometimes compared with the Judeo-Christian idea of apocalypse. However, while the Christian tradition sees apocalypse as a linear and final end, Hinduism interprets it as part of a cycle. Kali Yuga is not the end, but the prelude to a new rebirth, which will take place with the arrival of Kalki, the tenth avatar of Vishnu.

Hindu cyclicity and modern physics.

The cyclical conception of time in Hinduism has also fascinated modern scientists and philosophers, especially in relation to quantum physics. According to some cosmological interpretations in contemporary physics, the universe itself may follow a continuous cycle of expansion and contraction, an idea that resonates with Hindu cosmology. Albert Einstein's "oscillating universe" model, taken up later by other cosmologists, seems to recall the kalpa doctrine.

The theme of impermanence, central to both Hinduism and quantum physics, also creates a cultural and conceptual bridge between East and West. In the subatomic world described by modern physics, particles emerge and vanish in a continuous cycle of creation and destruction, similar to the process of birth, death, and rebirth that drives the cosmic cycles of Hinduism.

The idea of cyclical time is not only metaphysical or cosmological: it has an ethical and spiritual message. It invites human beings to reflect on their own impermanence, on the eternal rhythm of existence of which we are a part. In today's world, dominated by the relentless pursuit of linear progress, this vision offers an alternative perspective:

everything is connected, everything comes back. Dissolution is never the end, but an opportunity for rebirth.

Hinduism, with its elaborate conception of time, speaks to our age of uncertainty with a wisdom that is ancient but incredibly relevant today. Under the starry sky, in a universe larger than we can imagine, every end is a new beginning.

Cultural symbolism: places and texts.

Time, in Hindu philosophy, is not just an abstract phenomenon or a linear measure. It is life itself, a perpetual flow that repeats and transforms. This conception of time as a cyclic and infinite reality permeates sacred texts, architecture and cultural symbols, constituting one of the most fascinating pillars of Indian thought.

One of the most emblematic examples of this vision is found in Surya's Chariot of the Sun, carved in the 13th century in the Konarak temple in Odisha. This architectural masterpiece depicts the Sun god crossing the celestial vault aboard a chariot drawn by seven horses. Each wheel of the chariot serves as a sundial. The 24 wheels not only mark the rhythm of the days, but also deeply embody the cyclical nature of cosmic time. This temple is not simply a work of art: it is a celebration of Vedic cosmology and the concept that time does not flow in a linear fashion.

Time, in Vedic tradition, is personified as Kala, the Lord of Time. Kala is not a benevolent or malevolent entity. He is neutral, ruthless and inevitable. In the Vedas, the oldest texts in the Hindu tradition, Kala takes the form of the universal force that determines the movement of events. It is not static and does not "pass away," as in the Western perspective; rather, it regenerates. It is circular, cyclical and perpetual.

Perhaps the most powerful manifestation of Kala emerges in the Bhagavad Gita, one of the central texts of Hindu spirituality and an inexhaustible source of philosophical reflections. In the famous dialogue between Krishna and Arjuna, Krishna reveals his cosmic form, declaring:

"I am time, the destroyer of worlds."

With these words, Krishna does not merely describe the destructive power of time; he deepens the concept of its impermanence. Everything that exists is inevitably destined to transform and dissolve.

This teaching goes beyond a simple reminder of mortality. It invites us to overcome the illusions of the material world, to recognize that time as we know it is an emanation of illusory reality (Maya). In the great cosmic cycle, destruction is only a prelude to rebirth.

In the Hindu concept of time, the cycle takes on a monumental dimension through the Yugas, the four great eras that form the Maha Yuga, a cosmic cycle of 4.32 million years. Each Yuga represents a phase of progressive decay, from the golden splendor of Satya Yuga to the debauchery and chaos of Kali Yuga, the present era. When Kali Yuga comes to an end, the cycle begins again. This perspective rejects the idea of a definite beginning and end. The universe is eternal and lives through a rhythm without pause.

The cyclical nature of time is also reflected in daily rituals and Hindu festivals. The celebrations are not just commemorations, but real participations in the repetition of the cosmic rhythm. The festival of Makara Sankranti, for example, commemorates the transition of the Sun on its cyclical journey through the zodiac, while daily rituals such as sandhyavandanam celebrate the eternal cycles of day and night.

Hinduism does not limit time to a symbolic or speculative dimension. The cyclical and profound conception of time finds reflection in numerous texts, images and anecdotes. One of the most famous myths is that of Samudra Manthana, the churning of the cosmic ocean. In this mythological epic, time is depicted as a serpent, Vasuki, coiling around the sacred mountain, which is transformed into a churning. Deities and demons pull the serpent back and forth, emphasizing the perpetual alternation between creation and destruction. This myth, immortalized in frescoes and bas-reliefs at sites such as Angkor Wat, emphasizes that time is as much a generative force as it is transformative.

In Indian temples, the structure itself often tells the time. The layout of shrines follows the movements of the Sun, with doors oriented toward sunrise or sunset. Drums and bells rung in daily rituals recall the incessant rhythm of Kala, while statues of the gods with multiple arms or faces symbolize their ability to dominate the past, present and future simultaneously.

Finally, how to interpret this view of time today? For Hindu culture, time is both a practical reality and a spiritual challenge. If the cosmic cycle is inevitable, then the real human task is to go beyond it. This is the philosophical message that emerges from meditating on time in Vedic texts and cultural symbols: to understand impermanence in order to transcend the cycle itself, approaching the eternity of being.

Time, for the Hindu tradition, is not only a measure of past and future. It is a mirror that invites reflection on one's own nature and ultimate reality, just as Vedic texts and ancient temples continue to do after millennia.

A famous episode from the Mahabharata reinforces the idea of impermanence: the dialogue between Yudhishthira and the mysterious Yaksha. When Yaksha asks what is the most surprising truth of existence, Yudhishthira replies:

> *"Every day men see death around them, yet they live as if they were immortal."*

This underscores the illusory nature of linear time and reminds us of the importance of accepting impermanence as part of the eternal cycle.

Modern Reflections

The cyclical conception of time, which permeates the sacred texts of Hinduism, has challenged and fascinated the human mind for millennia. The ancient rishis (sages) of the Vedas described a time imbued with eternity and repetition, punctuated by cosmic eras (the kalpas) that govern not only the universe, but also the soul and its incessant dance between birth, death and rebirth. This approach, which unties our perception of time from the linear principles dominant in the West, fits surprisingly well with some modern reflections of quantum physics and the spiritual experiences of today's global culture.

In India, the cyclic vision is not only theological but also deeply cultural. An emblematic example is the Kumbh Mela, the festival repeated every twelve years in sacred locations such as Haridwar, Prayagraj, Nashik, and Ujjain. The choice of these dates is not random: it is based on complex astronomical calculations that reinterpret the cosmic cyclicality handed down in Vedic texts. During the Kumbh Mela, millions of devotees immerse themselves in the sacred waters, celebrating a time that is renewed and connected to the eternal. This particular ritual demonstrates how Hinduism has interwoven cosmic time with the everyday, moving beyond the division between sacred and profane.

The Hindu vision has not remained confined to the subcontinent. Western thinkers such as Carl Jung explored the similarities between the cyclical conception of time found in Vedic texts and the archetypes of the collective unconscious. Jung, in particular, found in ancient Indian symbols, such as the Mandala, a pattern of cyclical order that is reflected in the human psyche. The Mandala is not only a spiritual design, but also a visual representation of how time can be understood as recursive and infinite.

Hermann Hesse and the eternality of time.

European literature has also been able to capture the essence of this cyclical vision. Hermann Hesse, in his famous novel "*Siddhartha*" (1922), recounts the spiritual journey of a man in search of truth and enlightenment. In a key scene in the book, Siddhartha contemplates a river, understanding that time is not linear but simultaneous: past, present and future coexist in an eternal "now." This insight reveals obvious references to the Hindu philosophy of time, with its emphasis on impermanence and the interconnectedness of all things. Not surprisingly, the river, a fundamental symbol in Indian spirituality (like the Ganges), represents the continuous and infinite flow of life.

Hesse, inspired by Eastern texts, interprets the cyclical nature of time not only as a cosmic fact but as a psychological reality. Several scholars see in works such as *Siddhartha* an attempt to integrate Eastern wisdom with Western thought, creating a bridge between two seemingly distant worlds.

Modern science and nonlinear time.

Surprisingly, the insights of the ancient Hindu texts find an echo in modern science, particularly quantum physics. The concepts of time described by the rishis can be seen as a metaphor for some of the findings of contemporary physicists. For example, Einstein's theory of relativity destabilized the idea of absolute, linear time, suggesting instead that time itself is relational and depends on observational conditions. Similarly, in the subatomic world, particles do not behave predictably but exist in superposed states, similar to the perennial impermanence described by ancient Indian texts.

The comparison between Hindu cyclical time and the concept of "*eternal present*," suggested by quantum physics, has prompted well-known researchers, such as physicist Fritjof Capra, to explore philosophical similarities between quantum mechanics and Eastern spiritual traditions.

From Konarak to Stonehenge: sacred time in ancient cultures.

Time as a concept has been central to the great philosophical and religious traditions. In Hinduism, time is cyclical and eternal, a cosmic rhythm governing life, death, and rebirth. This view contrasts with the linear approach to time in Western traditions, which identify past, present and future as an irreversible sequence. Exploring the Sun Temple of Konarak, the Mayan cycles, and the megaliths of Stonehenge, a surprising convergence of cultures distant in time and space emerges.

In the Vedic texts and Puranas, time (Kala) is described as an endless circle. The ancient sages conceived the cosmos divided into yugas, great cyclic eras. This cycle regenerates incessantly, analogous to the alternation of day and night or the rhythm of the seasons.

The Konarak Sun Temple, built in the 13th century in Odisha, embodies this vision. The entire complex, shaped like a chariot, symbolizes the solar disk pulled by seven horses. Its wheels, carved in detail, represent perfect sundials that mark the passing of the hours. The structure celebrates the power of the sun as a symbol of cyclical regeneration. Visiting Konarak at the solstice, the sun's rays illuminate precise architectural details, demonstrating sophisticated astronomical understanding. This Temple is a stone poem about cyclical time that interweaves architecture and cosmology.

On the other side of the world, the ancient Maya developed a similar concept. Their famous Long Calendar measured time through cycles of 5,125 years, culminating in cosmic regeneration. For the Maya, the end of a cycle did not represent a catastrophe, but an opportunity for renewal. Chichen Itza, with the pyramid of Kukulkan, is an architectural symbol of this. During the equinox, the play of light and shadow recreates the image of a serpent descending down the steps, connecting heaven, earth and human time in a cosmic choreography.

In Europe, sacred time finds voice in the Stonehenge megaliths. Built about 4,500 years ago, this stone circle is aligned with the summer and winter solstices. In Neolithic times, Stonehenge served as an astronomical observatory and place of worship. At the summer solstice, the sun rises over the Heel-Stone, marking the change of seasons. Ancient Europeans saw time through natural cycles, an approach that shares common elements with Hinduism and the Maya.

These cultures indicate that ancient man understood time not only as a measurement, but as a spiritual experience. The summer Stonehenge recalls the symbolism of Konarak's solar chariot and Kukulkan's serpent. Each site establishes a deep connection between man, the cosmos and the sacredness of time.

The contrast with the modern West.

The advent of rational Greek thought, with philosophers such as Aristotle and Plato, constructed a different image of time as a progressive line. This view was further reinforced by the European Enlightenment, which placed man at the center of a process of continuous advancement into the future. Modern science, dominated by Newton's time equations, became detached from the sacred experience of time. Contrary to this, Hinduism and other ancient cultures shaped time as an eternal return, avoiding artificial segmentation between past, present and future.

Today, quantum physics comes surprisingly close to the cyclic view of time. In the subatomic world, the traditional concept of "before" and "after" dissolves. Physicist David Bohm described time as a wave, a cycle of interconnected moments. This perspective is reminiscent of the Hindu doctrine of Kala, suggesting that the "ancient visions" are more modern than we think.

Through the carved wheels of Konarak, the Mayan prophetic cycles or the alignment of Stonehenge, humans have celebrated time as sacred. These traditions offer a valuable lesson: time is not just a measuring tool, but a mysticism that connects us to the cosmos. Today, rediscovering it could help us mend a rift between science, spirituality and nature.

What can Hinduism teach us about modernity and impermanence?

We live in an age when time is a tyrant. Agendas are busy, goals incessant, and anxiety about the future seems a constant companion. However, Hinduism offers a completely different lens for looking at time. Far from Western linearity, Hindu philosophy invites us to reflect on the cyclical nature of existence and impermanence.

In this model, time is not an arrow flowing inexorably toward an unknown future. On the contrary, it is cyclical, like the turn of the seasons or the movement of the stars. This view has deep roots in ancient texts such as the Vedas and finds extraordinary expression in the Bhagavad Gita. Here Krishna, the embodiment of the supreme deity, guides the warrior prince Arjuna toward understanding Dharma, linking conscious action to awareness of time.

According to Hindu texts, time unfolds in great cosmic cycles called kalpas. A kalpa, equivalent to 4.32 billion Earth years, represents one day in the life of Brahma, the creator of the universe. At the end of each kalpa, the universe dissolves only to be reborn. This concept, known as *pralaya*, emphasizes the impermanence of things and the continuous transformation of reality. There is no absolute end, only a constant becoming.

Modernity looks at time as a resource to be managed, while Hinduism teaches us to observe it as part of a natural flow. If everything is transient, even the difficulties of life should be seen as passing moments. Krishna explains this to Arjuna in the midst of the battle of Kurukshetra:

> *"The non-manifest has a beginning and an end; the manifested is only a temporary interval."*

The message is clear. Nothing is eternal, neither suffering nor pleasure, and it is precisely this impermanence that gives meaning to existence.

The concept of impermanence (*anitya*) occupies a central place in Hinduism. To recognize it is not to fall into nihilism, but to learn to appreciate the present. Modern life, with its worship of productivity and progress, often ignores this principle. It chases after a better tomorrow, piling stress on the present. The Hindu perspective, on the other hand, invites us to slow down and find balance. It is not about abandoning ambitions, but managing them without being enslaved by them.

From a practical perspective, impermanence can help us reevaluate our priorities. The great masters of India such as Swami Vivekananda have reiterated the importance of acting in the present with awareness. Action, or Karma, must be selfless, without obsession with future outcomes. This concept, known as *nishkama Karma*, offers a way out of chronic anxiety. As Vivekananda said:

> *"Focus on an idea, live it, dream it, act on it. But don't get attached to the results."*

The Bhagavad Gita represents one of the most universal texts on the management of time and impermanence. In Chapter II, Krishna urges Arjuna to fulfill his Dharma as a warrior without fear of the outcome.

"You don't have to worry about success or failure. Just take action."

This advice is still valid today. Krishna does not incite immobility, but an understanding that the outcome of actions is beyond our control. This acceptance reduces anxiety and allows us to live with greater peace of mind.

Another significant passage concerns the cyclical nature of time. Krishna describes the alternation between creation and dissolution as an eternal dance, the divine *Lila*. This conception dismantles the modern idea of time as an enemy, turning it into an ally. If everything is constantly renewing itself, there is no reason to fear change.

Hindu philosophies resonate deeply in today's context. The realization that every event-positive or negative-is temporary can be an antidote to the fear that paralyzes many people. This call for balance also finds parallels in disciplines such as mindfulness, which has its origins in Eastern thought. Meditation and yoga practices, based on Vedic principles, help to ground oneself in the present and reduce anxiety.

Moreover, the idea of the cyclical nature of time makes us think about global problems such as climate change. If the world experiences cycles of destruction and rebirth, the importance of our actions becomes even more apparent. Every gesture, no matter how small, fits into this flow, transforming the individual into a janitor of this great harmony.

Hinduism, with its view of time and impermanence, offers us valuable tools for dealing with modernity. In an age that measures personal worth by productivity and speed, stopping to reflect on the cyclical nature of existence can be liberating. As Krishna suggests, acting with detachment and living fully in the present allows us to become attuned to the natural rhythm of the universe. Perhaps, understanding that nothing lasts forever is not cause for anguish, but liberation. And in this, Hinduism still has much to teach us.

From Kali Yuga to the Singularity: cyclical time meets technological future.

At the heart of Hindu philosophy, time is not linear. It is not a straight line projected toward infinity, as the dominant Western idea would have it, but a circle, an eternal return. In Vedic texts and epic poems such as the Mahabharata, time flows through vast cosmic cycles called yugas. These cycles define the eras of the world, in a continuous alternation of creation, decay and renewal. Among these eras, the Kali Yuga, in which we are believed to be living, represents the height of decadence. It is the dark age, marked by conflict, spiritual bewilderment and loss of balance between humanity and nature.

But what can the Kali Yuga, conceived thousands of years ago, tell us about the challenges of today? Is it possible that this ancient vision of cyclical time can help us interpret the historical moment we live in, marked by unprecedented technological acceleration? Most importantly, what might be the link between the Indian idea of a cyclical return and the modern perspective of the technological Singularity, the time when it is assumed that artificial intelligence will surpass human intelligence?

For Hindu texts, Kali Yuga remains a unique era characterized by the gradual loss of virtue and cosmic harmony. Its duration spans 432,000 years, according to the Shrimad Bhagavatam system, and culminates in a dissolution that prepares for the return to the golden age, the Satya Yuga, in which purity will again be restored. During Kali Yuga, time is perceived not as progress but as degeneration. Scriptures such as the Vishnu Purana describe the decline of human institutions, the rise of corruption and the fragmentation of spirituality.

Yet, there is a disturbing resonance with contemporary reality. We live in the age when acceleration seems to dominate every aspect of existence: technology, production, communication. Rhythms compress, distances seem to dissolve in the immediacy of the Internet and the digital. But, as in Kali Yuga, we face a growing sense of alienation, a sense of loss of core values that transcends cultures and eras.

A powerful image from Kali Yuga speaks of "*Shravana without Dharma*," or listening devoid of wisdom: a civilization inundated with information but unable to transform it into knowledge. Is this not a reflection of what we see today in today's hyper-connected society, immersed in constant noise and increasingly distant from the essential?

The *technological Singularity*, a concept made famous by visionary Ray Kurzweil, represents a point of no return. When artificial intelligence surpasses human cognitive capacity, evolution will take an unpredictable turn. It is unclear whether such a moment will lead to the unraveling of society, a transformation of human forms of existence, or a new heavenly era of progress.

But what happens if we look at this possibility through the prism of Hindu thought? The idea of cyclical time suggests that every climax of decay-social or technological-is only a prelude to regeneration. If Kali Yuga represents the ultimate distance from spirituality, then every human endeavor, even technological endeavor, can be seen as part of an evolutionary process that does not escape return, the fulfillment of a cosmic dance.

Those who believe in the return of the golden ages might interpret the *Singularity* not only as a threat, but as a transition. Can technology, with its dizzying pace, be a vehicle for a rebirth? Or will it itself - like the destructive power of Shiva, the great transformative of the Hindu triad - be the catalyst for a necessary demolition?

A central element of the Hindu view of time is impermanence. Nothing is fixed, nothing lasts. Not even the Kali Yuga, however long, is eternal. The god Kalki, Vishnu's ultimate avatar, will end this darkness, restoring cosmic order. Technology itself, viewed through this lens, is not final. In the Hindu world, the digital, with its transience, is yet another form of Maya, illusion. It cannot exist as an autonomous reality outside the cycle of time.

This approach blends perfectly with modern theories of quantum physics. Impermanence as a fundamental principle of the universe finds a foothold in scientific discoveries: subatomic particles appear and disappear, as if emerging from a quantum vacuum that seems reminiscent of the matrix of reality described in the Vedas.

Physics itself, through its questioning of the profound nature of time, offers a surprising point of contact with ancient thought. If time is relative, as Einstein demonstrated, and if there is no objective distinction between past and future, the cyclical model of the Hindus takes on an unexpected depth. It is not an anachronism, but a sophisticated metaphor capable of approaching the insights of modern science.

In a world that looks to the *Singularity* as a promise or, for many, a threat, Hindu texts invite us to deeper reflection. Speed and change are not a problem in themselves. Being unaware of the cycle, losing touch with the spiritual principle that moves it, is.

Through Kali Yuga and impermanence we can read a universal message: even in times of utmost darkness there is a call to renewal. Civilization can rediscover meaning, not by denying change, but by integrating it with wisdom. Hindu texts invite us to understand that time is neither enemy nor master. It is simply the stage on which every form of existence, from civilizations to artificial intelligences, dances its own ephemeral but eternal choreography.

From Kali Yuga to the Singularity, then, the common thread remains the same: to remember that everything-even technology-is part of the great breath of the universe.

Time in quantum physics.

Time, as familiar as it is mysterious. In its perceived linearity, it punctuates our lives. However, quantum physics has taken it into entirely new territory. There it is no longer a continuously flowing river, but a concept that can fragment and dissolve into an elusive dance. This scientific revelation puts us in a fascinating position: we are close to ideas that incredibly echo the ancient insights of Hindu philosophy.

In the quantum world, time behaves in a strikingly different way than it does in classical physics. In the models introduced by Albert Einstein's theory of relativity, time and space are linked into a single entity: spacetime. This structure complicates the traditional perception of a distinct past, present and future. Quantum physics, going further, suggests that time is not even a fundamental constant.

For example, experiments on entanglement question the uniqueness of time and space. When two "entangled" particles interact, they seem to synchronize instantaneously, ignoring traditional temporal notions. Albert Einstein, to describe this connection, provocatively called it "*phantom action at a distance.*"

Even more intriguing is the "delayed choice" experiment, originally proposed in 1978 by American physicist John Wheeler. This experiment shows how present events can theoretically influence the past, even if it breaks our usual understanding of causality. The concept of temporal linearity is, in this way, completely disrupted.

This reversal of our perception of time finds surprising affinities with Hinduism. In Hindu philosophy, time is not linear but cyclical-a constant repetition of creation, destruction and rebirth. The conception of the four Yugas-the great cosmic cycles of Satya, Treta, Dvapara and Kali Yuga-reflects an idea of time that has no definite beginning and no absolute end. We speak of billions of years in a continuous, unbroken flow.

According to the texts of the Upanishads, what we consider "past" and "future" are nothing more than relative manifestations. In the "Bṛhadāraṇyaka Upaniṣhad" we read:

"The eternal is beyond time, and is the silent witness of all things."

Quantum physics, with its overcoming of traditional temporal categories, seems to return to just that: time is no longer the absolute ruler, but a convention tied to a particular perspective of the world.

The similarities between quantum physics and Hinduism do not arise by accident. Modern physics and Eastern philosophies share a key point: both are attempts to answer fundamental questions about the universe and our existence. Among the proponents of this cultural bridge is Fritjof Capra, author of the celebrated book "*The Tao of Physics*" In this work, Capra points out how the discoveries of quantum physics have deep philosophical resonances with the sacred texts of ancient India.

Capra suggests that the cyclical perception of time, described for millennia in the East, is remarkably close to the conception of time within subatomic physics. The very idea of "Maya," that is, the impermanence and illusory nature of the material world, recalls the uncertain and probabilistic behavior of quantum particles. Nothing is really stationary; everything is constantly changing.

An interesting parallel between quantum physics and Hinduism can be seen in the Ganges River, revered in India as the "*Divine Mother*" and a symbol of life itself. Its waters are never the same: they move, evaporate, and return in the form of rain. The river seems to be an entity that traverses time, yet always remains impermanent. In quantum physics and Hindu philosophy, this metaphor is beautifully embodied: the Ganges is never the same, just like our ever-changing universe.

Discoveries in quantum physics and reflections on impermanence in Hindu thought bring us a common message: reality is not what it appears to be. Time is not rigid, nor is it something absolute. We live immersed in a flux in which present, past and future are constantly intertwined, like waves in an ocean.

This kind of understanding is not just academic or theoretical. It invites one to look at life with a sense of wonder at a universe whose mystery never abates. From an ancient temple in India to the futuristic laboratory of a physicist, the question remains the same: What is time? And perhaps, as the ancient Hindu sages suggested, time is just an illusion.

Time in quantum physics: a dialogue with Hinduism.

Time is a mystery that has always fascinated the human mind. Philosophers, scientists and mystics have tried to understand it, to unravel it, or at least to intuit how it works. When we look at the dialogue between quantum physics and Hindu philosophy, we find that the two approaches, though seemingly distant, seem to meet on common ground. Both offer a vision of time that challenges our ordinary, linear perception, opening us to deeper and, perhaps, truer dimensions.

In the traditional Western view, time is a stretch of history: a straight line that starts in the past, passes through the present and is lost in the unknown of the future. Hinduism, however, invites us to consider another perspective: time as a cycle. According to the Hindu sacred texts (the Vedas, the Upanishads and the Bhagavad Gita), the universe moves in an endless dance of creation, destruction and rebirth. This movement is represented by the concept of *kalpa*, a cosmic unity of enormous temporal proportions.

For the Hindu tradition, time is never fixed or linear. It flows, expands, dissolves, but then cyclically returns. Faced with this ancient conception, one wonders: might modern physics have something to say about this view?

The flexible time of quantum physics.

With the advent of quantum physics in the 20th century, science abandoned many of its certainties. Time, once considered absolute and immutable, has turned into a more fluid and elusive concept. Already Albert Einstein, with relativity, had laid the groundwork for an idea of time as something relative, influenced by speed and gravity. But it is thanks to figures such as Werner Heisenberg, Niels Bohr and John Wheeler that quantum time has taken on an even more enigmatic dimension.

One of the most fascinating aspects of quantum physics concerns entanglement phenomena. Two interconnected particles can affect each other instantaneously, regardless of the distance separating them. This implies that events do not always follow a linear time sequence. It breaks, in a sense, the arrow of time. In more subtle but relevant ways, quantum mechanics also allows the concept of "*delayed* measurements," where the result of a present measurement appears to influence the past of the observed particle, as demonstrated in experiments such as those devised by Wheeler.

These findings seem to lend credence to the idea that time is not unique and objective. But what does this mean for our deeper understanding?

Hinduism, like many Eastern traditions, invites the practitioner to transcend the illusion of linear time. According to the Upanishads, time is nothing more than a projection of the human mind. The ultimate reality, Brahman, exists beyond time. This principle finds expression in the experience of Moksha, liberation, a state in which the past and future vanish, leaving room only for the eternal present.

Even in quantum physics one finds a curious reminder of this timeless reality. Many scientists, beginning with the reflections of Erwin Schrödinger, have been struck by the convergences between quantum mechanics and Eastern philosophies. Schrödinger, fascinated by the Hindu vision, wrote:

> *"This separate life that each of us leads is an illusion. The life of every living being, past, present and future, is in the eternal present."*

Another scientist, David Bohm, known for his philosophical approach to physics, proposed the idea of the universe as an "implicit order," where past, present and future exist simultaneously as one reality. This perspective is reminiscent of the Hindu idea of a time that is not linear but enveloping, where all things occur in a higher, invisible dimension.

The eternal return: from physics to spirit.

If quantum physics has brought a revolutionary change in our understanding of time, Hinduism offers a spiritual key to the phenomenon. On the one hand, science tries to explain time with laws and formulas; on the other hand, Hindu philosophy invites us to experience it differently, accepting its impermanent nature and not fearing its cyclical nature.

In all this, one question remains suspended. Niels Bohr, Nobel laureate and one of the most influential voices in contemporary physics, once said:

> *"I am not asking that you understand quantum mechanics, but that you learn to live with it."*

Similarly, Hinduism does not ask us to dominate time, but to harmonize with it, to feel it as part of a larger cosmic design.

And perhaps it is here, in the brief moment when science and spirituality intertwine, that we can find an answer to our deepest questions about time. Not a definitive solution, but an invitation to observe, with open eyes and a curious mind, that eternal cosmic game of which, willingly or unwillingly, we are a part.

Time in quantum physics: a dimension that is no longer linear.

Classical physics - Newton's, to be clear - described time as a linear arrow, an inescapable stream flowing from the past into the future. This conception, which informs the Western view of time, underlies much of the way we live: we plan for the future, study the past and measure the present in terms of seconds, minutes, hours.

Quantum physics, however, has begun to question this linearity. The concept of quantum entanglement, for example, shows that two particles, once connected, remain interdependent at potentially infinite distances. Their behavior appears immediate and simultaneous, violating the classical concept of "cause and effect" typical of linear time. Albert Einstein, however, reacted with skepticism, and remained uneasy before a universe where simultaneity seemed possible.

Years later, increasingly advanced experiments, such as those of physicist Alain Aspect in the 1980s, confirmed that quantum entanglement is real. Today we know that time, at least in the quantum world, does not follow the rules of an ordered calendar. At the level of the fundamental particles, the time flow becomes flaky and complicated.

Time, from the perspective of quantum physics, turns out to be a much more fluid concept than understood by classical physics. Whereas the Newtonian model described time as a linear sequence, an "arrow" advancing from the past to the future, the quantum world breaks this illusion of simplicity.

Eugene Wigner and the role of consciousness.

One of the most fascinating aspects of quantum physics is its relationship to time and consciousness. On this issue, Hungarian physicist Eugene Wigner (1902-1995), Nobel laureate in 1963, left an indelible mark with his revolutionary idea. The consciousness of the observer, according to Wigner, is crucial for defining the events of quantum reality. And, consequently, to structure our experience of time.

Eugene Wigner belonged to the generation of scientists who transformed physics in the 1920s and 1930s. The birth of quantum theory quickly challenged concepts that had dominated Western scientific thought for centuries. These included the concept of time as a continuous and absolute line, an idea inherited from Isaac Newton's classical physics.

In quantum physics, time turns out to be elusive. When we observe the behavior of subatomic particles, we find that there is no obvious time sequence until a measurement is made. Before that time, the nature of particles is a superposition of possibilities, a dance of states that exist simultaneously. This is where Wigner's contribution comes in: without a conscious observer "watching," the collapse of the wave function (and with it the temporal sequentiality of what happens) may not occur at all.

Wigner wondered: what does it mean to observe? And, more importantly, who or what observes? This scientist proposed a bold hypothesis: the conscious intervention of the human mind is not just an accessory, but is essential to transform quantum potentials into tangible events. In short, for Wigner, time and reality exist only in relation to the observer.

Wigner's insight provokes wonder because of its consonance with certain philosophies of ancient India, particularly Advaita Vedanta, a non-dualist school of Hinduism. According to Advaita, the reality perceived by our intellect (the realm of forms, events and time) is Maya, that is, an illusion. This illusion is not a lie, but rather the veil that prevents us from seeing the ultimate truth: reality is nondual and beyond time.

In Vedantic thought, consciousness, Brahman, is at the heart of everything. Even the concept of time (past, present and future) is intrinsically linked to this supreme consciousness. For the Indian sages, events are not marked by the movement of clocks, but are created in the mind of the observer. Strikingly similar, Wigner suggested that the timeline is not an objective reality, but a construction of conscious observation.

To better understand these ideas, what, then, is the role of the observer in determining time? Here Wigner offers a suggestive image.

He likened the observer to a mirror: the quantum world is like a reflection that appears only when a conscious mirror comes into view. Without the latter, there is neither a before nor an after, but only a formless potentiality.

In Vedanta, the cosmic mirror is the universal awareness, the Supreme Self. Wigner and the Vedantic sages, though separated by centuries and disciplines, seem to answer the same basic question: is time real or a product of perception? Both, though using different languages, arrive at a similar answer: time emerges only in relation to mind and consciousness.

To support his ideas, Wigner often cited Erwin Schrödinger's famous "cat paradox." According to this thought experiment, a cat locked in a sealed box is in a state of "indefiniteness," alive and dead at the same time, until the box is opened to check. Quantum physics states that it is precisely the act of observation that decides the outcome. But who is the observer? For Wigner, a conscious observer with a mind is the subject who introduces the boundary between "potential" and "real."

This convergence of physics and philosophy is not accidental and continues to attract scholars and scientists. Many contemporary physicists, from Fritjof Capra to David Bohm, have explored the links between the insights of quantum physics and Eastern sapiential traditions. Although the scientific community has not universally accepted ideas such as Wigner's, his work has opened horizons of reflection that push beyond the boundaries of the laboratory.

Conclusions like Wigner's, although speculative, invite us to look at science not just as a technique, but as a discipline that also addresses the deepest questions of human beings. And perhaps, as the ancient Indian sages suggested, the key to understanding the mystery of time is to understand the mystery of the consciousness that observes it.

Anecdotes and cultural connections.

There is a place in India, Kashi, now known as Varanasi, that bears the mysterious weight of time upon itself. According to Hindu tradition, this thousand-year-old city exists "*outside of time*," suspended between becoming and eternity. It is here that pilgrims come in search of something extraordinary: to abandon the illusion of linear time and immerse themselves in the cyclical nature of existence.

This cyclical conception, deeply rooted in Eastern philosophies, surprisingly finds a reflection in modern quantum physics.

In many quantum physics experiments, time seems to work in ways that defy ordinary understanding. The traditional linear view of a past leading to a future is challenged by the interaction between subatomic particles. An emblematic example is John Wheeler's "*delayed choice experiment*," conducted in the 1970s. Wheeler demonstrated that a particle, such as a photon, can "decide" its state only when the outcome is observed, even if this occurs after the photon has already traveled or interacted. In other words, the future seems to influence the past. This paradoxical idea brings us closer to a nonlinear view of time, subtly reminiscent of that of Hinduism.

According to Indian sacred texts, linear time-the "Kala"-is only a superficial pattern, an illusion, known as "Maya." In contrast, deep time, cyclical time, is eternal, marked by the cosmic rhythm of the great universal cycles called "yuga." Quantum physics does not go so far as to support this eternity, but it destabilizes the certainties of time: it invites us to consider the fact that our linear perception may be only an approximation of a more complex reality.

At the heart of Hindu philosophy, the city of Varanasi becomes a living symbol of this vision of time. Built on the banks of the sacred Ganges River, the city is considered immortal. It is said that Shiva himself founded Kashi and that those who die there are freed from the cycle of rebirth, entering the realm of eternity. Here, every sunrise and sunset mark a time that does not flow but returns upon itself. Pilgrims, immersed in rituals, perform mantra repetitions and ablutions to reconnect with this endless dimension.

Quantum physics, with its revolutionary insights, seems to touch on such a perception. Experiments on time, quantum chaos and superposition of states -- in which a particle can be in multiple places or times simultaneously -- ignore the rigid definitions of linear time. Time itself becomes a malleable variable, not the fixed entity that governs our daily lives.

Perhaps the most fascinating point of contact between Eastern philosophy and quantum physics lies in their ability to push beyond appearances. Just as Hindu sages invite us to look beyond Maya-the illusion of the phenomenal world-physicists also seek to explore the hidden substratum of reality. Varanasi, with its cyclical nature and promise of eternity, is a powerful and metaphorical image for both of these perspectives.

In the 20th century, Albert Einstein himself helped rethink the concept of time. His theory of relativity showed that time does not flow uniformly. The idea that time can expand or contract depending on speed or gravity effectively expanded the boundary of physics, bringing it closer to ancient philosophical views. Einstein, while not explicitly influenced by Hinduism, demonstrated that Newton's absolute time-so dear to the Western tradition-was an illusion, just like the concept of Maya..

If time is indeed an illusion, then the dialogue between physics and philosophy is not just an intellectual curiosity but a path of deep inquiry. In a quantum particle that defies the past there is perhaps an echo of the eternal cycle of Kalachakra. On a misty morning in Varanasi, when the sun laps the Ganges, one can really wonder if time is our companion or just a reflection of something larger.

Discontinuous weather

Time, this seemingly familiar dimension that punctuates our lives, has always fascinated human beings. But what happens if we go beyond appearance? If we interrogate the depths of quantum physics and Hindu philosophy, we discover that time is not at all what it seems. Both of these disciplines, so different on the surface, come to a surprising conclusion: time may be an illusion, a construct that deceives perception.

Quantum physics, the study of matter and energy on an infinitesimal scale, has brought to light a puzzling fact: time is not a straight line, but a fragmented and fluid reality. In the famous Wheeler "*delayed choice*" experiment, photons, particles of light, seem to choose whether to behave as waves or particles depending on human observation. What is puzzling is that the observation does not depend on the present moment, but can also be influenced by events that will happen "in the future." Such observation makes time nonlinear, but something malleable, like a tape that can be rolled up and unrolled at will.

"Time is not what you believe."

so stated Richard Feynman, one of the greatest physicists of the 20th century. According to Feynman, the behavior of subatomic particles seems to suggest that future and past are not separate entities but intertwined. The sequence of events, as we perceive it, is not an absolute given, but a manifestation of deeper, invisible but powerful conditions.

This view finds a curious parallel in Hindu philosophy, which considers time a form of Maya, or illusion. According to Advaita Vedanta, one of the most profound schools of Hinduism, time and space have no existence of their own. They are transitory manifestations of Brahman, the Absolute, the only permanent reality. The human perception of time, with the past slipping away and the future stretching to the horizon, is a limited construction, a narrative that the being uses to navigate the incomprehensible.

An ancient allegory found in the Puranas describes Brahma, the Creator, experiencing time cycles so vast that a single day of his existence corresponds to billions of earth years. This immense temporality, called *kalpa*, far exceeds human understanding, reminding us how relative and elusive time can be. Interestingly, Brahma's relativity of time has an echo in Einstein's theory of relativity. For Einstein, as well as for the ancient Hindu texts, time and space are intertwined, and their perception varies depending on the perspective of the observer.

If quantum physics presents us with fragmented and nonlinear time, Hinduism invites us to consider time as an eternal cycle. The great yugas of Indian tradition (cosmic eras that alternate in long periods of creation, maintenance and dissolution) call to mind the quantum concept of "*uncoordinated time.*" There is no absolute beginning and end, but a perpetual becoming, a dance of the universe.

Wheeler's quantum experiment and Brahma cycles enlighten us on a central point: the present, as well as the past and future, is imbued with a fluidity that we cannot comprehend with ordinary senses. In Vedanta, this awareness is called "*Kaivalya,*" a liberating experience in which the tyranny of linear time ceases. The same principle may emerge from future developments in quantum physics, when we finally unveil the deeper essence of this dimension.

While quantum physics is based on scientific observation and complex mathematical calculations, Hindu philosophy walks a path of introspection and myth. However, both share a crucial insight: time is not a solid entity but a deceptive illusion.

We can thus regard quantum physics and Hindu philosophy as two explorers of the same mystery. Isn't it fascinating to think that the insights of ancient Indian scriptures - handed down over millennia by sages who meditated on the banks of the Ganges - echo in the discoveries of advanced laboratories, where light itself seems to play with our perception?

Time, in the light of these reflections, is no longer the companion that punctuates the hours with regular ticks. It is a wave that bends, a circle that closes, an illusion that invites human beings to awaken to a deeper reality.

"Time: isn't it time to stop measuring it, and start observing it?"

Feynman wondered. And perhaps time itself, in its eternal quantum game, is waiting for us for an answer.

Time that does not flow.

Imagine two particles. One particle is located in Milan, the other in Tokyo. No communication would be possible under the ordinary rules of space and time. Yet their properties remain mysteriously connected. Changing something about the first particle instantly changes the second, even thousands of miles away. This is quantum entanglement, a phenomenon that defies common-sense intuitions and challenges our understanding of time.

Quantum physics, however, has not stopped there. The phenomenon has been verified in the laboratory, for the first time since the 1980s, by physicists such as John Clauser and Alain Aspect, and it continues to be studied. But what does entanglement tell us about time or, rather, its possible illusion? And why do some juxtapose it with concepts from Hinduism?

According to Hinduism, Brahman is the ultimate reality, the absolute principle that transcends all distinctions of space and time. All existence is united in a single fundamental matrix, of which the visible universe is only an illusory manifestation, called Maya.

This view, elaborated in the Upanishad texts, invites us to overcome the superficial perception that divides the "before" from the "after," or the "here" from the "there." At this point the parallel with entanglement becomes fascinating: even in quantum physics, entangled particles seem to ignore not only distance but also temporal sequentiality. It is as if they operate in an underlying, deeper dimension where these categories do not exist.

David Bohm, fascinated as much by science as by the great spiritual traditions, attempted to propose an alternative explanation for the mysteries of the quantum universe. Bohm spoke of an "implicit order," a kind of "wrapped-up" reality that contains within itself all the information of the cosmos. In the implicit order, every particle and every event do not exist as isolated entities, but are all part of a vast interconnected whole. Bohm likened this implicit order to a hologram: a three-dimensional representation where each individual fragment contains the complete image.

Bohm's idea is strikingly reminiscent of the Vedantic concept of Brahman. Like Bohm's hologram, Brahman is beyond the apparent division among the elements of reality. Everything is one, and this unity is manifested not in linear time but in an eternal present.

In this perspective, time loses its absolute character, becoming a kind of functional illusion, necessary for our limited mind, but not fundamental in the ultimate understanding of reality.

Mandala and cosmic interconnectedness.

The concept of timeless interconnectedness also recurs in the symbols of Hinduism. One of the most fascinating examples is the mandala. Mandalas, so central to Eastern philosophy, represent the universe as a perfect unity, articulated in harmonious circles converging toward a center. Every line and figure in the mandala hints at the whole, suggesting that each point somehow contains the whole design.

For a quantum physicist, a mandala could almost graphically represent the idea of entanglement or a holographic reality. No point is isolated. Everything is connected, in an invisible but profound way.

In Hinduism, Brahman is not only a cosmological concept. It is also an inner experience. Atman, defined as the deep Self of every human being, is considered a direct expression of Brahman. To know the Atman is to transcend the illusion of time and space to perceive pure, timeless awareness.

A similar intuitive experience has also fascinated some physicists such as Erwin Schrödinger, one of the fathers of quantum mechanics. Schrödinger, an admirer of the Upanishads, wrote that

> *"The distinction between past, present and future is only an illusion, however persistent."*

In these words we glimpse the echo of that timeless awareness that Vedantic sages for millennia have described as "enlightenment."

What really links quantum entanglement to Brahman? Both seem to indicate that separation and sequentiality are but a superficial level of reality. If quantum physics investigates the deep unity of the universe with the tools of science, Hinduism has been exploring it for centuries through philosophical and spiritual insight.

Both confront us with a question that shakes our way of seeing the world: what if time, which we perceive as a flowing river, is actually an illusion? In this mysterious and interconnected universe, perhaps we are just at the beginning of an understanding that may one day reconcile science with ancient wisdom.

Time and Non-Time: Hindu philosophy and Schrödinger's cat.

Quantum physics and Eastern philosophies might seem, at first glance, worlds apart and unable to converse. Yet bringing them closer allows for the exploration of fascinating concepts about the nature of time. The investigation begins with one of the most famous mental experiments in modern physics: the "*Schrödinger's Cat*" paradox. This, compared with the Hindu cosmology described in the Vedas, transports us into a surprising reflection on time as a simultaneous experience and the superposition of temporal states.

In 1935, Austrian physicist Erwin Schrödinger presented the theoretical experiment of the cat locked in a box. Inside is a complex quantum mode: an unstable atom, a Geiger counter tied to a vial of poison, and a system that condemns the cat to an unusual fate. Until the box is opened, the cat exists in a state of quantum superposition: alive and dead simultaneously. It is the observer, by his act of observing, who collapses these superposed states into a single reality.

In this disturbing scenario, time takes on an unprecedented role. Before observation, there is no clear and definable "before" or "after." The temporal dimension appears fluid, nonlinear and often inaccessible to traditional human conception. This is where Hindu philosophy comes in, where the idea of simultaneous time finds ancient roots.

Hindu philosophy does not perceive time as a linear sequence, but rather as an interweaving of past, present and future coexisting in the "*Trikala*." In Vedic texts, it is stated that the universe itself is governed by cyclical and cosmic processes, in which time takes a circular form, accessible in simultaneity.

A significant example is the avatar of Vishnu in his manifestation as Narayanan. Narayanan is described as one who traverses the ages without being affected by the passage of time. In this divine dimension, the past is never gone, the present is never bound to a definite "now," and the future is already enshrined in the folds of the eternal.

The observer: the focal point between physics and metaphysics.

Schrödinger's Cat paradox poses an intriguing question: does the observer affect reality? What happens in the box, in fact, depends on the direct intervention of the human observer. The role of the central observer is not only a theme of quantum physics, but also a cornerstone of Hindu philosophy. In the Vedic texts and Upanishads, the Self, or "*drashta*," is defined as the supreme witness. He is the one who sees but does not act. It is the still, out-of-time point that observes the world, differentiating itself from the mutability of sensory experience.

The Upanishads speak of this observer with a powerful metaphor: the Self is the sun that illuminates everything that happens on earth. However, the sun does not participate directly in events. Similarly, the observer in the quantum context does not enter the box but, by the mere act of observation, transforms reality into a specific condition. This bridge between Hindu philosophy and modern physics creates an extraordinary dialogue.

A powerful image of the simultaneity of time, according to Hindu philosophy, is the sacred river Ganges. Indians believe that immersing oneself in the Ganges today is like participating in an eternally present act. Brahma, Vishnu and Shiva, representing creation, maintenance and destruction respectively, exist in every epoch of the cosmos, just as their influence is reflected in the flow of the river. Similarly, in quantum physics, the superposition of quantum states suggests an equivalent dynamic: the present already contains all the potential of what is to come.

The question of time, then, unites seemingly distant worlds. Science, with its precise mathematics and Schrödinger's mental experiments, opens doors that Eastern philosophies have long contemplated on the metaphysical plane. Both point to a reality that defies the limits of ordinary perception.

As physicist and philosopher Fritjof Capra wrote in *The Tao of Physics*:

> *"the mystical conceptions of the East and modern physics no longer appear as separate universes."*

So it is not surprising that a quantum experiment devised in the heart of twentieth-century Europe succeeds in giving new voice to ancient Hindu teachings.

In the fundamental questions of reality, we find ourselves suspended between science and spirituality. Schrödinger and Narayanan both tell us about a world where time turns out to be, in the end, only an illusion. An illusion that the observer can pass through with an act of pure awareness.

Chapter VI. A bridge between physics and spirituality.

Meditative practices in Hinduism.

The art of meditation in temples of the Hindu tradition.

Walking barefoot on a polished granite floor, under an archway decorated with intricate carvings, is not just an architectural experience. It is a spiritual act. In the sacred temples of India, Hinduism manifests itself in a silent dialogue between man, space and the divine. These places are not just centers of worship. They are true cathedrals of silence, built to facilitate *dhyana*, deep meditative practice.

Among these temples, the one at Chidambaram in Tamil Nadu shines as a beacon of cosmic significance. Dedicated to Nataraja, the dancing god, Chidambaram celebrates more than just one aspect of divinity. It represents the eternal dance of reality-a movement that, according to ancient Hindu texts, reverberates through all things. It is in this setting that devotees gather in meditation, seeking to achieve inner silence.

The Chidambaram temple is an example of how sacred space is designed to lead to transcendence. At the center of the complex is the *sancta sanctorum*, the innermost chamber, where the image of Nataraja resides. However, there is something unusual: part of the sancta *sanctorum* is empty. This space, known as *Chidambaram Rahasya*, is a symbol of the Supreme Self, the Absolute that cannot be described or seen.

The entire design of the temple seems to invite a meditative experience. The tall columns that rhythm the courtyards evoke the flow of thoughts, while the dancing statues evoke the resonance of eternal movement. The architectural aspect is never merely decorative. It is a tool that guides the individual to a state of concentration. It is believed that even the sound of devotees' chants and bells, amplified by the geometries of the walls, stimulates specific aspects of the mind.

Meditation in Hindu temples does not only involve physical silence. It often involves sound, which becomes a portal to the transcendent. The practice of *Nada Yoga*, described in the Chandogya Upanishad, is a fundamental tradition. In this form of meditation, sound is used to become attuned to the cosmos.

Nada Yoga is divided into two main sound paths: external sound (Ahata Nada) and internal sound (Anahata Nada). In the former, devotees focus on the sound of mantras, such as the famous "Om," considered the primordial vibration of the universe. In temples, this sound is recited in groups, creating a collective resonance that also involves the bodies.

The second, more subtle path requires absolute silence. Sitting on a cold stone floor, immersed in the shadows of the arches, meditators try to listen to the inner sound, the subtle echo of their own existence. This sound, according to tradition, resides in the heart of every human being and is a gateway to connect with Brahman, the Absolute.

Sound waves and quantum physics: a modern dialogue.

But why is sound so important? Modern physics offers us a curious analogy. Quantum mechanics describes matter not as solid and unchanging, but as a collection of vibrational waves. Just as in Nada Yoga, where everything is perceived as vibrating energy, physics teaches us that subatomic particles are characterized by an associated wave, a continuous "dance" in space-time.

It could be said that the Hindu tradition, through sound rituals, intuitively perceived what science has only discovered in modern times. In the Chidambaram temple, during solemn celebrations, mantras and the sound of bells fill the spaces. These sounds, amplified by the temple structure, seem to resonate with the same primordial vibration that quantum physics associates with the universe.

Suman, an elderly devotee who has been visiting Chidambaram regularly for decades, describes his own meditative experience with clear eyes:

> *"When I close my eyes and hear the rhythmic drumming during the ceremony, it is as if everything inside me begins to dance. It is an energy that cannot be explained in words. The temple is not just a place, but an experience."*

Many similar descriptions can be found in the diaries of nineteenth-century Western travelers, such as the British Francis Buchanan, who visited the temples of Tamil Nadu. Buchanan described the silence of the nights in the temples, broken at intervals by the chanting of Brahmins. He, too, noted the profound psychological effect of those repeated, cadenced sounds, which seemed to take the mind to another place.

The real purpose of meditation in Hindu sacred spaces is not just to find calm. It is to rediscover that fundamental Oneness that runs through everything. The Chandogya Upanishad states:

> *"This whole universe is Brahman. In silence, enter the heart, and discover that you are One with it."*

Devotees who gather in sacred temples, from the Kashi Vishwanath in Benares to the remote Arunachaleswarar in Tamil Nadu, know that the common thread of the meditative experience is the inner journey. Whether in silence or in the vibration of mantras, they seek to understand that universal truth, which the temple itself embodies: the outside reflects the inside.

In the sacred temples of Hinduism, where every stone seems to resonate with life, profound wisdom is revealed. Sacred silence is interwoven with sound, the visible with the invisible, the physical with the spiritual. The temple then becomes much more than a place. It is a bridge between the mind and the cosmos, where meditation and Nada Yoga remind us that, after all, everything is vibration. Everything is dance, just as the eternal Nataraja, cosmic dancer, teaches us from the heart of the Chidambaram temple.

Bringing the mind into total silence.

Hinduism, in its many philosophical and spiritual strands, has always regarded meditation as a fundamental tool for exploring the essence of reality. Among these practices, *dhyana* occupies a central space. It is not simply a relaxation technique, but a profound process of transcending the limitations of the mind and recognizing the spiritual nature of existence.

The word dhyana is derived from the Sanskrit root "dhī," meaning "to think" or "to reflect." However, the meaning transcends any intellectual reflection. Dhyana is the art of "*stopping*" and bringing the mind into total silence. As the fifth-century Indian philosopher Patanjali states in his Yoga Sutras:

> *"Meditation is the uninterrupted flow of consciousness toward a single object" (Yoga Sutra 3:2).*

This "uninterrupted flow" promotes self-awareness, creating a bridge between the personal self and Atman, the universal self.

Mind, silence and the universe.

The concept of "total silence" is more complex than it may seem. In dhyana, silence is not simply the absence of sound or the avoidance of thought. It is the subtle ability to suspend mental activity, to go beyond the turbulence of the mind and emotions, like a still lake reflecting the sky. For the Indian masters, mental silence leads to higher levels of awareness, bringing us closer to the vision of Atman, the universal Self.

An illuminating example comes from the Vedic texts. In the ancient Mandukya Upanishad, it is stated that silence is the purest sound, represented by the fourth state of consciousness (*turīya*), a stage beyond sleep, wakefulness and dreaming. This state of pure silence is the prelude to union with the divine, a dimension reminiscent of the intangible and mysterious nature of quantum particles, which seem to "exist" only when observed, but maintain an intrinsic connection with the entire universe.

Hindu temples are designed with a specific purpose: to guide the mind to inner silence. As tradition explains, each architectural element has a symbolic function. The central location of the garbhagriha, the *sanctum sanctorum* where the deity is enshrined, represents the Atman. At the heart of these spaces, silence reigns supreme. The sound of the mantra, as well as the silence that follows its repetition, allow the practitioner to enter a deep meditative state.

Modern science and meditation: an unexpected convergence.

Interesting is the parallel between the silence sought in dhyana and the principles of quantum physics. Quantum theories show how matter, once observed in its subatomic form, defies all traditional conceptualization. Particles appear and disappear, manifesting a kind of "*silent potentiality*," as if the whole emerged from the void.

This conception connects deeply with the Hindu view. Dhyana invites the practitioner to stop mental activity, entering the silence that creates and sustains everything. And just as in quantum particles, the meditator's inner void is rich in potentiality. Beneath the silence, the same universal energy pulses.

A modern example is provided by the work of physicist David Bohm, who proposed the concept of implicit order. Bohm hypothesized that visible reality emerged from a deep, invisible, and potentially infinite state. This "implicit order" is very close to the concept of Brahman in Hinduism: the ultimate, indescribable reality manifested in the dynamism of the universe. The philosopher Erwin Schrödinger, one of the fathers of quantum physics, also claimed to be influenced by Hinduism, speaking of a deep unity that connects all reality.

To achieve this state of mental silence, Hinduism proposes specific tools. One of the most significant is Nada Yoga, the yoga of sound. Through the chanting of mantras, such as Om, one guides the mind to focus on a single point, gradually dropping all thought. The initial sound becomes an echo that slowly fades away, leading the practitioner into deeper silence.

Another example is the Trataka technique of staring at a flame or other form of visible energy. Constant attention helps calm inner turbulence, bringing the practitioner closer to the mind without movement.

Total silence, in dhyana, is not just a personal experience. It is the key to understanding universal truth, rooted in both particle physics and the metaphysics of the Vedic scriptures. At a time when science and spirituality seem distant, dhyana reminds us that these two disciplines can meet in the same aspiration: to understand a world beyond the senses, to arrive at a state of absolute stillness, in which the self dissolves and only being remains.

Purpose and meaning of meditation.

In Hinduism, dhyana is much more than mere stillness of mind. It is an inner journey, a crossing of the veil of illusion (Maya) to discover the true nature of consciousness. It is the heart of meditative practices in the Indian tradition, a path leading to Moksha, the ultimate liberation from Samsara, the cycle of birth and death.

The word dhyana, often translated as "meditation," emerges as early as ancient texts such as the Upanishads and finds full systematization in Patanjali's Yoga Sutra. It means absorption or contemplation without distraction. It is not just mental retreat, but a gradual extinguishing of identification with body and mind.

According to the Advaita Vedanta school, founded on the teachings of Adi Shankaracharya (8th century CE), the material world is illusory. Maya, the appearance of reality, deceives the individual into believing that he or she is separate from the Whole. Dhyana has the task of dissolving this perceptual error. Having overcome the illusion, what remains is Brahman, the Absolute, the infinite Self that is everyone's true nature.

One master who fully embodied dhyana was Ramana Maharshi (1879-1950), a key figure in the modern Indian spiritual landscape. Ramana did not teach elaborate techniques. Summit of his doctrine was the direct method of self-inquiry. He invited his disciples to ask themselves a seemingly simple question, "*Who am I*?" This question is not a mental exercise, but a plunge into the inner void, a tool to free oneself from mind and conditioning.

Arunachala, a sacred mountain in Tamil Nadu, was for Ramana a living symbol of the Supreme Self. Near the Arunachala is the Ramana Ashram, a spiritual center still visited by thousands of seekers from around the world. Here, bathed in silence, men and women sit in contemplation. Many recount that it was enough to gaze into his eyes to feel a deep peace. He himself meditated, motionless, in states of absorption so intense that even the body was forgotten.

For many Hindu schools of thought, dhyana is the means of "really seeing." Maya is like a dream that makes us mistake the unreal for the real. When the mind calms down, the separation between subject and object vanishes. The insight then emerges that the individual Self (Atman) is one with the universal Self (Brahman).

A link to quantum physics becomes interesting here. Modern physicists, such as Erwin Schrödinger, were deeply influenced by Advaita Vedanta. Schrödinger wrote:

> *"Multiplicity is only appearance. In truth, there is only one mind."*

This principle strongly evokes the idea that dhyana can lead to a unified view of reality, similar to the quantum concept of "entanglement," where everything is connected.

Dhyana is not something remote or reserved only for masters. The practice has been preserved in Indian temples and *ashrams*. It is alive in the silent meditation sessions in Ellora caves, in the teachings of the monks at Sringeri Monastery or in the attenuated breath of pranayama practitioners.

A famous anecdote brings us to Swami Vivekananda (1863-1902), another spiritual giant of India. During an intense meditative state, he described the experience of his mind as "*a drop falling into the infinite ocean*." This image encapsulates dhyana: the loss of individuality to embrace that which is eternal.

Dhyana represents a millennia-old path that connects the depths of the spirit with the frontiers of science. It is no accident that modern theorists find parallels between the Indian idea of Atman and some insights from quantum physics on the nature of reality. The practice of meditation, which touches the body, mind and Self, invites the dissolving of any apparent division.

Following dhyana is like learning to slip into the present. Those who meditate place themselves beyond time, beyond space. It is not a theoretical process, but a personal experience. Through it, the individual recognizes himself not as a separate body, but as pure consciousness, infinite and formless.

The path of dhyana remains, through the ages, an open way for anyone who has the courage to look within himself or herself and face the greatest question, "Who am I?" And from this answer, found in silence, the real is illuminated.

The seven chakras: gateways to energy and meditation.

In Hindu tradition, the seven chakras are defined as energy centers located along the spine, considered key elements for spiritual awakening. These points are not only mystical symbols, but represent a bridge between body, mind and consciousness. Through meditation and yogic discipline, the chakras offer a path to enlightenment and allow one to connect with a higher level of universal energy.

The term "chakra" comes from Sanskrit and means "wheel" or "circle." This image suggests a continuous movement of energy. Hindu philosophy describes seven main chakras: starting from Muladhara (at the base of the spine) to Sahasrara (at the top of the head). Each chakra is associated with specific colors, natural elements, and personality qualities. For example, Muladhara represents stability and connection to the Earth, while Sahasrara is the symbol of spiritual transcendence.

Ancient texts such as the Tantras and Upanishads describe chakras not only as spiritual elements, but also as points where vital energy, called *prana*, converges. They are associated with "*nadis*," or energy channels, which carry energy throughout the body. Their activation is seen as essential to achieving psychophysical balance and, ultimately, enlightenment.

Meditation and awakening of Kundalini energy.

One of the most common practices related to chakras is Kundalini yoga. This discipline combines breathing techniques, physical postures (*asanas*), mantra repetition and meditation to awaken the dormant energy known as kundalini. This energy is traditionally represented as a coiled serpent at the base of the spine, in the Muladhara chakra.

During practice, energy awakens and rises along the Sushumna Nadi, passing through each chakra. As the energy moves upward, it opens inner "doors," leading the practitioner to experience higher states of consciousness. The ultimate goal is to reach Sahasrara, the crown chakra, where union with the absolute is achieved.

Sri Aurobindo, an important figure in modern Indian philosophy, described kundalini awakening as a crucial moment in the process of spiritual transformation. In his writings, he stressed the importance of integrating ancient practices with a broader understanding of human nature. Contemplating the chakras, according to Aurobindo, also means exploring the development of consciousness in its entirety.

Chakras and quantum physics.

The traditional explanation of chakras is based on an energetic and spiritual understanding of the human body. In recent times, some scholars and practitioners have tried to connect the principles of chakras with discoveries in quantum physics. The basis of this analogy lies in the vibrational nature of the universe.

Chakras can be seen as nodes of energy or electromagnetic fields that interact with their surroundings. Quantum physics, which explores subatomic particles, highlights the connection between energy and matter. This parallel suggests that chakras may not only be spiritual realities, but also physical phenomena, anchored in a universal energy intelligence.

A quote from physicist Niels Bohr, known for his work on quantum theory, poses an interesting reflection:

"The opposite of one deep truth can be another deep truth."

This principle, which invites the consideration of seemingly opposite realities as complementary, can apply to the relationship between science and spirituality.

Chakra meditation represents an ancient discipline capable of remaining relevant even in the modern world. Through techniques such as Kundalini yoga, practitioners can not only achieve greater self-awareness, but also understand the holistic nature of the universe.

Sacred places such as Varanasi continue to offer inspiration to spiritual seekers, while teachings of masters such as Sri Aurobindo guide toward an integration of tradition and modernity. At the same time, analogies with quantum physics stimulate a dialogue that could lead to new insights on both spiritual and scientific levels.

After all, the chakras remind us that we are more than just physical bodies. We are, in every sense, energy in motion.

Meditating on Time: The Maya Illusion and Quantum Reality.

In the depths of Indian philosophy, time is not real at all. The ancient Upanishads, among the most mystical and poetic Hindu texts, describe it as *kaala*, a construct of the mind. According to these scriptures, time is an illusion, a creation of consciousness itself. This concept merges with the idea of maya, the veil of deception that presents the world as separate and fragmented. Curiously, modern physics-particularly quantum physics-seems to be quietly approaching similar views.

Meditations on time, central to many Hindu practices, aim to deconstruct this illusion. In the fields of *dhyana* (deep meditation), yogis investigate time not as a sequence of moments, but as an alterable, subjective perception. This same paradigm fascinates theoretical physicists, who today speak of reversible time or even the possibility that time exists only in relation to the observer.

The concept of time in the Upanishads is not linear. In the Vedic tradition, everything exists in an eternal present, which the mind, through its limitations, divides into past, present and future. This crucial fragment emerges, among others, in the Brihadaranyaka Upanishad, where it states:

"He who knows time knows Brahman."

Brahman, in the Hindu view, is the absolute, the whole, which transcends all human definition. Time, therefore, separates the individual from universal truth, anchoring him or her in the material dimension.

Meditating on time allows one to overcome these limitations. Traditional yogis engage in practices that take them beyond their perceptions.

The experience of time dissolution is often connected with a state of deep meditation. Thought is removed from the constraints of memory or expectations. One reaches a dimension that Hindus call *turiya*, the "fourth state," beyond wakefulness, dream and sleep. This practice, thousands of years old, is surprisingly echoed by contemporary quantum physics paradigms.

In quantum physics, the conception of time is anything but rigid. The fundamental equations governing the subatomic world are reversible. In theory, this means that the past and the future could be interchangeable. Traditional science does not yet know how to explain this peculiarity, since it contradicts the arrow of time defined by thermodynamics (entropy). However, if time is a mental construct-as the Vedic doctrines suggest-this apparent contradiction dissolves.

The link between the human mind and the flow of time has been explored by physicists such as John Wheeler, who introduced the concept of "*participatory univers.*" According to this view, the observer collaborates in the creation of physical reality. This theory, although complex, resonates with the ancient Vedic insight: consciousness is the basis of reality. Nothing exists separate from the observing mind.

Physicist David Bohm, known for his theory of "implicit order," also emphasized how time could be a projection in which the entire universe manifests itself. At a deeper level, Bohm suggested, time and space continuously reform as holographic images, as a single "unified field." Within this perspective, time loses its character of absoluteness, just as Hinduism claims.

In many ashrams of India, meditative practices on time have been handed down for centuries. One of the best-known techniques involves focusing on the breath, observing its flow without identifying it with mental processes. Once the mind calms down, spontaneous insights surface aimed at showing the unreality of time. Rishikesh, sacred country at the foot of the Himalayas, is a place where such practices are cultivated by modern yogis and international spiritual seekers.

These subjective experiments also have implications for neuroscience. Recent studies on advanced meditators show that medical practice can alter perception of time, expanding the present moment. Brain MRIs reveal that certain areas of the brain related to time measurement "switch off" during deep states of meditation. This suggests that the state of consciousness may directly intervene on the experiential level of time.

Although rooted in traditions far apart in time and space, Hinduism and quantum physics open up fascinating conversations about our relationship with time. Both seem to suggest that what we perceive as separate is actually part of a single, vast, interconnected reality.

Meditating on time is thus not only a spiritual exploration but also a challenge to modern science. Ancient Indian rishis and quantum physicists share a bold vision: that the key to understanding reality is hidden in consciousness and in overcoming temporal constraints. It is up to us, observers and participants, to cross the bridge between physics and spirituality.

The breath of the universe. Pranayama and the cosmic connection.

The word pranayama comes from Sanskrit: "*prana*" means vital energy or universal breath, while "*ayama*" means expansion or control. In the ancient texts of Hinduism, such as the Upanishads and Patanjali's Yoga Sutra, prana is described as the breath that permeates all existence. It is not just air entering the lungs, but a subtle energy that flows through every living being and connects it to the whole.

Paramahansa Yogananda, the celebrated author of Autobiography of a Yogi, emphasized this cosmic connection. In his teachings, Yogananda explained that the breath is not just a physiological action, but a portal to universal awareness. He likened breath control to a "switch" that turns on the understanding of our deep connection to the absolute, or Brahman.

Hinduism has been talking for millennia about a universal energy that animates everything. The idea appears strikingly akin to modern field theory in quantum physics. Albert Einstein, with his unified field theory, sought to prove that all matter and energy share a common root. Although the term "prana" does not appear in Einstein's work, the parallels are fascinating. Just as prana spreads through every being, so in modern theories quantum waves vibrate and permeate the universe.

The ancient Vedic sages, centuries before the advent of modern science, already sensed this interconnectedness. For them, controlled breathing was a means of becoming attuned to these cosmic vibrations. Through pranayama, the individual becomes an active participant in the great game of the universe, just as an observer modifies a quantum system in modern physics.

Pranayama is much more than a spiritual metaphor. On a practical level, breath regulation disciplines the mind and calms the nervous system. During meditation, techniques such as *Anulom Vilom* (alternate breathing) or *Kapalabhati* (purifying breathing) establish a rhythm that harmonizes body and mind. This synchrony allows the individual to experience deep states of awareness.

A modern example comes from the "*Isha Foundation*" of Sadhguru, a contemporary Indian master. In his advanced meditation programs, pranayama is used to align individual energy with universal energy. Sadhguru describes the breath as a messenger:

"Through the breath we can dialogue with existence itself."

Ancient Hindu wisdom holds that, through breath control, one can overcome the duality between the self and the whole. The monistic philosophy of advaita, promoted by Adi Shankaracharya in the 8th century CE, states that the individual (Atman) is not distinct from the absolute (Brahman). Pranayama thus becomes a practice for dissolving the barriers between the individual self and the universe.

Paramahansa Yogananda used to tell how this communion could be experienced through deep meditation. He often quoted his teacher, Swami Sri Yukteswar, who described the breath as the "*gateway movement*" to higher states of awareness. In the experience of advanced meditation, the breath slows to a stop, leaving the practitioner in a state of perfect oneness with the cosmos.

Many scholars see pranayama as a meeting point between tradition and science. On the one hand, the breath represents a biological function that can be studied and measured. On the other, it is a profound symbol of the spiritual connection between the human being and the universe. This duality perfectly reflects the dialogue between Hinduism and quantum physics: two seemingly distinct perspectives that converge into a broader understanding of existence.

In the 21st century, meditation and breath are finding application beyond religion. Pranayama-inspired techniques are integrated into mindfulness, psychotherapy and stress-relieving practices around the world. The message remains the same: when you calm your breath, you can perceive the deeper order of things.

The breath, invisible but omnipresent, is the "*thread between the visible and the invisible*," as Yogananda put it. Through pranayama, Hinduism invites us not to be mere observers of the universe, but active participants. In an age when science is pushing toward understanding the quantum foundations of existence, this ancient wisdom seems more relevant than ever. Breathing is no longer just a necessary act for life. It becomes the key to connecting with the eternal rhythm of the cosmos.

The mind as a quantum field in the yoga tradition.

Between the symbolic peaks of the Himalayas and the mysteries of quantum mechanics, a link emerges that fascinates scientists and philosophers: observation. Hinduism, with its millennia-old spiritual tradition, offers a unique perspective on the role of the observer, a concept that is echoed in modern physics. In both views, observation is not passive. It is the act that generates reality. This parallelism, which might seem esoteric, delves into one of the deepest territories of yogic philosophy: the "*purusha*," the Silent Observer.

In yoga philosophy, "*purusha*" represents the eternal, unchanging and conscious principle. According to Patanjali's Yoga Sutra, human experience is divided between the purusha and prakriti, that is, the material, changing world. The purusha is the pure, detached and still observer, watching without judgment the flow of thoughts, emotions and sensations. The mind, with its waves, is like a lake in which these experiences are reflected. But who is it that observes these reflections? It is the purusha.

Through dhyana, deep meditation, the yogi learns to identify with this Silent Observer. Achieving this awareness means going beyond the ego and recognizing that perceived reality is nothing but a construct. Swami Vivekananda, during his famous address to the 1893 World Parliament of Religions in Chicago, forcefully explained this idea:

> *"The soul is neither the body nor the mind; it is the eternal witness. Everything else changes, but the soul remains."*

Practicing dhyana means sitting in silence and watching one's thoughts as one watches the clouds pass by in the sky. This detachment guides the yogi to realize that thoughts arise from a deeper "field," a field of pure energy, just as in the subatomic world what we see emerges from a complex web of particles and waves. In this sense, purusha could be understood as man's "quantum mind," where every possible reality exists as potentiality until observed.

Hinduism, for millennia, has offered a teaching that quantum physics is only recently recognizing. In the famous double-slit experiment, the choice of whether or not to measure a particle affects its manifestation. It is the role of the observer that determines whether the particle behaves as a wave or as a particle. An unobserved reality remains a possibility. This principle is strikingly similar to the yogic view: what is observed through consciousness determines the structure of reality.

Swami Vivekananda, during his stay in the United States, outlined a crucial concept:

> *"All matter is only a secondary manifestation of consciousness."*

This seemingly mystical statement speaks of a universe in which matter does not exist in the conventional sense. Modern physics, with quantum theory, agrees. For example, Werner Heisenberg, father of the uncertainty principle, observed in turn:

> *"The world is just a web of potentialities that becomes concrete only through observation."*

These parallels between science and spirituality would not surprise a yoga practitioner. The ancient wisdom of the Vedas already sensed that the separation of observer and observed is an illusion. The human spirit, according to the ancient texts, does not just create its own reality; it is itself reality. During dhyana, the yogi does not seek to change the world, but to know who really observes.

Interconnections between science and meditation.

Contemporary science is beginning to probe the mysteries that the meditative practices of Hinduism have addressed for millennia. In particular, quantum physics now offers new insights into the nature of reality, proposing visions that find surprising echoes in the spiritual insights of Indian sages.

For centuries, meditation techniques such as dhyana and pranayama have indicated a path to explore consciousness and the interconnectedness of the individual and the universe. The ancient Vedic texts, including the Upanishads, state that what we perceive as separate is, in fact, one indivisible reality. This reality is called

Brahman, the supreme, absolute and omnipresent principle, the source of all that exists.

In the 20th century, pioneers in quantum physics began to question the traditional boundaries between matter and consciousness. Eugene Wigner, a physicist and Nobel laureate, emphasized that the observer was not a neutral figure in the measurement of quantum phenomena. According to Wigner, human consciousness plays an active role in determining the behavior of physical reality. This insight is strikingly close to the Vedantic view that the reality we perceive is partly constructed by our consciousness.

David Bohm, another great innovator in quantum physics, proposed the concept of "implicit order." For Bohm, the universe is not a collection of separate elements, but an undivided whole in which each part is connected to all the others. This view clearly reflects Vedantic philosophy, in which the phenomenal universe (what we perceive) is only an appearance, like the veil of Maya. The true fabric of reality is invisible to the eyes, but accessible through meditation. Bohm himself was influenced by his dialogue with Jiddu Krishnamurti, a leading Indian philosopher who often spoke of the need to transcend conceptual thinking in order to experience unity.

Meditation as an "observer" of reality.

Hindu meditative practices are based on the direct experience of consciousness. Through techniques such as mantra repetition (japa), breath control (pranayama) and concentration (dharana), practitioners seek to overcome the illusion of the individual self. Meditation leads to experiencing a state of deep connection. The mystic no longer feels separate from the world, but part of an interconnected whole, just as the wave is not distinct from the ocean.

There are similarities between this experience and the quantum concept of "entanglement," or quantum entanglement. In laboratories, physicists have shown that two particles, once connected, remain in a state of instantaneous correlation, even if they are far apart. This fundamental interconnection, which transcends time and space, seems to echo the Vedic insight that everything is one under the veil of illusion.

An emblematic example of the bridge between meditation and science is found in the experiences of Nikola Tesla. Tesla, while not a quantum physicist, became interested in Eastern philosophies and meditated regularly. He declared that many of his inventions had "come to him in visions," states of deep insight that could be compared to the meditative states described in Indian tradition.

The figure of Erwin Schrödinger, famous physicist and father of quantum mechanics, is another significant example. Schrödinger was deeply influenced by the Upanishads. He wrote that the concept of universal consciousness described in the Vedic texts matched what was emerging in his understanding of physics. In one of his letters, Schrödinger stated that the distinction between subject and object is an illusory barrier.

India offers many sacred places that relate scientific and spiritual concepts. Rishikesh, in the foothills of the Himalayas, is known as the world center of meditation. Here, millennia-old traditions continue to be practiced alongside modern seminars that explore the relationship between mind and quantum physics. Each year, scholars from around the world gather to discuss science and spirituality, combining the sutras of the sages with the equations of physicists.

Hinduism has always perceived meditation as a tool for accessing ultimate truth. Quantum physics, with its paradoxes and revolutionary discoveries, is opening the door to a deeper understanding of the universe around us and ourselves. If Brahman is the ultimate essence of being, perhaps modern science, just like the Indian sages, is discovering that investigating external reality also means exploring the inner realm of consciousness. A bridge, that between physics and spirituality, that seems increasingly firm and promising.

Meditation and quantum neuroscience.

Meditation, an ancient spiritual practice explored in depth by Hinduism, is attracting increasing interest from modern science. This interest is no accident: neuroscientists and quantum physicists are discovering surprising connections between the effects of meditation on the mind and the mysteries of reality described by quantum physics. But what is the connection?

Meditation: Break from reality and inner journey.

Quantum physics has revolutionized our understanding of the world. In the famous double-slit experiment, scientists discovered that subatomic particles behave as both waves and particles, but only when observed. Before the observation occurs, the particle exists in a superposition state, a set of possibilities.

This behavior seems analogous to the state of the mind during deep meditation. When the practitioner enters *dhyāna*, ordinary thought dissolves. The mind ceases to chase a defined reality. It loses its habitual boundaries and enters a realm of fluid and infinite possibilities.

This link between mind and reality was also explored in the famous meeting between Albert Einstein and Rabindranath Tagore in 1930. Tagore, an Indian poet and philosopher, argued that reality was not independent of human observation. Einstein, an advocate of an "objective" universe, was initially skeptical but was impressed by Tagore's argument, which brought together philosophical and spiritual insights. Their dialogues highlighted what is apparent today: the boundaries between science and spirituality are less defined than previously believed.

In recent years, studies on meditation have revealed surprising connections with brain functioning. During deep meditation, brain activity is significantly altered. Meditative states are associated with increased coherence of brain waves, particularly in the gamma wave range. These states reflect greater "connectedness" between brain regions and seem to support the perception of unity with the whole described by many meditators.

But where does quantum physics come in? Some neuroscientists, such as Stuart Hameroff, have hypothesized that brain processes might depend on quantum phenomena. Hameroff, together with physicist Roger Penrose, has proposed the "quantum mind" model. According to this theory, microtubules in brain cells could facilitate quantum action in consciousness. Although controversial, this model offers a possible bridge between meditation, quantum physics and neuroscience.

The Indian mystic Ramana Maharshi (1879-1950), known for his practice of silent meditation, described the meditative experience as letting go of the "*small self.*" Ramana taught that silence is not empty but full of awareness. In one of his reflections, he said:

> *"What remains after the mind quiets down is absolute reality."*

This idea seems to be reflected in the quantum concept of "wave function collapse," in which observation collapses a range of possibilities into a single reality.

If quantum physics shows that reality is plastic and shaped by observation, Hinduism suggests that the mind can "observe" a deeper dimension through meditation. Both disciplines, while taking different paths, open up a universal question: who, or what, observes? Meditation and physics seem to whisper the same truth, each in its own language.

Perhaps, as Erwin Schrödinger, the father of quantum mechanics, argued,

> *"Mind is not a product of nature; mind is the matrix of all things."*

It is in this mystery that Eastern philosophy and quantum physics meet, reminding us that the frontiers between science and spirituality are meant to be explored.

The brain in meditation: the quantum dance between waves and consciousness.

For centuries, Hinduism has explored the nature of consciousness through the tools of meditation and introspection. Today, modern science, with the support of quantum physics and neuroscience, allows us to investigate how meditative states can affect the brain and potentially alter our understanding of reality and consciousness.

Meditation, practiced in the Indian Vedic tradition for thousands of years, acts profoundly on the brain. Neuroscience studies have shown that during deep meditation -- as in the *samadhi* described in Patanjali's Yoga Sutras -- brain waves follow a unique dynamic. Experiments by Richard Davidson, professor of neuroscience at the University of Wisconsin, on Tibetan monks trained in transcendental meditation have produced surprising results.

Davidson, in his laboratory, recorded a significant increase in gamma waves in the brains of monks. These waves, which oscillate at very high frequencies, are associated with states of expanded awareness, deep attention and mental plasticity. The study, conducted in the early 2000s, showed that such waves can synchronize different brain regions. The monks themselves reported feelings of oneness with the universe, experiences vaguely reminiscent of the principles of non-locality described in quantum physics.

In quantum physics, the concept of non-locality implies that two particles, once "entangled," can affect each other, regardless of the distance between them. This reflects the sense of universal connectedness that emerges from meditative experiences. According to the monks interviewed, in deep meditation "there is no separation between self and universe." A description curiously akin to what quantum mechanical scholars observe in the behavior of particles.

Meditation and quantum decoherence.

Meditation also affects the brain through a phenomenon called "*neural coherence*." During deep meditation, brain waves align, creating a kind of harmony or "order." This concept has parallels in the field of quantum decoherence. According to quantum physics, quantum systems coexist in overlapping states until an interaction causes "decoherence," or collapse into a single observable state.

Neuroscientist Bruce Alan Wallace, an expert on Buddhist meditation and Indian spiritual references, has suggested that meditation might promote a kind of internal coherence of the mind, similar to maintaining an efficient quantum state. When the meditative brain enters a state of order, it may facilitate a higher perception of reality, perhaps even reducing interference between the inner and outer worlds, as if the mind synchronizes with the "quantum field" that drives the universe.

Hindu texts, particularly the Yoga Sutras, often speak of non-ordinary states of consciousness. Patanjali defines dhyana as uninterrupted concentration on the nature of the Self or the absolute (*Purusha*). This state, described as a union between subject and object, has strong parallels with the perception of interconnectedness that emerges from quantum entanglement.

An example of these experiences can be found in the lives of Indian mystics. Ramana Maharshi, one of the most influential spiritual teachers of the 20th century, told of meditative states in which the individual self disappeared, leaving him "immersed in the eternal." This is reminiscent of what quantum physicist David Bohm called "implicate order," a hidden reality in which everything is connected.

The neuroscientific discoveries of Davidson and others, combined with striking similarities to quantum principles, are building a bridge between spirituality and science. In 2014, the Dalai Lama, during a dialogue with physicists and neuroscientists, stated:

> *"The mind has an unlimited capacity to develop, just like the universe."*

This spirit of curiosity reflects the heart of the Vedic tradition, which has been investigating the inner universe for centuries.

Meditation thus emerges not only as a spiritual practice, but as a means of exploring a complex territory situated on the border between consciousness and quantum mechanics. Perhaps, as Hindu sacred texts and modern studies suggest, the human mind is a microcosm reflecting the secrets of the universe itself.

Hinduism, with its ancient practice of meditation, offers an extraordinary lens for examining the mysteries of consciousness. Modern science, through brain studies and quantum physics, is beginning to illuminate how these practices can affect brain waves and our perception of reality. Meditating, in this sense, is no longer just a spiritual act. It is a quantum dance between brain waves, consciousness and the universe.

From Samadhi to quantum consciousness.

Meditation is not just a method of relaxation. Numerous neuroscientific studies are beginning to show that sustained, focused attention during meditative practice can produce significant changes in brain structure and functioning. Studies by Italian neuroscientist Giuseppe Vitiello, for example, propose an intriguing hypothesis: the mind, through focused attention, could influence the quantum fields that permeate our physical reality.

Samadhi (to "*put together*," "*unite with*") is a Sanskrit noun specific to Buddhist and Hindu religious cultures. This term defines the meditator's cosmic union with a deity, or with the object of meditation, and his or her consequent liberation (Moksha) from earthly bonds.

Vitiello, who has explored connections between brain systems and quantum phenomena, argues that the human brain can be read as an open quantum system. During a state of intense concentration-such as that which occurs in Samadhi-a form of coherence could be created between neural networks and the subtle energy of quantum fields, similar to what occurs in superconductivity phenomena.

British physicist Roger Penrose has delved into these topics, focusing on the microtubules found in brain cells. These microscopic filaments, according to Penrose and his collaborator Stuart Hameroff, could act as quantum circuits, capable of processing information not only at the classical level, but also at the quantum level. In a state of deep meditation, then, microtubules could act as a bridge between mind and matter, between individual consciousness and cosmic reality.

In Hinduism, Brahman is the absolute reality, the ultimate essence from which everything emerges and to which everything returns. It is described as eternal, infinite and unmanifest. This view finds striking similarities with the physical concept of the quantum vacuum, which is not "nothingness" but rather a pulsating field of energy in which particles and antiparticles continually emerge and dissolve.

The Indian philosopher Adi Shankaracharya, in the 8th century, defined Brahman as:

> *"the unique reality that permeates everything, beyond the illusion of the manifold."*

Modern quantum physics, although from a completely different point of view, describes the quantum vacuum in not so distant language. Both systems of thought speak of a fundamental dimension hidden behind phenomenal multiplicity.

In a contemporary interpretive key, meditation-and particularly the attainment of Samadhi-could be seen as a tool for "tuning in" to this ultimate reality. The human mind, then, would not just be a passive observer, but an agent interacting with the universe on a deeply creative and quantum level.

Concrete examples are not lacking. Studies of experienced meditators, such as Tibetan monks analyzed by teams of neuroscientists since the 2000s, have shown that meditation produces states of unique neural coherence, with synchronized brain waves and patterns that defy conventional neurophysiological understanding. Some researchers speculate that this coherence may reflect an interaction with levels of reality that are not accessible in ordinary states of consciousness.

It is worth mentioning the great physicist Erwin Schrödinger, who had a deep interest in Indian philosophy. Schrödinger, famous for his "cat" paradox, declared:

> *"There is undoubtedly one universal mind. Plurality is only an illusion."*

This thought, which resonates with the Hindu idea that Atman is actually a manifestation of Brahman, suggests a profound convergence between Eastern philosophy and modern physics.

From Samadhi to quantum consciousness, meditation is revealed as an extraordinary tool for exploring the ultimate nature of reality. Science, particularly quantum physics, is beginning to offer a modern language for understanding phenomena described for millennia in spiritual traditions. Perhaps, the real bridge between physics and spirituality is not an abstract concept, but the direct experience of consciousness itself. And this experience, as taught by the millennia-old wisdom of Hinduism, may hold the key to awakening the ultimate truth about our place in the universe.

What does neuroscience say?

In recent decades, neuroscience has offered new insights into the relationship between meditation and consciousness. Using advanced technologies such as functional magnetic resonance imaging (fMRI) and electroencephalography (EEG), scientists have begun to explore how meditative practices affect the brain. The results of these studies not only evoke parallels with principles of quantum physics, but also solidify the millennia-old link between Eastern spirituality and modern science.

A crucial example is the 2011 study conducted at the *University of Wisconsin-Madison* by neuroscientist Richard Davidson. Davidson, a longtime collaborator with the Dalai Lama, analyzed the brains of Tibetan Buddhist monks, known for their intense meditation practices. Through electroencephalography, the researchers found a significant increase in gamma waves, a type of brain activity associated with advanced processes such as concentration, higher perception and creativity. During deep meditation, the intensity of these waves far exceeds the levels found in the so-called "ordinary mind." According to Davidson, meditative practice could enable individuals to access states of awareness that transcend normal everyday experience.

These findings are revolutionary, not only for their implications for neuroscience, but also for what they suggest about the concept of mind. The intensification of gamma waves would highlight hitherto underestimated brain plasticity. Meditation, according to these studies, is not simply an art of the mind, but a real tool for reshaping the brain.

Richard Davidson's work finds resonance in advanced theories of quantum physics. Hungarian physicist Eugene Wigner, Nobel laureate in 1963, theorized that consciousness is fundamental to the collapse of the quantum wave function. In simple terms, Wigner suggested that physical reality depends on the observer. Without consciousness, reality would remain in a state of potentiality. This viewpoint relates to the role of meditation. If the state of consciousness affects observation, and thus reality, then a mind trained in higher meditative states could exert a direct impact on the structure of reality itself.

A significant example of these ideas is found in the spiritual tradition of the Advaita Vedānta of Hinduism. Here it is taught that ultimate reality (Brahman) is observed through conscious interaction with the individual self (Atman). The insight described by this philosophy-namely, that the observer and the observed are deeply interconnected-has striking similarities to the principles of quantum physics, which studies the interaction of subatomic particles and the observer.

Returning to neuroscientific work, it is worth mentioning studies conducted at Harvard Medical School. In 2012, researchers found that eight weeks of daily meditative practice physically changes the structure of the brain. They observed an increase in gray matter density in areas related to memory, learning and emotion management. This leads to deep thinking: meditative practice influences not only subjective experience, but also the biological substrate of the brain.

A fascinating parallel to this is found in the ancient Vedic texts. The teachings describe meditation as a tool for "purifying" the mind and awakening the latent potential of consciousness. In ancient India, practitioners were convinced that control of thought and breathing could transform both body and mind. Modern neuroscientific discoveries seem to validate these millennia-old insights.

Meditation represents that ideal bridge between scientific research and Eastern spirituality. Not only does it reveal the extraordinary capabilities of the brain, but it invites us to explore the role of consciousness in the construction of reality. As physicist Niels Bohr argued,

> *"When talking about atoms, language can only be used as poetry."*

This also applies to the dialogue between science and spiritual practices: two fields that meet to explain the unexplainable. Meditating, perhaps, is the poetic language that connects mind, universe and reality.

When the atom meets the soul: meditation, neuroplasticity and quantum potential.

Eastern philosophies, particularly Hinduism, have always been concerned with the connection between the mind, the body and what lies beyond: the infinite. Meditation, at the heart of Hindu spiritual practice, represents a way to transcend material limitations and reach a higher state of awareness. Today, thanks to modern science, we can better understand the mechanisms by which this happens. Neuroscientific studies and quantum theories shed new light on ancient insights, suggesting that meditative practices can not only change the brain, but also alter the way we perceive reality.

Concentration is the heart of meditation (Dhyana). In the Bhagavad Gita, one of the fundamental texts of Hinduism, Krishna teaches Arjuna that the mind of one who meditates steadily is "*like a flame that does not waver*." This state of concentration is more than just a mental exercise. According to modern neuroscience, meditation can physically change the brain through a process known as neuroplasticity.

One of the leading researchers in this field is Sara Lazar, a Harvard psychologist. In the 2000s, Lazar conducted several studies documenting how meditative practice, particularly mindfulness-based meditation, increases gray matter density in areas of the brain responsible for memory, empathy and emotional regulation. One of his most celebrated studies, published in 2011, showed that eight weeks of daily meditation can lead to significant structural changes.

This phenomenon shows that the human brain is not rigid, but can be shaped through practice. Through meditation, neural connections are strengthened, improving attention span and, often, changing the way we perceive our emotional and physical limitations. Hinduism has always emphasized this possibility of "*self-transcendence*"- Dhyana is not just a practice, but a transformation.

The intuition that consciousness is something more deeply rooted than individual experience paves the way for quantum physics. In recent decades, some scientists have speculated that the human brain may be the theater of quantum phenomena. Among them, British physicist Roger Penrose, Nobel Laureate in 2020, and physician Stuart Hameroff have proposed the theory of quantum neurodynamics. This theory suggests that human consciousness may depend on quantum processes occurring in the microtubules of brain cells.

Unlike the macroscopic phenomena observed in the classical world, the quantum realm is governed by uncertainties, superpositions and nonlocal correlations. This view opens the way to a fascinating idea. If quantum mechanisms exist in the brain, these dynamics could explain not only the coherence of conscious experience, but also the possibility of free will. Many scientists agree that consciousness is too complex a phenomenon to be explained linearly through neurochemistry alone.

According to some modern spiritual philosophers, this may connect to the Hindu concept of Atman, the individual soul that is part of a greater whole. Just as in the quantum realm particles cannot be understood separately from their energetic context, similarly the soul, according to the Vedic views, cannot be separated from the universe (Brahman).

The ability to transform the brain through meditation is a modern demonstration of what ancient Hindu philosophies had meant by the term "awakening." Through practice, according to the tradition, it is possible to reach states of consciousness in which the limits of space and time dissolve. This is reminiscent of some quantum theories that see reality as a fluid interconnectedness, where the observer plays a key role in determining what is perceived.

The bridge between physics and spirituality finds a point of contact here. Quantum physics teaches us that the world is not made up of separate objects, but of energies linked by deep, invisible relationships. Hinduism, through Dhyana, invites us to remove the mental barriers that prevent us from recognizing our unity with the whole.

Meditation and quantum physics seem to travel parallel paths toward the same goal: the rediscovery of the infinite. The former teaches how to transcend internal limits through awareness and discipline. The second shows us that what appears finite and separate is, in fact, interconnected and limitless.

The words of Maharishi Mahesh Yogi, father of Transcendental Meditation, seem to anticipate this connection:

> *"Meditation takes you into the realm of the infinite: what always was, always will be, and is here now."*

A message that many believe echoes in the mysteries of modern physics, where the atom meets the soul.

Meditation and time. "Time ceases to exist."

Many meditation practitioners, both modern and ancient, describe a profound and mysterious feeling: time seems to dissolve. Seconds, minutes, hours lose all meaning. This experience is not just a subjective idea, but is deeply rooted in the spiritual traditions of India and finds interesting parallels in quantum physics today.

In the philosophical system of Yoga, described in Patanjali's ancient Yoga Sutras, the state of samadhi represents the pinnacle of meditative practice. In this state, the boundary between the individual and the rest of the universe vanishes. There is no longer any "before" or "after." The mind accesses a dimension that transcends linear time. In samadhi, past and future become irrelevant because everything is perceived as eternally present.

This experience of meditation seems to have an interesting counterpart in a key concept in quantum physics: non-locality. In quantum mechanics, two particles that have become entangled can remain connected instantaneously, regardless of distance. This phenomenon, called quantum entanglement, suggests that spatial and temporal separation, so central to our daily lives, may not be a fundamental property of reality.

Physicist David Bohm developed this idea further. Bohm proposed the concept of "implicit order," a framework in which time and space are not absolute entities but emerge from a deeper, undivided reality. In this view, what we perceive as temporal separation is only an illusion of our mind. In the deepest level of reality--as in the experience of samadhi--everything is intrinsically connected.

The expansion of consciousness in samadhi is lyrically described in Vedic texts. In the Kaivalya Upanishad, an ancient Indian philosophical text, we read:

> *"When the mind dissolves into peace, time vanishes. The Supreme Self is all that remains."*

This description seems to resonate with the notion of an implicit order, in which every fragment of the universe is interconnected.

Modern meditation practitioners report similar experiences, confirming how much these insights transcend the boundaries between culture and time. Indian master Ramana Maharshi, active in the 20th century, explained that the Self, once realized, does not perceive time as fragmented. For him, past and future were condensed into an "eternal present." These words seem to anticipate the parallels emerging from science today.

Even in contemporary practice, people from different cultures have similar experiences. Chade-Meng Tan, a Google engineer and author of books on applied meditation, recounts how a deep practice enabled him to "*perceive the world beyond the thought of time*." These personal accounts reveal how universal the theme is, able to unite India's mystical past and the Western technological present.

Modern neuroscience is beginning to investigate how meditation can change the perception of time. Studies using brain scans show that, during deep states of meditation, activity in the parietal cortex-the region of the brain responsible for spatial and temporal awareness-is reduced. This may explain why many meditators report the feeling of no longer being bound by time.

Neurobiologist Francisco Varela, known for his work on the intersection of science and Buddhism, hypothesized that the mind in meditation accesses an "atemporal" vision. For Varela, meditation is a form of internal exploration of time, comparable on an experiential level to the theoretical challenges faced by quantum physics. Just as quantum equations challenge the idea of linear time, so meditation unhinges this conception at the subjective level.

The dissolution of time is not a foreign concept to Western philosophies. St. Augustine, in his famous text "*Confessions,*" wrote:

> *"There is no past, except as a memory, nor a future, except as an expectation. Only the present is real."*

This philosophical insight finds striking parallels with descriptions of samadhi: time as a mental construction rather than an objective reality.

However, it is in Hindu culture that we find the richest and most articulate descriptions of time as an illusion, or Maya. In the Bhagavad Gita, Krishna explains to Arjuna that the world perceived by ordinary man is an illusion, while the ultimate reality is timeless and unchanging. This theme is also explored in Indian cosmological concepts, with the Kalachakra, or "wheel of time," representing the eternal cyclical return as opposed to the absolute.

The bridge between physics and spirituality rests on common insights that remain remarkably relevant. Eastern sages described states of consciousness that seem to resonate with phenomena explored by modern physics, such as non-locality or implicitness. Meditation, with its ability to make us perceive eternity in the present moment, invites us to reconsider our relationship with time. In this encounter between ancient traditions and new scientific discoveries, a universal truth emerges: time, as we perceive it, may just be a doorway to a deeper reality.

The illusion of time: neurobiology of meditation and quantum theory.

Ancient Hindu Vedic texts, such as the Rig Veda and the Bhagavad Gita, invite us to reflect on a revolutionary concept: time is not real. It is an illusion, part of the deceptive plot called Maya. For centuries, Western thought has treated time as a straight line, a succession of evenly flowing instants. But today, quantum physics and neuroscience are beginning to converge on a view that is strikingly reminiscent of the ancient Vedic teaching. Time, as we perceive it, may be a construct of the mind.

Meditation, in particular, offers a tool for exploring and even transcending the perception of time. Harvard University neuroscientist Judson Brewer has studied the brains of experienced meditators such as Buddhist monks and yogis for years. Through

MRI, Brewer observed reduced activity in the "*default mode network*" (DMN), a neural network related to self-reference and self-perception. When this network "shuts down" during meditation, the boundaries between past, present and future seem to dissolve.

It is through this "turning off," or slowing down of the DMN, that meditators describe feelings of timelessness. For example, Matthieu Ricard, a well-known Buddhist monk and scientist, has often recounted how long meditation sessions can make hours feel like minutes. This neurobiological phenomenon is not a simple subjective effect. It is an open door to a deeper way of experiencing reality.

If we look at quantum physics, we find surprising parallels. Theoretical physicist Carlo Rovelli, in his book "*The Order of Time*," explained that time, as we conceive it, does not exist. Einstein's general relativity already shows how time is a relative event, influenced by gravity and speed. Rovelli goes further: in the quantum scale, time dissolves completely, existing only as a relation between events.

This physical idea reflects the Vedic view of time as a construct of the mind, not an objective entity. Ancient Indian rishis claimed that time is nothing more than a dance of Maya, and that the true purpose of existence is to go beyond this illusion. Similarly, Rovelli suggests that our perceptions are partial, limited by how our brains construct reality.

The human mind plays a central role in the creation of time. Neuroscientific studies indicate that temporal perception is largely related to our attention. When we meditate, we stop dividing reality into temporal fragments and access a more holistic experience in which past and future merge into an eternal present.

Indian yogis speak of *samadhi*, the highest state of meditation, as an experience in which barriers between the self and its surroundings, including temporal ones, vanish. This deep state of awareness coincides, according to some spiritual scholars, with insights gleaned from quantum physics. For example, the famous physicist David Bohm saw consciousness as fundamental to understanding quantum reality.

It is fascinating to see how cultures distant in time and space arrive at similar insights. India, a land of meditation and philosophical depth, provides lessons that resonate with modern science today. On the one hand, the Mahabharata and the Puranas describe time as cyclical, not linear, strikingly anticipating concepts of modern physics. On the other, modern neuroscientific research is revealing how millennia-old spiritual practices, such as meditation, profoundly alter human biology and perception.

Places, such as the ancient monastery of Rishikesh on the banks of the Ganges, or Kuśināgara, where the Buddha attained *parinirvana*, remind us that this wisdom arose in real and specific contexts. Today, neuroscience laboratories in the United States and Europe empirically explore a similar purpose: to free the mind from deceptive constructs.

Meditation and quantum physics seem to pursue a common intent. Both seek to strip time of its illusory linearity. Science, with its rigorous experimental methods, and spirituality, with its deep introspection, may appear distant. However, they converse, they meet. Meditation helps to experience directly what modern physics proposes: temporality as a result of the mind in action.

In this way, ancient spiritual masters and modern theoretical physicists unknowingly work together on the same question. What is time? Is it real or a construct? The answers are not yet definitive, but the path laid out is clear. To meditate is to explore the infinite, just like a quantum journey: to face the mystery of time and rediscover the eternal present.

A world changing with the mind: experiments and implications.

One particularly famous experiment links the quantum world to consciousness. The "*Double-Slit Experiment*," repeated several times from the 20th century to the present, showed that light can behave as both a wave and a particle, depending on the presence of an observer. This experiment, according to many theorists, implies that consciousness-or at least observation-plays a key role in defining states of reality.

So can meditation, which profoundly alters the mental state, influence the way we perceive (or even shape) reality? According to Amit Goswami, quantum physicist and author of the celebrated book "*The Self-Aware Universe*," the answer is yes. Goswami argues that through the enhancement of consciousness achieved through meditation, we can access deeper levels of reality.

One of the most curious experiments aimed at proving this hypothesis was conducted in the 1990s at *Princeton Engineering Anomalies Research* (PEAR). Participants, during states of intense concentration or meditation, tried to influence the random behavior of machines called *Random Event Generators*. The results suggested significant variations in the data, as if the mind could "*order chaos*" through intention alone. This type of research remains controversial, but for proponents it proves that consciousness can interact with physical reality.

The mind-matter connection explored by quantum physics has a deep echo in the sacred texts of Hinduism, particularly in the Vedas and Upanishads. In these scriptures, the concept of Brahman-the absolute, indivisible reality-is intimately connected with the concept of Atman, the individual self. According to these ancient teachings, the external universe is but a reflection of the inner consciousness. This view, however philosophical, blends surprisingly well with the modern assumption that observation, and not simply matter, is what shapes reality.

An evocative cultural example is found in the Sanskrit poem of the Mandukya Upanishad, which describes the universe as an immense dream state. Here, the tangible world is compared to a dream created by the mind. Similarly, physicist John Wheeler, one of the greatest explorers of the quantum world, described the universe as a "participatory universe," one in which consciousness plays a central role in shaping what we perceive.

These convergences between quantum physics and meditation do not yet answer the central question: does the mind create the world, or only interpret it? However, the convergences themselves suggest a conceptual revolution that could profoundly change the way we understand our relationship with reality....

This new alliance between physics and spirituality raises many questions. If consciousness affects reality, can meditative practices become concrete tools for dealing with psychological malaise and even some illnesses? Scientists are beginning to build bridges between quantum mechanics and spiritual research, but there is still a long way to go.

To date, models of study combining meditation and quantum science show that the world, after all, is more flexible and less determined than we thought. Perhaps the great change lies not only in understanding matter, but also in recognizing that through awareness reality becomes, as Goswami himself said, a creative opportunity.

In this vision, the ancient and the modern meet. Hinduism and quantum physics speak, in different but equally powerful ways, of the immense potential of human consciousness.

Meditation and the science of the future.

India, home to ancient spiritual traditions and meditation, is becoming a hub in the dialogue between modern science and Eastern philosophy. In particular, meditation, a millennia-old practice deeply rooted in Hinduism, is emerging as a key element in scientific research on neuroscience and quantum physics. This bridge between spirituality and modern rationality is opening new avenues for understanding human consciousness and the fundamental laws of the cosmos.

Research institutes in India play a key role in this evolution. *The Indian Institute of Science in Bangalore*, one of the country's most prestigious centers, is at the heart of innovative studies that seek to combine Western neuroscientific models with the wisdom of Eastern meditation. Indian scientists are investigating how contemplative practices, such as meditation, can affect the structure and function of the human brain. In this regard, research suggests that meditation may have measurable impacts on brain waves, promoting neuroplasticity and activation of neural networks often associated with states of deep attention and introspection.

A leading figure in this debate is Indian quantum physicist Subhash Kak, known for his studies on the Vedic model of consciousness. According to Kak, the Vedas offer a view of reality that could explain phenomena still unexplained in modern physics. He argues that while contemporary science explores subatomic particles and quantum fields, Vedic thought considers consciousness a fundamental principle of the cosmos, something that is not only subjective but also intrinsic to reality itself.

Kak explains that many theories in quantum physics, such as that of wave function collapse, raise questions about the role consciousness plays in determining reality. A question that is echoed in ancient Indian philosophy. In the Vedas, the concept of "Chitshakti" (*the force of consciousness*) suggests that the universe is the result of a conscious act. This idea could provide a new perspective for understanding the non-local dimensions of the quantum world, such as entanglement between particles.

An interesting example combining meditation and science comes from research conducted on Indian yogis and monks. In recent years, neuroscientific studies have used techniques such as magnetic resonance imaging to observe the minds of individuals who have been meditating for decades. One famous study, conducted in Dharamshala in northern India, found that experienced meditators exhibit greater coherence in gamma brain waves. This suggests that prolonged meditation can benefit not only spiritually, but also cognitively and neurologically.

However, these are not just scientific discoveries. The cultural and spiritual element of meditation remains central. In India, places like Varanasi and Rishikesh continue to attract scholars, mystics, and scientists from around the world. These sacred places serve as meeting points between the modernity of science and the antiquity of spirituality. In Varanasi, for example, the *ghats* on the Ganges River are not only spaces of religious devotion but also symbols of a holistic philosophy that sees the body, mind, and cosmos as one.

Meditation is also influencing modern medicine. Practices such as pranayama, an ancient yogic breathing system, are being studied for their impact on the autonomic nervous system. It has been found that techniques such as mindful breathing can reduce cortisol levels in the blood, thereby decreasing stress-one of the leading causes of illness in contemporary society. This highlights how ancient meditative techniques have the potential to complement modern models of mind-body care.

Finally, the dialogue between meditation and quantum neuroscience also has a deep philosophical side. If quantum physics has taught us that reality is very different from how it appears (with concepts such as the multiverse or quantum superposition), meditation invites us to explore this "invisible reality" within ourselves. In both cases, there is a search for something that transcends the senses, a greater truth that unites matter and mind.

In the future, India may become even more central to this dialogue. With a millennia-old tradition of spiritual practices and centers of research excellence, the country is a natural laboratory for bringing together physics, neuroscience, and meditation. As Subhash Kak himself argued:

> *"To understand the universe one must first understand the mind, because the mind is both: the means and the result."*

The science of the future, perhaps, will not only be a matter of formulas and theories. It will also be a matter of silence, introspection and awareness, combining the best of ancient wisdom with the light of modern knowledge.

Practical examples and an invitation for reflection.

In the modern era, science and spirituality seem to be dialoguing more and more. Two seemingly distant worlds, such as classical Hinduism and quantum physics, are finding surprising points of contact through an age-old tool: meditation. This bridge of connection reveals a new vision of man and the universe, capable of inspiring scholars and spiritual practitioners alike.

Hinduism has offered a unique approach to understanding reality for millennia. According to this tradition, meditation is the key to accessing the deeper dimensions of existence. An extraordinary example is offered by Paramahansa Yogananda in his famous "*Autobiography of a Yogi.*" In the book, Yogananda recounts experiences that defy the laws of space and time. Through meditation, he describes states of consciousness in which the universe appears as an interconnected unity.

Steve Jobs, famous founder of Apple, was deeply inspired by this book. A copy of "*Autobiography of a Yogi*" was the only book Jobs carried everywhere, giving it even to close friends. Jobs described this reading as a window into the nature of reality, highlighting how meditation was not only a spiritual practice but also a methodology for developing insights that could transform the way we think.

Meditation, according to Yogananda, allows one to explore not only the depths of the mind, but to connect with the "unified field" that pervades everything, a concept that has affinities with the language of quantum physics. Indeed, physics speaks of interconnections at the subatomic level, where the boundaries between space, time and matter seem to vanish. Meditative states of consciousness could mirror these realities, suggesting a profound harmony between science and personal experience.

In recent years, neuroscience has begun to investigate the extraordinary abilities of the brain during meditation. Studies conducted *by the University of California at Los Angeles* (UCLA) have shown that meditation can alter brain waves, stimulating a state of coherence between different areas of the brain. In this state, known as brain synchrony, the brain appears to act as an integrated whole.

Meditation, in neuroscientific terms, activates the prefrontal cortex and thalamus, improving concentration and emotion management. But there is an even more interesting aspect: some research suggests that the meditative experience may influence quantum reality.

According to physicist John Hagelin, a meditator and proponent of the "*unified field theory,*" consciousness is not a mere byproduct of the brain. Instead, it could be a fundamental element-a basic building block of the entire universe.

Practitioners of advanced meditation tell of experiences that seem to defy the laws of matter. A study conducted in India analyzed the body temperature of monks immersed in deep meditations of the "*Tummo*" type. During the practice, the monks were able to voluntarily increase skin temperature, defying the body's normal responses to cold. This example, in addition to confirming the power of mind over matter, invites reflection on the invisible connections between thought and physical reality.

Practical examples of transformation through meditation.

One important case concerns the *Transcendental Meditation* (TM) program devised by Maharishi Mahesh Yogi. This method, widely used in the West, has shown tangible effects not only on the well-being of individuals but also in the social context. In Fairfield, Iowa (United States), a meditation group was formed devoted to demonstrating the collective power of the mind. According to the data collected, when a significant number of people practiced TM at the same time, there were drops in crime rates, violence and even traffic accidents. This phenomenon has been called the "*Maharishi Effect*" and continues to be studied to highlight possible links between collective consciousness and physical reality.

Another impressive study comes from the *Bhutanese Mindfulness Movement.* In Bhutan, known for pursuing "*national happiness*" as an indicator of progress, meditation is frequently integrated into the educational system. Students who practice mindfulness daily demonstrate not only better academic achievement but also greater emotional and spiritual resilience. Bhutan thus represents a current example of how ancient practices can be integrated into a modern society to generate harmony.

Science, with quantum physics, tells us that reality is much more complex and elusive than our senses perceive. Hinduism, through its meditative practices, invites us to look within ourselves to overcome the illusions of the material world (Maya). Together, these perspectives pose us a fundamental question: what is our place in such an interconnected universe?

Meditation is no longer just the preserve of ascetics or mystics. It is an opportunity for anyone who seeks to explore the relationship between body, mind and cosmos. Paramahansa Yogananda encourages us to reflect on how thin the line is between what appears real and what is essence. Quantum physics teaches us that we are all part of a web of invisible connections.

The invitation, then, is to experiment. Try closing your eyes, meditating and embracing the silence. Perhaps, in that silence, you will discover not only yourself, but also the incredible vastness of the universe.

Meditating to observe. The parallel between the quantum observer and Hindu awareness.

A fascinating analogy emerges in quantum physics and Hindu philosophy: the role of the observer. For physics, observing is an act that alters subatomic reality. For the yogic tradition, on the other hand, observing the Self is the path to awakening to ultimate reality. In both cases, what happens has consequences: observation shapes realities and transforms it.

In the famous double-slit experiment, devised in the early 20th century and still a symbol of quantum complexity, scientists discovered that subatomic particles, such as electrons, behave differently depending on whether they are observed or not. When no one observes them, they move like waves, dispersing in an interference pattern. But, if they are observed, they behave as particles, choosing a specific position. This discovery raised a question that goes beyond the scientific level: what role does the observer play in the formation of reality?

Flipping the quantum microscope into an inward-facing instrument, Hinduism and particularly the yogic tradition propose a strikingly similar idea: human beings become what they observe about themselves. Meditative practices, as described in the ancient Vedic texts and Upanishads, are aimed at awakening full awareness of the Self (Atman). In Hinduism, the mind is compared to an ocean: agitated at the surface, but calm and deep at its core. Meditation allows one to cross the waves of the surface mind to reach that hidden depth, free of turbulence.

Observation of the mind during meditation is not a passive act. It is a dynamic process in which the meditator shifts awareness from outward thoughts, emotions and perceptions to recognizing a deeper level of reality: the very essence of consciousness. Just as the quantum physicist observes particles, the yogi observes the inner Self.

The relationship between quantum physics and Hinduism emerged strongly in the 19th and 20th centuries, thanks to the dialogue between Western scientists and Eastern philosophers. One of the most famous names is that of Erwin Schrödinger, creator of the wave equation and fervent scholar of Vedic texts. Schrödinger, fascinated by Hinduism's concepts of unity and permanence, wrote:

" There can be no true observer who is distinct from the Whole. He will in any case be an integral part of the observed reality. ".

Niels Bohr, one of the fathers of quantum mechanics, also integrated Eastern symbols such as taijitu (the Chinese yin-yang) into his coat of arms as a tribute to the complementarity of natural processes.

According to Vedic tradition, reality is the result of the oscillatory manifestation of Maya-the illusion that hides the true universal nature behind dependencies of perception and thought. Comparing this metaphor with quantum, the probability wave of particles-before being observed-can be read as a representation of fluid reality, similar to Maya's illusion. Observation "collapses" the wave itself, revealing a particular physical state. The yogi, similarly, through meditation and complete control of the mind, dissolves the illusion and comes to unity with the divine.

Andrew Newberg, a neuroscientist and pioneer in the study of the neural correlates of spirituality, has explored for decades how contemplative practices affect the brain. His studies have revealed that intense meditation causes changes in crucial brain areas, such as the parietal lobe, which regulates perception of space and time, and the prefrontal lobe, which is associated with attention and awareness. During deep meditation, the boundaries between the self and the external world seem to dissolve, leading to the feeling of union.

Yogis describe this experience with the term samadhi, the meditative state in which the boundaries of individual perception dissolve, allowing one to become attuned to the Whole. Newberg himself stated that meditation can measurably alter the perception of reality, something that brings it closer, at least conceptually, to the role of the quantum observer.

The centrality of the observer, shared by physics and Hinduism, prompts reflection on a fundamental question: are there boundaries between what is observer and what is observed? This question does not necessarily entail an answer, but opens a fruitful dialogue between different approaches to the mystery of reality.

In temples across India, from Varanasi to Rishikesh, bells ring out as echoes of mantras, a constant invitation to return to inner observation. In laboratories around the world, physicists study ever more refined experiments to understand the role of quantum "collapse." Yet, perhaps both paths lead to the same horizon: the search for what connects human experience to the fabric of reality.

At a time when science and spirituality are often considered opposites, the dialogue between quantum physics and Hinduism shows that the boundaries are never as rigid as they seem. As physicist David Bohm, another great innovator of quantum thought and a scholar of Eastern mysticism, said:

> *"Science and spirituality are not separate worlds. Like the observer and the universe, they constitute a single dance."*

Chapter VII Brahman: the ultimate reality

Brahman, the universal principle that permeates everything.

From the heart of Eastern philosophies emerges the concept of Brahman, the universal principle that permeates all that exists. In the Hindu tradition, Brahman is described as the indivisible essence from which everything comes and to which everything returns. A fascinating idea that today finds unexpected parallels in the world of quantum physics, particularly in the mysterious Higgs field.

The Higgs field, theorized in 1964 by British physicist Peter Higgs and experimentally confirmed in 2012 at CERN, is an invisible presence that permeates the entire universe. It is because of this field that particles gain mass, just as waves on the surface of a lake reveal the presence of water below. Similarly, according to Vedic texts such as the Upanishads, Brahman is the ultimate reality, the invisible "fabric" that gives meaning and substance to the cosmos.

An interesting, almost prophetic anecdote can be found in the Hindu idea of Maya, the illusion that hides the true nature of reality. Similarly, modern physics reminds us that what we perceive as solid or tangible is, in fact, largely empty: atoms, with their tiny nuclei and orbiting particles, are 99.9999% empty space, and their cohesion depends on fundamental forces such as the Higgs field or quantum bonds.

In both fields-spiritual and scientific-an impressive insight emerges: everything is interconnected. The Higgs field and Brahman suggest that there is no real separation between what we see and what is. This brings to mind what Erwin Schrödinger, father of quantum physics, said; in his book "*What Is Life?*" he wrote that the discoveries of modern science had brought him closer to the philosophical traditions of India.

Places like Benares, for millennia the cradle of Brahman research, and Geneva, home of CERN, today represent two poles of a universal quest: one spiritual, the other physical. Yet both seem to converge on a single truth: the universe is an indivisible "whole," where knowledge and experience, science and meditation, guide us toward a deeper understanding of our place in reality.

Brahman, the Higgs Field and the fundamental structures of physics.

For millennia, man has wondered about ultimate reality, that profound essence that holds the cosmos together. The Vedic sages of ancient India identified this universal principle with the term *Brahman:* the infinite, the all, the indivisible unity that permeates all things. In their meditations, encapsulated in the Upanishads (philosophical texts composed between the 7th and 2nd centuries BCE), they described Brahman as the invisible and unchanging basis of reality. Not a personal god, but an omnipresent and fundamental force.

In a surprising convergence, modern physics, through concepts such as the Higgs field, seems to touch on similar ideas. The Higgs field, concretely discovered in an experiment at CERN in 2012, is a kind of "invisible fabric" that allows particles to acquire mass. The similar search for "grounding" has led Indian scientists and mystics on parallel paths, albeit with different tools: mathematics and experiments on the one hand, philosophical intuition on the other.

To understand Brahman, the Upanishad texts use evocative metaphors.

> *"Just as oil permeates the sesame seed or the essence of butter is hidden in milk, so Brahman is everywhere, invisible but omnipresent."*

recites the Chandogya Upanishad. Similarly, the Higgs field is everywhere, but invisible to the senses. Physics needs giant accelerators to find the "trace" of a force that, paradoxically, is always there.

The universal interconnectedness described in Hinduism, through Brahman, is reflected in the behavior of subatomic particles. People like Erwin Schrödinger, one of the fathers of quantum mechanics, were fascinated by Vedic philosophy. Schrödinger studied Hinduism and repeatedly cited the "*striking coincidence*" between his scientific insights and the ancient doctrine that "*every being is one with the universe*."

Inside the *Large Hadron Collider* near Geneva, physicists pursue truths that the ancients sought in caves and forests of the Indian subcontinent. Despite the technical distance, the spirit of the journey is similar. For example, Peter Higgs, before the scientific confirmation of "his" boson, was moved by the same wonderment described by Yajnavalkya, a Vedic sage who lived in the 1st millennium B.C., who stated that Brahman cannot be known with the senses, but only intuitively.

Finally, the analogy between the path of Hinduism and that of quantum physics reveals something fundamental. The human mind-both that of the mystic and that of the scientist-has always yearned for one thing: to understand the mosaic of reality, uncovering each piece of it. And, perhaps, by uniting Brahman with the Higgs boson, we intuit that the answers are not as far apart as they seem.

In a passage from the Upanishads, ancient sacred texts of Hinduism, the sage Yajnavalkya states:

> *"Everything that exists is enveloped by Brahman. Nothing exists outside of it."*

This concept, which defines Brahman as the universal principle and ultimate reality, now finds striking parallels with modern quantum physics, particularly the Higgs field.

Brahman, described as the essence that permeates all that exists, is often depicted not in concrete terms, but as an indefinable, omnipresent and invisible force. Similarly, the Higgs field, theorized in the 1960s by British physicist Peter Higgs, is a subtle, invisible network that runs throughout the universe. It imparts mass to elementary particles through an interaction that conditions their behavior. In a sense, everything with substance in our universe exists and interacts only because of the Higgs field, just as Brahman is responsible for the existence and coherence of the whole of creation according to the Vedic view.

Quantum physics has revealed to us that reality is not made up of solid, separate objects, but of waves, fields and deep connections. This insight echoes the Vedic concept that all the multiplicity of the world is illusory (Maya): a projection of the one underlying reality, namely Brahman. The Upanishads describe Brahman as "*the thread that holds the pearls of a necklace together*." This poetic image suggests a unifying principle that connects everything we see. Similarly, the Higgs field represents an invisible fabric that unites matter and energy at a fundamental level.

Yajnavalkya, in the Brihadaranyaka Upanishad, invites us to contemplate an interconnected universe, dominated by Brahman, and proposes a path of inner knowing to perceive it. Similarly, modern physics invites us to explore the mysteries of the universe through its smallest components. The discovery of the Higgs boson occurred in 2012.. This particle, sometimes known as "*the God particle*," represents a breakthrough comparable in terms of understanding to the revelation that everything is not separate but part of a whole.

Physicist Fritjof Capra has explored these connections between modern physics and ancient Eastern philosophies. Capra asserts that quantum theories and Vedic texts converge on a single point: the universe as a network of relationships and interactions. Parallel paths, according to Capra, that demonstrate how science and spirituality can offer us complementary images of reality.

Brahman and the Higgs field are not identical concepts; their nature, metaphysical and scientific respectively, places them on different planes. However, both invite reflection on an intrinsically connected universe in which the individual self is never truly separate from the whole. Perhaps, as the ancient Vedic sages and modern scientists suggest, our purpose is not so much to understand the universe as an external entity, but to perceive our deep unity with it.

Maya, Brahman and quantum interconnectedness.

Hinduism describes the concept of Maya as the veil of illusion. This term refers to the perception of a seemingly separate and fragmented reality. According to the Vedic texts, the world we observe-with its variety of forms, colors and phenomena-is only an illusory manifestation. Behind this multiple appearance lies Brahman, the one ultimate and indivisible reality.

Quantum physics, surprisingly, seems to suggest something similar. The quantum superposition principle states that particles exist in multiple states simultaneously until they are observed. The act of observing, therefore, does not merely record reality but defines it. This feature recalls the mystery of Maya, which, until "transcended" through knowledge of Brahman, maintains the illusion of a separate and discrete world.

A modern parallel with Brahman emerges in the concept of the Higgs field. This fundamental quantum field, theorized in 1964 and confirmed in 2012 with the discovery of the Higgs boson at CERN, permeates the entire universe. Just like Brahman, it is invisible in our everyday senses but fundamental to the structure of reality. Without the Higgs field, matter itself would have no mass.

The Indian philosopher Adi Shankaracharya, who lived in the 8th century, explained that understanding Brahman means "*seeing beyond illusion*." Similarly, increasingly refined scientific observation has enabled us to glimpse previously inaccessible dimensions of reality. However, each new discovery opens up deeper questions. Quantum physics itself prompts many to question whether the universe is really separate from the observer or whether, on the contrary, it is deeply interconnected, just as Vedic wisdom suggested.

For both the physicist and the meditator, the question remains the same: Is the world we perceive real or is it an illusion? Discovering the answer might mean, as Hinduism teaches, reconciling appearance with truth, the visible with the invisible. Or, perhaps, simply accepting that reality and illusion are two faces of the same universal wave.

Brahman and the wave nature of particles.

Hinduism, with its fascinating and complex cosmological vision, defines Brahman as the ultimate reality, the universal principle that permeates everything. Brahman is neither a form nor a concrete entity. Instead, it is the unchanging essence beyond time and space, the source of the entire universe. It is what lies behind every manifestation of visible reality, binding everything to that invisible "fabric" that unites the Whole. It might sound like a purely philosophical or spiritual concept, but some parallels with modern quantum physics make it surprisingly close to scientific language.

A fascinating bridge between science and spirituality emerges when we talk about the wave nature of particles. The double-slit experiment, first conducted by Thomas Young in 1801 and later reinterpreted in the 1920s as the basis of quantum mechanics, is a cornerstone of modern physics. In the experiment, electrons -- fundamental building blocks of matter -- exhibit dual behavior: they can behave either as particles or as waves, depending on human observation. This dual nature has left scientists and philosophers facing profound questions about the ultimate nature of reality.

Brahman, in Vedic philosophy, is strikingly similar to this idea. In Indian thought, the cosmos appears as a play on Maya, the illusion of material reality, supported, however, by a fundamental "wave" that is Brahman. Like the quantum wave, Brahman is invisible and immaterial, but it is essential for the existence of forms. Therefore, what we perceive as "reality" is only a transient manifestation of something deeper and more fundamental.

This parallel also leads us to reflect on the relationship between Atman and Brahman. According to the Bhagavad Gita and the

Upanishads, Atman-the individual soul-is nothing but a fragment of Brahman. Here, too, a duality emerges, somewhat like in the behavior of particles in the quantum world. When we meditate on the concept of Brahman, the dualism between the individual and the Whole dissolves: just as the wave becomes a particle when observed, Atman is recognized in Brahman when we move away from the illusion of separation.

The concept of Brahman invites us to reflect that the distinctions between science and spirituality may be less rigid than we think. Perhaps, as the great physicist David Bohm suggested, the universe is not fragmented into separate parts but a huge "implicit order," an infinite network of interconnections. Bohm wrote:

> *"The absence of fragmentation is the most fundamental order of nature."*

It is not difficult to discern, in this, an echo of Vedic thought.

Today, quantum physics continues to explore the secrets of space, time and particles. But the ancient wisdom of Hinduism, with the concept of Brahman, reminds us that the explanation of who we are may not lie only in scientific laboratories. The ultimate reality may ultimately be a single, mysterious wave that connects everything.

The Dance of Shiva in the quantum field. Brahman as the union of matter and energy.

The figure of dancing Shiva, the Nataraja, is one of the most powerful symbols in Hinduism. His dance represents the eternal cycle of creation and destruction, both movements necessary for the balance of the universe. In the philosophical context, Shiva is not merely a god, but an expression of the ceaseless dynamic of Brahman, the ultimate reality that permeates everything.

Brahman is conceived by Vedic texts as the universal principle that sustains existence. He is beyond any concrete definition, an invisible force that unites matter and energy. This view also fascinates many theoretical physicists today, as it is reminiscent of the modern description of fluctuations in the quantum field. The quantum vacuum, seemingly empty, is actually full of energy. Particles arise and disappear from this basic oscillatory state in a continuous dance of creation and destruction.

Shiva's dance offers an extraordinary image to describe the behavior of the quantum field. In 1931, the great physicist and mathematician Fritjof Capra wrote that seeing the depiction of Nataraja led him to imagine the dance of subatomic particles. Capra emphasized how, at moments of peak concentration during his studies of quantum physics, the connection with Eastern symbology became natural and almost inevitable.

Similarly, the Higgs field seems to evoke the essence of the concept of Brahman. The Higgs field is what allows particles to have mass, connecting them into a single invisible network that governs the entire universe. Everything is interconnected, everything shares a common matrix. This is also the idea at the heart of the Vedic concept of Brahman: an indivisible unity that manifests itself in infinite forms.

In southern India, the Chidambaram temple is home to one of the most celebrated representations of dancing Shiva. Here, the symbolism of Nataraja finds tangible expression, marrying art, religion and cosmology. The position of the Nataraja statue's hands and feet are said to represent not only philosophical principles but also natural laws. The raised right hand symbolizes protection, while the raised left foot represents liberation from the cycle of birth and death.

The parallel between Brahman and discoveries about the quantum field suggests that ancient Eastern wisdom and modern science are not separate but reflect each other. Shiva, the one who destroys in order to create, recalls the quantum void, which is never truly "empty," but a fertile field of infinite possibilities. A profound message emerges from Shiva's dance: matter and energy are one, closely connected in an eternal cycle.

Brahman and the Higgs Field: A meeting of East and West.

In the universe of human thought, modern science and ancient philosophies have always moved on seemingly separate paths. At certain moments, however, these paths seem to cross, offering fascinating parallels that demonstrate how universal reflection on reality is. One such point of convergence is the relationship between Brahman, the fundamental principle described in the texts of Hinduism, and the Higgs field, a key discovery of modern physics.

In Hinduism, Brahman is the ultimate reality: a universal essence that permeates all that exists. He is not a personal deity but an abstract principle that transcends time, space and matter. In the Vedas and Upanishads, ancient Sanskrit texts, Brahman is defined as "*sat-chit-ananda*" (*being, consciousness and bliss*): the origin and sustainer of everything. A description that, even in its philosophical complexity, suggests that everything in the universe is connected by an invisible substratum.

On the other side of the world, CERN in Geneva has announced the discovery of the Higgs boson. This tiny, elusive fragment of matter is part of a universal field, the Higgs field, an invisible energy that permeates the entire universe. According to quantum physics and the Standard Model, this field is what gives mass to particles, allowing the cosmos as we know it to form.

The parallel is striking. Just as Brahman is seen as the fabric that unites all things, the Higgs field is the infrastructure on which the universe stands. Both suggest the existence of a fundamental, ubiquitous, interconnected reality. Physicist Fritjof Capra was among the pioneers in suggesting this analogy. Capra noted how quantum physics, in its embrace of indeterminacy and interconnectedness, came close to the mystical insights of Eastern philosophies.

Another interesting example comes from the research of Erwin Schrödinger, one of the fathers of quantum mechanics and, surprisingly, a passionate scholar of the Upanishads. Schrödinger, in his journal, wrote that the idea of a "universal self" described in Hindu thought was strikingly similar to the unitary view proposed by contemporary physics.

India itself offers cultural cues that evoke this interconnectedness. For example, the concept of "Maya" (illusion) in the Vedas suggests that what we see is but a superficial manifestation of reality, hidden beneath deeper layers. In a parallel way, modern physics reminds us that solid matter is, in fact, composed of particles that behave both as waves and as corpuscles, with no absolute materiality.

In Geneva, CERN is home to scientists of different nationalities and cultures. This research site invokes the idea of universal knowledge, just as Brahman represents the underlying unity of humanity. In Geneva, in the tunnel of the LHC, the Higgs boson was identified, poetically referred to as "the God particle" (although physicist Leon Lederman, who coined this expression, meant it in an ironic spirit).

The idealized encounter between Brahman and the Higgs field is not a scientific demonstration, but an invitation to reflect. Can modern science, with its tools, approach the understanding that Eastern sages explored through introspection millennia ago?

Brahman, quantum vacuum and cosmic creation.

For millennia, Indian sages have contemplated Brahman, the universal principle of Hinduism. According to tradition, Brahman is the ultimate reality. He has no form, exists beyond time and space, and permeates the entire universe. Sacred texts, such as the Upanishads, present him as the source from which everything emerges and to which everything returns-at once invisible and omnipresent. In the space of modern physics, some scholars have seen a fascinating parallel between Brahman and the quantum vacuum. Both, for Indian mystics and physicists alike, represent a fundamental background from which reality manifests.

In the context of contemporary physics, the "quantum vacuum" is not simply empty space, but a state of fundamental energy that permeates the universe. This has been paralleled by the concept of Brahman as a universal substratum in which all phenomenal manifestations find their origin.

The quantum vacuum is not a "nothingness." It is a kind of "cosmic soup" vibrating with potential energy, a primordial state that holds the secret of creation. Some physicists, such as David Bohm, have proposed the idea of an implicit reality, a deep order from which particles and all the complexity of the universe emerge. Similarly, Hinduism recounts the cosmic creation myth through Brahma's dream. In the myth, the universe is born and dies cyclically, just as the elementary particles manifest and dissolve in the quantum vacuum.

Interestingly, places and times differ, but human aspirations are similar. Sages in ancient Indian ashrams, meditators on the banks of the Ganges millennia ago, sought to grasp the eternal through the language of spirituality. Today, in Geneva laboratories or among space telescopes, scientists pursue the same eternal through mathematics and physics. Quantum vacuum and Brahman, while coming from very different cultural orbits, both offer a powerful symbol: an undivided reality that transcends the chaotic and fragmented appearance of the everyday.

In this convergence, something new is reflected. The myth of Hinduism is not only "spiritual." It can inspire the scientific mind, just as modern science can make visible the intuition of that universal Brahman that permeates everything. The universe, after all, always remains a mirror. It is up to us to decide how to look at it.

In the context of contemporary physics, the "quantum vacuum" is not simply empty space, but a state of fundamental energy that permeates the universe. This has been paralleled by the concept of Brahman as a universal substratum in which all phenomenal manifestations find their origin.

Brahman and the quantum potential.

Since the dawn of civilization, man has sought to understand the deep fabric of reality. In India, Vedic texts developed the concept of Brahman, the universal, eternal and indivisible principle that permeates all that exists. In modern quantum physics, we find striking parallels in phenomena such as non-locality and particle behavior, which seem to suggest a world of invisible connections and deep interactions.

An ancient Vedic metaphor likened the universe to a spider web: each strand is separate, yet each point is connected to the rest, an image that remarkably mirrors the quantum concept of interwoven reality.

Over the centuries, Vedic masters have described Brahman as the ocean of which everything is wave or echo. The Chandogya Upanishad, one of the oldest Vedic texts, states:

"All that is, is Brahman."

One of the main teachings of Adi Shankara, an 8th century Indian philosopher, explains that all apparent diversity in the sensible world is only an illusion (Maya), while everything is an expression of the one indivisible reality of Brahman.

In the world of modern physics, the phenomenon of non-locality offers an unexpected echo to ancient thoughts. Quantum particle experiments show that two entangled particles, although separated by immense distances, communicate instantaneously. This "instantaneous" connection defies our intuitive boundaries of space and time. Non-locality seems to affirm the existence of a level of reality in which everything is interconnected, beyond physical distances.

An evocative parallel can be drawn with the Higgs field, discovered in 2012, which permeates the entire universe and imparts mass to particles, seemingly unifying the microcosm and macrocosm. The idea that the universe is traversed by a universal "field" reveals striking similarities with the Vedic concept of Brahman as the basis of all visible and invisible phenomena.

A classic anecdote in Hinduism speaks of salt dissolved in water, used by sages to describe Brahman:

"You can't see the salt, but you can taste it everywhere."

This recalls the nature of quantum particles, which, being in a wave state, are everywhere and nowhere at the same time.

For ancient Indian rishis, Brahman could not be "seen" or "measured" but only perceived in one's inner self, through meditation and intuition. Similarly, in modern physics, many of the quantum dynamics manifest only indirectly, leaving room for immense theoretical complexity. The idea that everything is interconnected, as Niels Bohr, one of the fathers of quantum mechanics, said, is central to both physics and Eastern philosophy:

"Everything we observe is intimately connected to who we are."

Einstein himself, despite being skeptical of some implications of quantum physics, proposed a unified universe where space and time were inseparably intertwined. This view might be reminiscent of the concept of Maya in Hinduism: the world as an appearance concealing an overarching, unchanging and timeless reality, namely Brahman.

In the Chandogya Upanishad, an ancient sacred text, we read:

"As all pearls are joined by an invisible thread, so everything is connected in Brahman."

This thread might evoke, in the modern mind, the image of quantum fields binding particles and forces into one sublime cosmic dance.

At base, both quantum physics and Vedic philosophy raise a common dilemma: Is there something beyond space and time? If Brahman is what exists beyond perceived reality, it could represent a prescientific insight into what modern physics attempts to develop with multiverse or string theories. Both, in different ways, seek to penetrate the invisible, to seek the fundamental unity.

Perhaps the greatest message that both Brahman and quantum fields give us is that the universe is not just matter but an eternal interplay of connections, where everything participates in something greater. For the mystic poet Rabindranath Tagore, who won the Nobel Prize in Literature in 1913, this idea is clear:

"Infinity is not something found somewhere else. The infinite is here, behind the screen of visible reality."

It was an allusion to Brahman, perhaps, but also to that invisible universe that we explore in science laboratories today.

The striking similarities between the ancient concept of Brahman and modern quantum physics phenomena certainly do not prove a definitive connection. However, they do prompt us to reflect on a common insight shared by science and spirituality: the world in which we live is more complex, subtle and connected than we can perceive. Brahman and quantum potential both seem to indicate that, beneath apparent divisions, there is a single reality that unites everything. Perhaps, as ancient Vedic wisdom suggests, we are already part of an indivisible whole, although we often forget this.

Brahman and the mystery of quantum entanglement

According to Vedic philosophy, everything that exists is a manifestation of Brahman: there is no separation between what appears distinct, but only different modes of the same unified reality.

Curiously, this insight seems to find a surprising echo in one of the most enigmatic phenomena in quantum physics: entanglement. This is the mysterious correlation that binds two particles separated in space, to the point that what happens to one of them seems to instantaneously affect the other, regardless of distance. It is as if the two particles were threads of the same invisible fabric, woven together by what physicists call "non-locality."

Entanglement may seem to be a phenomenon exclusive to modern physics, but the parallels with Brahman suggest that similar insights were already part of the philosophical musings of ancient cultures. Swami Vivekananda, one of India's most celebrated spiritual masters, spoke of Brahman as:

> *"the rope on which the entire illusion of the universe is projected."*

An image that might recall the idea of a universal field in which everything is connected, just like two particles connected at the quantum level.

Another interesting example emerges from the writings of physicist David Bohm, a central figure in the study of quantum mechanics and implicit order. Bohm, fascinated by the holistic nature of the universe, proposed that all visible reality is the manifestation of a larger, indivisible pattern. A view that aligns with Vedanta, the philosophical school of Hinduism that explores the concept of Brahman and its relationship to the phenomenal world.

Whether from the texts of the Upanishads or the laboratories of a physicist, the message seems clear. Quantum entanglement and Brahman offer a vision of reality as an inextricable web, in which everything is connected and nothing is truly isolated. This idea, while coming from such different contexts, leads to a powerful reflection: the universe may not be made of separate objects, but of deep relationships, of invisible threads that unite the whole.

Perhaps, as speculated by Niels Bohr, one of the fathers of quantum physics

> *"deep truth can only be expressed by the complementarity of opposites."*

Eastern philosophy and Western science together could tell the same story with different languages.

From the Big Bang to Vedic knowledge. Brahman and the origin of the universe.

> *"Reality is one, even if the wise call it by different names."*

This verse from the Rigveda, one of the oldest texts in the Vedic tradition, might seem like a distant echo. Yet its resonance spans the centuries, even going so far as to dialogue with some of the most revolutionary ideas in modern physics. The concept of Brahman, the universal principle and foundation of all that exists, surprisingly recalls some of contemporary science's insights into the origin and nature of the universe.

More than thirteen billion years ago, according to the Big Bang theory, the entire universe was enclosed in an infinitely dense point. Then, a violent expansion gave rise to time, space and energy, laying the foundation for what we know today as reality. In the Hindu tradition, however, the universe originates from Brahman, an indivisible, eternal and unfathomable reality that permeates everything. It is not a god in the personal sense, but an omnipresent force that in itself contains every possibility of existence. The comparison with the primordial energy of the Big Bang is not accidental.

The Chandogya Upanishad, (circa 6th century B.C.) states, "*All this is Brahman.*" This concept reiterates the fundamental interconnectedness of reality, an idea many scientists now attribute to the Higgs field. Introduced in 1964 by physicist Peter Higgs, this invisible field permeates the entire universe. It is because of it that elementary particles gain mass, transforming into what we see or perceive. Again, as with Brahman, it is impossible to separate the particle from its context. Everything is plastic, fluid, but interconnected.

An ancient Hindu myth tells that the visible universe is the breath of Brahman. When Brahman exhales, universes are born in a jubilation of creation. When he inhales, everything returns to itself, dissolving. This cosmic cycle of expansion and contraction is not entirely dissimilar to the scientific hypothesis of the oscillating universe, according to which the Big Bang may not have been the only creative episode, but one among infinitely many.

Places such as Varanasi, on the banks of the Ganges, have been the cradle of meditations on these issues for millennia. In this sacred city, sages reflected on the nature of the soul (Atman), which, according to tradition, is intimately connected to Brahman. Here, too, we find an echo in quantum physics: the correlation between distant subatomic particles (entanglement) suggests that the universe is less separate than it appears.

The Big Bang and Brahman, separated by millennia and disciplines, seem to converge toward a single truth: the universe arises from a common source, which is energy, substance and spirit together. Perhaps, at the basis of science and spirituality is not rivalry, but the search for the same answer to the deepest questions about existence.

Is there a universal "All"?

For millennia, the spiritual traditions of the East have questioned the ultimate nature of reality, seeking answers that go beyond what is visible. In Hinduism, the concept of Brahman represents the heart of these reflections. Brahman is the universal principle, the essence that pervades all that exists. It is an indivisible, infinite and invisible reality beyond time and space. For Indian sages, such as the rishis of the Vedas, Brahman was not just an abstract idea, but the fundamental truth that connects every fragment of the universe.

Surprisingly, a similar idea seems to emerge millennia later in the field of modern physics. Advances in quantum mechanics show us that what we perceive as "solid" is, in fact, a complex and mysterious web of waves and particles. The Higgs field, theorized in the 1960s and concretely discovered in 2012, has brought to light a profoundly fascinating idea: a kind of invisible energetic "fabric" permeates the entire universe, imparting mass to particles and, thus, giving rise to what we call reality. This elusive field recalls, in an almost poetic way, the concept of Brahman.

If we go back more than three thousand years, in the Vedas-the sacred texts of Hinduism-Brahman is described as that from which everything arises and to which everything returns. For the Indian mystics, every form, every sound, every thought is a manifestation of that one reality. There is no separation. This philosophy is summed up by the ancient Sanskrit phrase "*Tat Tvam Asi*" (You are That), which invites every human being to recognize his or her own deep connection with the cosmic whole.

Similarly, modern physicists, observing subatomic phenomena, have discovered that at the fundamental level there are no sharp boundaries between one particle and another. Electrons are not static objects, but waves of probability, existing only within an interconnected network of relationships. Here, the separation between observer and observed seems to dissolve, recalling the undifferentiated unity of Brahman.

Professor Fritjof Capra, with his famous book "*The Tao of Physics*," explored this dialogue between science and Eastern spirituality. Capra writes:

> *"Modern physicists are reconnecting with worldviews that ancient sages knew thousands of years ago: the idea of an indivisible reality that is both dynamic and interconnected."*

It is no accident that many theoretical physicists, such as Erwin Schrödinger, the Austrian scientist and father of quantum mechanics, have been deeply influenced by Eastern thought. Schrödinger himself, reading the Upanishads, declared that their concept of universal unity bore a striking resemblance to what physics was trying to describe.

Imagine a symbolic image: on one side, the Rishikesh Ashram on the banks of the Ganges, where spiritual masters meditated on Brahman. On the other, the *Large Hadron Collider* in Geneva, where scientists detected the Higgs boson. Although the tools used are radically different (the contemplative mind of a rishi and the particle accelerators of a physicist) the goal seems similar: to understand the profound nature of what connects all that exists.

The concept of Brahman and that of the Higgs field both raise a decisive question: is there a One that underlies and transcends all diversity? In Hinduism, this question takes on metaphysical and spiritual overtones. In physics, on the other hand, we explore a mathematical and experimental language. But the challenge remains the same. The insights of mystics and the discoveries of physicists like Peter Higgs seem to suggest that some form of "universal whole" may indeed exist. Perhaps, in the end, the journey of science and spirituality converges in a single quest to know the invisible.

Brahman and Einstein's search for the unified field.

Albert Einstein spent the last years of his life searching for something extraordinary: a single equation, almost divine in its simplicity, capable of unifying all the forces of nature. This dream, often called the "*Theory of Everything*," represents the desire to find a universal principle underlying the entire universe. Yet, ironically, the idea of a unifying principle is not new. Philosophies such as Hinduism have pondered it for millennia through the concept of Brahman: the indivisible and all-pervading essence of the cosmos.

Einstein was a staunch rationalist, but he is said to have admired the mysterious. In a 1929 interview he stated:

> *"I want to know how God created this world. The rest is details."*

His "God" was not necessarily a religious deity, but a higher cosmic order. In a surprising way, Brahman is mirrored in this idea. For the ancient Vedic texts, Brahman is the ultimate reality, the invisible and incomprehensible substratum that sustains the universe. According to Hinduism, all that exists (matter, energy, space and time is a single manifestation of Brahman.

A modern parallel emerges with the physical concept of the Higgs field. The Higgs field is a kind of "invisible fabric" that permeates the entire universe, imparting mass to particles. Similarly, Brahman is not a separate thing or entity, but an immanent principle that sustains every manifestation of reality. Like the Higgs field, Brahman is everywhere, though invisible to our ordinary senses.

The sacred texts speak of Brahman as the source not only of destruction but also of creation and transformation, an eternal cycle reminiscent of the quantum behavior of particles, in constant motion between states of existence.

This vision of an interconnected and unified universe has fascinated the Western mind for centuries. In the 19th century, writer and poet Henry David Thoreau, immersed in reading ancient Vedic texts, imagined a cosmos without divisions, governed by a universal harmony. Other 20th-century thinkers, such as Erwin Schrödinger, father of quantum mechanics, found similarities in the Eastern texts with their discoveries about the probabilistic nature of reality.

Einstein himself, although he did not explicitly refer to Eastern teachings, seemed to allude to a similar view. For him, the differences between space and time, matter and energy, were superficial. Like Brahman, all apparent reality dissolves into a single essence, an ultimate truth yet to be known. Although Einstein never found the "*Theory of Everything*," his vision remains an extraordinary modern echo of an ancient wisdom.

Perhaps, in the tension between science and spirituality, we will find an eternal question rather than an answer. Is it possible for the universe to speak the two languages? Einstein, ideally, might have wondered the same. And Brahman, imagined by Vedic sages millennia earlier, still seems to whisper it to our most sophisticated laboratories and particle accelerators.

Brahman and the hidden symmetry of the universe.

Vedic texts define Brahman:

> *"that from which everything is born, through which everything lives and into which everything returns."*

a description that evokes a unifying idea capable of challenging ordinary understanding of the world.

But what does Brahman have to do with modern physics? Contemporary cosmology offers us an extraordinary parallel: the Higgs field. This field permeates the entire universe by imparting mass to particles. Without the Higgs field, the universe could not have form, substance or coherence. Doesn't this almost sound like talking about Brahman in scientific terms? The invisible, the fundamental, becomes the principle that gives substance to the whole.

Then there is the wave nature of particles, described by quantum mechanics. Niels Bohr, one of the founding fathers of quantum physics, stated that:

> *"everything we call real is made of things we cannot consider real."*

Such a statement suggests a level of reality suspended between what we see and what we intuit. This concept is close to Vedic thought, which sees Brahman as that which lies beyond phenomenal appearance.

A further connection can be found in the words of Erwin Schrödinger, physicist and author of the famous quantum equation. Schrödinger was often inspired by Eastern texts. He was convinced that the unity of the universe was intrinsically linked to a great hidden coherence. He wrote:

"The total number of minds in the universe is one. In reality, the distinction between individual minds and Brahman is an illusion."

Schrödinger recognized the power of Brahman as a theoretical symbol of an interconnected universe.

It is no accident that many modern physicists have found inspiration in Eastern philosophy. The analogy between Brahman and the symmetrical nature of the cosmos, where the visible and invisible coexist in perfect balance, recalls the notion of symmetry hidden in the standard model of particle physics. In both cases, one perceives an underlying unity that transcends all fragmentation.

Imagine, then, this universe as an immense symphony. Each sound stands out, but it is through the coherence of the notes that harmony emerges. Similarly, Brahman represents the whole. He is not just the sum of the parts, but the principle that gives structure and order to the whole.

On the one hand, the millennia of Vedic wisdom. On the other, the equations of physics. These two seemingly distant worlds seem to suggest the same fundamental truth: the universe, in its complexity, speaks a single language. Or, as Indian mystics would say, everything is Brahman.

Brahman and the fate of quantum physics.

"Brahman is all that truly exists. The rest is illusion."

So goes one of the most famous phrases in the Upanishads, fundamental texts of Hindu philosophy. This concept guides millennia of spiritual reflection: everything in the cosmos is united by an ultimate, invisible but fundamental reality that transcends sensory perception. But what happens when this ancient idea meets the language of modern quantum physics?

In recent decades, scientists and philosophers have found surprising similarities between Brahman, the absolute principle of the

Hindu tradition, and some discoveries in contemporary physics. One intriguing analogy is that between Brahman and the Higgs field, the mysterious invisible web that, according to particle physics, permeates every corner of the universe and gives mass to elementary particles. British physicist Peter Higgs hypothesized its existence in 1964, but it wasn't until 2012 that the boson associated with the field was finally observed at CERN in Geneva, sparking a huge worldwide stir.

Like Brahman, the Higgs field is omnipresent, though not visible to our eyes. Only through its interaction with "visible" particles does the field become experiential, but what animates it remains deeply enigmatic. Similarly, the ancient Indian rishis described Brahman as the origin and sustainer of all that exists, a unique force that manifests itself in everything without ever losing its unity.

Another interesting affinity lies in the interconnectedness of reality at the quantum level. Modern physics has shown us that subatomic particles are more like waves than solid objects. They exist in a constant state of fluctuation and interaction. In this sense, the idea of a world in which everything is deeply intertwined recalls the Vedantic vision of a cosmic unity, subtle but omnipresent.

Vittorio Marchi, one of Italy's best-known popularizers of quantum physics, often emphasized these connections. According to him:

> *"quantum physics is not discovering anything new, it is just rewriting in mathematical terms what the sacred texts of humanity have been telling for centuries."*

However, the dialogue between science and spirituality is not without debate and, often, controversy. More orthodox scientists, such as Richard Feynman, have cautioned against the superficial use of mystical metaphors in science, preferring to focus on experimental data.

The philosophical convergence between Brahman and quantum physics, however, stimulates some profound questions. Is science really getting closer to understanding a universal Brahman-like principle, or is this an interpretive coincidence due to limitations of our language? And, again, can the ancient idea of a fundamental unity of the universe suggest new directions for scientific inquiry?

The Upanishads often closed with an affirmation of hope and awareness, "*You are That*." This statement is an invitation to recognize that each individual is an inseparable part of the cosmic unity. Perhaps, as Einstein suggested when he spoke of the "cosmic mystery," modern physics is also searching for a similar universal sense. And who knows, in the future, science and spirituality may discover that they are not only complementary, but intimately connected, just like the waves of an invisible ocean guiding the fate of the universe.

Chapter VIII. The Atman: The Immortal Self -

Atman and the superposition.

In the ancient Vedantic philosophy of Hinduism, the concept of Atman represents the immortal self, the eternal, indivisible spark that lives in every individual. This essence is considered a direct manifestation of Brahman, the universal cosmic principle, a formless entity that permeates everything. It is one of the most profound ideas in Hinduism, and it lends itself, surprisingly, to an intriguing connection with modern quantum physics: the superposition of states.

Superposition, in quantum mechanics, describes the extraordinary ability of a subatomic particle to exist simultaneously in multiple states. The image of "Schrödinger's cat," simultaneously alive and dead until observed, is famous. Similarly, in Vedanta, Atman appears as a reality that transcends the limitations of time, space and stable identities. It manifests itself in an infinite number of ways through individual perception, but, beyond appearances, it always remains unchanged and indivisible.

Adi Shankara, one of Hinduism's greatest philosophers who lived in the 8th century, taught that the phenomenal world as we see it is an illusion (*Maya*). He argued that the essence of the Atman is unique and universal, despite presenting itself as fragmented in the individuality of each human being.

"When a man knows the universal self, he becomes immortal."

Shankara wrote in his commentaries on the Vedic sacred texts. This "universal self," invisible but omnipresent, can be seen as the key to a multidimensional reality, an idea that quantum physics explores through the notion of "superposed states."

Let us now imagine a parallel: just as a single particle can exist simultaneously in multiple positions and conditions, so Atman, being a manifestation of Brahman, simultaneously inhabits infinite levels of existence. In the human context, this could mean that the self, by virtue of its connection with the universal self, has access to multiple potentials. Every choice, every possibility, could exist in a kind of "*quantum lattice*" of consciousness.

One striking example is found in the Vedantic cosmology of ancient India: the competing worlds or "*lokas*" described in sacred texts. These lokas, parallel planes of existence, echo the possibility of many coexisting realities, principles that physics today hypothesizes with multiverses. Interestingly, archaic India already envisioned the complexity of reality with a perspective that today seems remarkably modern.

Let us return again to Schrödinger's cat. Until the box is opened, the cat is both alive and dead: it is an entity that exists in a condition of fundamental ambiguity. Is this not reminiscent of the human experience of the self? Often, we find ourselves oscillating between identities, roles and possibilities coexisting simultaneously in our consciousness. The Atman, as the Vedantic texts suggest, is not bound to any one reality. It is reflected in every mask we wear, never being fully defined by any of them.

On a famous trip through India, Shankara told a parable to clarify the concept. He saw a man frightened by a rope, mistaken for a snake. This misperception of the snake, Shankara explained, is Maya: appearance deceives, while the "rope" (the underlying reality) is the Atman. Something similar happens in physics. Particles are never exactly what they appear during observation: they hide behind equally valid dualities, such as wave and particle.

This overlap between Vedanta and physics could be accidental or depend on a universal intuition of the human mind. Nevertheless, the dialogue between the two disciplines remains one of the most fascinating frontiers of contemporary philosophy and science. Perhaps, just where the Atman and quantum particles meet, we can glimpse a new vision of our role in the universe: finite beings dancing with infinite possibilities.

Atman and the Quantum Field.

In Hinduism, the concept of Atman, the immortal self, represents the deepest essence of the individual, a reflection of the absolute divine, Brahman. Atman is not related to body or mind, but is that entity eternal that transcends time and space. In the world of quantum physics, this idea finds a curious resonance. Quantum superposition, a phenomenon in which particles exist simultaneously in multiple states, seems to enter into dialogue with the manifold and illusory nature of the self (Maya) taught by Vedic texts.

Werner Heisenberg, one of the fathers of quantum mechanics, visited India in 1929. During his meeting with Rabindranath Tagore, a poet and philosopher deeply rooted in Vedic thought, Heisenberg said:

"I found in Eastern philosophy a way to understand quantum uncertainty beyond rigid Western categories."

This dialogue between science and spirituality recognizes a common point: ultimate reality is not fixed, but probabilistic and interconnected.

Physics describes the quantum field as an infinite field of possibilities. Subatomic particles, such as electrons, do not occupy a definite position but are "*probability waves.*" Only observation collapses these waves into a definite state. Similarly, in the Vedic view, Atman lives in the illusory play of multiplicity, but always remains a witness and point of unity. Self-awareness emerges, as in the collapse of quantum waves, only when the observer becomes aware of his or her true essence.

The parallel between Atman and the quantum field enriches both science and spirituality. While for science the self might be a point of convergence in a probabilistic universe, for Hinduism Atman is the absolute reality waiting to be recognized. Both fields, with different languages, invite us to consider a world in which the whole is interconnected.

According to the sacred texts, Atman is not a separate entity, but a direct manifestation of Brahman, the universal absolute. It matters little how many forms or identities it may take: its essence remains unchanged, like a mirror that reflects without ever changing. This concept, elaborated millennia ago, seems to find a correspondence in modern quantum physics and the idea of superposition of states, where one particle can exist in several conditions simultaneously.

The Mahabharata, the great epic poem of ancient India, offers an illuminating narrative. In one iconic episode, Arjuna, a warrior prince, looks inside the Vishvarupa, the universal form of the god Krishna. In that moment, he sees all manifestations of the universe coexisting simultaneously: his past lives, his future battles and the infinite possibilities of existence. Krishna, in his commentary on this divine vision, reminds Arjuna that everything-including his fears and desires-is a reflection of a single truth: the Atman, which lies beneath the multiplicity of appearances. This reflection of multiple identities is a symbolic representation of quantum superposition.

In quantum mechanics, superposition explains how a particle can be in contradictory states, such as being both wave and particle at the same time. Similarly, Atman is the unique essence that can be reflected in multiple forms and identities, never being limited to any one of them. Swami Vivekananda, one of the greatest Indian philosophers of the 19th century, wrote:

"Atman is not present in anything, but everything is in Atman."

This phrase captures the idea that, as in a quantum particle, the ego's identity is not locked into a definite form, but lives in an open field of possibilities.

The practice of meditation also offers a mirror of this idea. In the highest meditative states, described in the Upanishad, the boundary between self and universe dissolves. Those who meditate experience their consciousness as an infinite field in which all personal identities coexist simultaneously: parent, child, student, universal being-all are real and unreal at the same time. Vedic rishis likened this insight to a calm body of water in which each wave is distinct, but the essence always remains that of water.

Hinduism also offers a poetic image of this fluid identity structure. Mount Meru, the cosmic axis described in sacred texts, is the center around which infinite worlds revolve. Each world reflects a different perspective, but the center remains still. Similarly, Atman is the eternal, unchanging core that observes the endless dance of becoming.

The parallels between Vedanta and quantum physics are never completely overlapping, but they open up a profound reflection on how we conceive of identity and reality. Both disciplines seem to suggest that we are much more than what appears on the surface. Atman, like the quantum particle, might inhabit the mirror of consciousness, reflecting infinite possibilities that we, sometimes consciously, fail to see. Perhaps, the answer to the mystery of who we are lies in the ability to recognize that plurality and harmonize it, knowing that the ultimate essence that defines us is never altered.

Atman and uncertainty: Werner Heisenberg and the invisible self.

In 1929, German physicist Werner Heisenberg, father of the uncertainty principle, visited India. There he met the poet and philosopher Rabindranath Tagore and became acquainted with the Bhagavad Gita, the sacred text of Hinduism. Heisenberg, famous for his discovery of the fundamental limit in the knowledge of subatomic particles, was surprised to discover that the concepts of indeterminacy and relativity had been present, in philosophical form, in Eastern culture for millennia.

The heart of Indian thought, expressed in the Bhagavad Gita, revolves around the idea of Atman, the immortal self. Atman is the invisible essence that every individual possesses, and it is a manifestation of Brahman, the universal absolute. But this "self" is not definable. Like the electron in the quantum world that cannot be precisely observed at a point and time, Atman eludes description. It is infinite and intangible; one senses it, but does not measure it.

The connection between Heisenberg and the Bhagavad Gita, according to some biographers, deeply influenced his view of science. He found a similarity between the behavior of particles and Eastern thought.

> *"Quantum physics is not so far from the spiritual reality described in Indian texts,"*

affirmed Heisenberg, explaining that science was moving toward a space where uncertainty and complexity reign, just as in the sacred texts.

An overlapping electron can be in two states at once, moving from being to nonbeing, just like the philosophical idea of Atman that makes the self both universal and personal, tangible and transcendent. Uncertainty, then, is not a defect, either of science or spirituality, but their strength.

Heisenberg left India impressed by the depth of those teachings. He returned to work in Europe with a new element: the realization that the laws of the microcosm need a philosophical sensibility to be understood.

In a universe where observer and observed are inextricably intertwined, physics and spirituality, distant only in appearance, meet on the brink of the impossible.

The Illusion of the Self: Atman and the quantum multiverse.

The idea that reality is an illusion is a fascinating theme that links ancient Eastern philosophies and modern discoveries in quantum physics. In Hinduism, the concept of Maya describes the visible world as a deceptive projection divorced from ultimate truth. At the same time, quantum physics, with its multiverse theory, proposes the coexistence of infinite parallel worlds, each with its own alternate versions of events, people and identities.

Atman, according to Vedic texts, is the immortal and unique essence that lives beyond all illusory appearances. Connecting it to the quantum multiverse, we could imagine the Atman as the thread that runs through and connects all versions of the self that exist in the multiple hypothetical realities. Just as any subatomic particle can be in a state of superposition-that is, existing simultaneously in multiple states-so the idea of a single self but fragmented into infinite experiences seems to resonate with both views, religious and scientific.

The Bhagavad Gita, one of Hinduism's most celebrated sacred texts, states:

> *"Just as a person puts on new clothes, leaving aside old ones, so the Atman takes on new bodies."*

This quote can be seen not only as a metaphor for reincarnation but as an allusion to the multidimensional reality of the universe. The self could, therefore, have "parallel" experiences in several simultaneous incarnations, just as the multiverse theory suggests.

Adi Shankara, the great 8th-century philosopher, also argued that discernment of Atman was the key step in breaking the illusion of the separate self created by Maya. In a sense, his thought resonates with the modern scientific insight that different observers can influence and shape distinct realities. Just as the role of the observer is crucial in quantum science, it is also crucial for those seeking spiritual enlightenment: recognizing the true nature of Atman means transcending the illusions of this fragmented reality.

A suggestive image may arise by imagining the multiverse as an infinite range of possibilities, each represented by a different world. However, at the center of it all, like a motionless and invisible axis, is the Atman. This immortal self is the one constant that runs through the various "I's" scattered throughout the parallel worlds described by science.

The West began to approach these issues thanks to people like Werner Heisenberg, one of the fathers of quantum mechanics. Heisenberg himself studied Eastern thought in depth and recognized surprising similarities between the insights of modern physics and those of Vedic texts. Again, the famous quote from Schopenhauer, the German philosopher fascinated by Hinduism, sounds prophetic in this context:

> *"The world is my representation of the world."*

In a reality constructed by observation-whether of the mind or of inner consciousness-who we really are might just be the manifestation of that immortal thread that holds everything together.

Finally, the link between Atman and multiverse is an invitation to reflect on the concept of identity. If every decision, every action, and every possibility exists in some alternate branch of quantum reality, then what we are in this life is only a partial reflection of our eternal self. To acknowledge this perspective is not simply to embrace a speculative view of the cosmos but also to rediscover a deeper awareness of who we are: not passive spectators but integral and inseparable parts of an interconnected universe.

In a world where science and spirituality are increasingly mixing, the ancient concept of Atman is finding new life in light of recent quantum discoveries. Or, perhaps, it is just a reminder of how far humanity has always been able to look beyond appearances.

Atman and the "Schrödinger's Cat" paradox.

Atman represents the immortal self, a spark of Brahman, the ultimate and all-pervading reality. Hinduism holds that Atman is not subject to the constraints of space, time or illusory phenomena. This self does not change, observes everything, but remains unchanging. It is a silent witness to becoming.

In the world of modern science, a strikingly similar idea emerges in quantum mechanics. The phenomenon of superposition allows subatomic particles to exist in multiple states simultaneously, up to an "observation." And it is here that an interesting parallelism emerges between Atman and Schrödinger's famous cat paradox.

In 1935, physicist Erwin Schrödinger described a thought experiment to explain the uncanny nature of quantum systems. Imagine a cat locked in a box with a mechanism related to the decay of a radioactive particle. If the particle decays, a mechanism kills the cat. The particle, according to the rules of quantum physics, exists at the same time in two states: decayed and undecayed. Consequently, until we open the box, the cat is simultaneously alive and dead.

This paradox reflects the idea that reality is indeterminate until the act of observation. Atman, according to Hindu philosophy, could be considered the "witness" of this ambiguity. The observer does not modify states, but makes them manifest.

A deep connection lies in the idea of Maya, the illusion in which the phenomenal world exists. Maya wraps reality like a deceptive canvas. However, Atman remains detached, pure, and unadulterated, observing events without being involved.

Similarly, the role of the observer in quantum physics has a fundamental function: to define what is real. What we see emerges only when we observe. Without the observer, reality seems to remain suspended, like the infinite possibilities of the particle in superposition. The Bhagavad Gita, the Hindu sacred text, states:

> *"For the soul there is neither birth nor death. It exists and never ceases to exist. It is not born, it does not die, it is eternal, original, never began and will never end. It does not die when the body dies."*

This connects intimately with Schrödinger's notion, in which the system is suspended in eternal possibility until observation defines it.

Hindu sages such as Adi Shankara, a great proponent of Advaita Vedanta, also described Atman as the mirror that reflects the phenomenal world without ever being altered by it. Just as the mirror remains unchanged in the face of images, Atman observes multiple states without changing.

Quantum meditation: accessing Atman and superposition of states.

In many spiritual traditions of India, Atman is considered the deepest core of the human being. It is the eternal, immortal essence that, according to Vedic texts, represents a direct manifestation of Brahman, the universal cosmic principle. In a sense, the Atman can be said to be a divine spark present in every individual. But what happens if we bring this millennia-old concept into dialogue with the findings of quantum physics?

Modern physics has shown us that subatomic particles can be in superposition states. This means that particles exist simultaneously in multiple states or places until an external observation intervenes. This phenomenon, at the heart of Schrödinger's experiments and the intricate quantum world, has fascinated scientists and philosophers alike. And surprisingly, one can see a connection with the concept of Atman and meditation practices.

Mindfulness meditation, extensively described in the Vedas and later texts such as the Upanishads, is much more than a relaxation technique. Some scholars and practitioners consider it a method of "tuning in" to the deepest levels of consciousness, allowing one to perceive one's oneness with the universe. Anecdotes from great spiritual masters, such as Adi Shankaracharya, tell how meditation leads to the realization that the Atman is not limited to the body or mind, but is a multidimensional entity, beyond time and space.

On the neuroscientific level, recent studies have explored how meditation alters brain circuits and creates states of "connectedness" between different areas of the brain. Neuroscientist Sir Roger Penrose, with physicist Stuart Hameroff, proposed the theory of "quantum mind." According to this theory, microtubules in brain cells could play a role similar to that of quantum particles, with states of superposition and quantum collapse within the human mind. These phenomena could explain not only consciousness, but also deeper meditative states.

Imagine, then, meditation as a tool for accessing the "superposition" of the self. A practitioner, in mental stillness, could simultaneously perceive the multiple aspects of his or her identity: the individual, relational and cosmic self. As Nobel laureate Erwin Schrödinger, heavily influenced by Vedic philosophy, wrote ,

> *"The distinction between the self and the universe is ultimately an illusion."*

Sacred places, such as Varanasi on the banks of the Ganges, have been places of meditation and spiritual quest for millennia. But today, even in scientific laboratories, the possibility that deep connections exist between mind, universe and subatomic particles continues to inspire questions. We are part of a unique reality, where ancient wisdom and more modern physics, each in its own way, bring us closer to the same insight: we exist as a web of states, visible and invisible, tangible and transcendent.

Through meditation, perhaps, we can glimpse and even experience this mystery. A mystery already described by the word Atman, the immortal soul.

Kalidasa and the Suspended Particle: Poetry and Physics of Atman.

The work of Kalidasa, considered one of the greatest poets and dramatists of classical India (4th-5th centuries CE), evokes with extraordinary sensitivity the union of individual soul (Atman) and infinite cosmos (Brahman). In his verses, the universe and the soul are not separate entities, but two sides of the same fluid, changing, but intimately connected reality. This ancient thought seems to be in dialogue with one of the most revolutionary ideas in modern physics: quantum superposition.

In quantum physics, a particle can exist in multiple states simultaneously until an observer causes its "collapse," that is, determines its "position" or "state." This principle, seemingly foreign to ordinary human understanding, poetically recalls the Vedic idea of the multiform self. Atman, according to Hinduism, is the immortal self, capable of transcending individual identity to merge with universal reality. It is both a unique and multiple entity.

Kalidasa, in his poems, describes the self as a "drop" capable of dissolving into the universal ocean while retaining its essence.

(Remember David Bohm's image of the drop of ink in glycerin?)

In the epic masterpiece "*Meghadūta*" (The Messenger Cloud), an exiled "yaksha," a symbol of the suffering soul, invokes a cloud as a conduit to carry a message to the distant beloved. The cloud becomes the bridge between the individual self and the vast expanse of the cosmos. This symbolism echoes the principles of the quantum world, where the individual particle is one with the cosmic probability network.

An interesting parallel can also be found in European poetry. John Keats, in his "Ode to a Nightingale," speaks of a soul suspended between different worlds, recalling a theme akin to that of the Indian poet: the tension between individuality and cosmic union. However, it is Kalidasa who defines, with his refined Sanskrit aesthetic, a language that accommodates this duality as a harmonious union, not a conflict.

Modern science, represented by pioneers such as Schrödinger and Heisenberg, recognized in Eastern philosophy an interesting prenatal insight into quantum discoveries. Schrödinger, in particular, explicitly mentioned the idea of the Atman as a model for conceiving of a reality in which the boundaries between observer and observed dissolve. Kalidasa, in many verses devoted to nature and the transcendent, anticipates a sense of "indefiniteness" that quantum mechanics unveiled with its equations.

Rethinking Kalidasa through the lens of quantum physics is not just an intellectual exercise. It is an emotional and aesthetic journey that reminds us how poetry and science can converge. Atman, like the suspended particle, teaches us that identity and reality are larger concepts than we seem to perceive. In both cases, what seems separate is, after all, indivisible.

Atman and the quantum observer. Who are you when you look?

The idea that observation can influence reality is one of the most fascinating concepts in quantum physics. Particles that are in states of "superposition" collapse into a single observable reality when an observer intervenes. This apparent scientific magic finds a striking parallel in Hindu thought, where the Atman plays a similar role as an observer of universal reality.

According to Hinduism, Atman is the individual essence, the unchanging core that observes the world through the veil of Maya, the illusion of materialistic reality. In the texts of the Upanishads, we read that only the Atman makes perception of the world possible. It is the ultimate observer, a silent witness that transcends body and mind. The sages in the texts of the Vedas write:

> *"The Self is the witness, eternal and formless."*

This concept can be compared to the quantum double-slit experiment. In this experiment, a particle - for example, a photon - passes through two slits simultaneously, revealing superposition of states. However, when an observer measures the particle's path, this particle chooses a single trajectory: the simple act of observation changes the result.

The implications are extraordinary. The observer, therefore, is himself part of the reality he observes. Werner Heisenberg, one of the fathers of quantum mechanics, emphasized this relationship by saying:

> *"What we observe is not nature itself, but nature exposed to our method of questioning."*

Hinduism further expands this concept. In the dialogues of the Bhagavad Gita, Krishna explains to Arjuna that the Atman, being timeless and unchanging; nevertheless, watches the drama of earthly life even though he does so with a divine detachment. In other words, the observer (Atman) not only watches, but is also part of the creation of what he watches. This is reminiscent of the philosophical interaction between the human observer and quantum reality.

How do these ideas help us better understand who we are? When I observe the world, who am I really? Is it the body and mind that churns or is it the Atman, the self that calms the confusion of the visible world? As both quantum physics and Hindu philosophy point out, perhaps the real answer lies in the very act of observing. We are not just spectators: we are also participants. In this mirror between East and West, the observer and the observed reflect each other, revealing a hidden link between science and spirit.

The dance of the multidimensional self. Shiva, Atman, and quantum superposition.

At the heart of Hindu philosophy, the Atman, or immortal self, manifests as the eternal essence within each individual. Atman is at once a fragment of the infinite Brahman, the ultimate and indivisible reality. From another perspective, quantum physicists have discovered an intriguing phenomenon called "*superposition.*" According to quantum mechanics, a particle can be in several states simultaneously, until observation "decides" one state among the possible ones. Surprisingly, this condition seems to echo the concept of Atman as a multidimensional presence, in perpetual transformation.

Shiva, one of the most iconic figures in the Hindu pantheon, perfectly embodies this dance of change. Known as the destroyer and the transformer, Shiva represents the continuous movement between the possibilities of being. In Indian mythology, his famous "cosmic dance" (the *Tandava*) symbolizes the creation, destruction, and regeneration of the universe. This dance, which originates in myth and resonates in temples dedicated to Shiva such as the one at Chidambaram in Tamil Nadu, becomes a powerful metaphor for the way the Atman himself moves between his different states.

Werner Heisenberg, during his stay, conversed with Rabindranath Tagore, the Bengali poet and philosopher. The two discussed the interconnection between modern physics and Eastern philosophies. Heisenberg, impressed by the holistic view of reality typical of Hinduism, found resonances between the probabilistic nature of quantum phenomena and the idea of a fluid, interconnected self. The superposition of states, for example, recalls the idea that Atman is neither rigid nor monolithic, but can vibrate between different perspectives and forms, in dialogue with Brahman.

When one thinks of Shiva's dance, one can imagine the Atman immersed in a "quantum dance" among infinite possibilities. Like a particle, the self can oscillate between different states, waiting to manifest a definite form. This concept is reinforced by the Bhagavad Gita, in which Krishna states:

"There is neither birth nor death for him who is immortal. He has no beginning and no end. Although the body perishes, the Atman never dies."

Atman is thus not only in a position of unity, but also in an eternal interplay of multiplicities, just like the particles described by quantum physics.

What makes this parallel fascinating is the modern convergence of two seemingly opposite worlds: ancient spirituality and contemporary science. Shiva dances in mythology, but he also dances in the microcosm of quantum reality. Both point to a universe in constant motion, where identity-both of the Atman and of particles-forms and dissolves in an infinite web of possibilities.

Atman, Déjà Vu and quantum intersections.

Déjà vu is an almost mystical experience. Millions of people around the world describe the strange feeling of "having already experienced" a present moment. But where does this unexplained phenomenon originate? In the depths of Indian philosophy, the concept of Atman, the immortal soul, may offer a surprising explanation. And, incredibly, even quantum physics may help decipher this mystery.

In Hinduism, the Atman is the eternal essence of the individual, a fragment of the divine Brahman. This immortal soul exists beyond the constraints of time and space. In modern physics, something similar is observed in the principle of quantum superposition. In this state, a particle can be simultaneously in multiple positions, or assume multiple conditions, until it is observed. The idea is elegant and disorienting: a "divided" microcosm that exists in multiple realities.

What if déjà vu is something like that? What if the Atman floats, like a quantum particle, between different dimensions of space and time? The human mind, in fact, could perceive overlapping instants of a parallel reality. These once separate moments become for consciousness a single trace of meeting memories.

The Bhagavad Gita, one of the sacred texts of Hinduism, states:

"The body changes, but the Atman remains unchanging. It is birthless and deathless."

This description seems to resonate with the idea of superposition: the Atman traverses the different states of existence, preserving itself beyond the physical world.

A modern example might be Hugh Everett III, the U.S. physicist who developed the many-worlds interpretation of quantum mechanics in the 1950s. Everett suggested that every decision or event generates a parallel universe. Today, scholars of the subject such as Michio Kaku speculate that consciousness can also "remember" these alternate universes, creating déjà vu-like experiences.

Déjà vu, then, may represent a "*glitch*" in perception. The Atman, at that point, experiences all its possible lives, if only for a fragment. Like a grain of sand falling through the folds of time, this immortal spirit retains within itself the echo of parallel universes.

Ancient Indian mystics meditated on the deep nature of the soul, believing that every human being contained a fragment of the eternal divine. Similarly, physicists today meditate on subatomic particles, wondering if the universe itself is not animated by a similar logic. Could it be that quantum mechanics, rather than contradicting mysticism, is the most concrete proof of it?

The experience of déjà vu reminds us of the mysterious connection between what we think we understand and what remains beyond our control. The intersections between the Atman and quantum theories open the door to a new way of seeing reality. A way in which the divine and the subatomic coexist, perhaps, in the eternal now.

Ocean of Self: Atman as wave in quantum physics.

The ancient sages of the Vedas described Atman as a wave in the vast ocean of existence. Similarly, quantum physics offers us a fascinating parallel: the wave-like behavior of subatomic particles. Where metaphysics seeks the meaning of consciousness, science explores matter; both meet in an image that unites the microcosm with the infinite.

Quantum physics has revealed to us the idea of an undefined reality, where a particle can exist in superposition states. In this superposition of states, where being expands beyond a single identity, we can glimpse an affinity with Atman. Atman is also not confined. It is undulating, fluid, connected to the ocean of total existence.

A striking picture emerges by studying the Schrödinger equation, which describes the "probability wave" of particles. We can never know for sure where an electron is at any given time, but we can imagine this particle as a wave stretching through space. Similarly, Hinduism invites us to consider Atman as a spark of the absolute that transcends individuality and expands beyond the boundaries of our everyday perception.

An enlightening example comes to us from the aphorisms of the Upanishads, those sacred texts written about 3,000 years ago. In the Isha Upanishad we read:

"He who sees everything in himself and himself in everything feels no fear."

This passage resonates with the quantum view: the interconnection between what exists inside and what exists outside. Modern physics, with quantum entanglement experiments, has shown us that two particles separated by sidereal distances can influence each other. Similarly, the Atman is both unique and interconnected with every element in the universe.

The sacred city of Varanasi, still the meeting point of spirituality and knowledge, can be seen as the "living laboratory" of this philosophy. It is the place where generations of Hindu scholars have reflected on the universal self, perhaps contemplating the very flow of the Ganges, which, like a wave, symbolizes transformation and continuity.

Let us think for a moment about the wave-particle dualism of quantum physics and connect it to this vision. We can imagine Atman, the Self, as the wave: undefined, expanding, elusive with rational thought. And we can imagine the physical body as the particle: a definite point in space, visible, but still a manifestation of the infinite quantum ocean. Both realities coexist without being contradictory.

Today, as the scientific world attempts to observe particles in ever greater depth, we are approaching fundamental questions about identity and presence. What is the nature of the self? Where does our consciousness begin and where does it end? These questions, which Galileo and Einstein could hardly imagine as fields of scientific study, echo with extraordinary modernity in the simultaneous study of Atman and the subatomic world.

The wave would not exist without the sea. Atman would not exist without Brahman. The particle would not manifest without the wave. In this fascinating play of metaphors, philosophy and science brush up against each other, reminding us of what the German poet Novalis dared to suggest centuries ago:

"Philosophy is nostalgia: the desire to be everywhere at home."

And perhaps, in the concept of Atman understood as a wave, and the quantum idea of a constantly expanding reality, we can find a harbor that brings us a little closer to home.

Atman and quantum entanglement.

The invisible thread that binds every element of the universe is one of the most profound insights emerging from both Indian philosophy and quantum physics. In Hinduism, the concept of Atman represents the immortal essence, the individual soul that is a direct manifestation of the great all: Brahman. This eternal self is not isolated. It is interwoven with every other fragment of life in a web that transcends space and time.

Quantum physics offers a strikingly similar view. The phenomenon of quantum entanglement shows that two particles, once connected, continue to influence each other at any distance, as if they were two sides of the same coin. Albert Einstein called it, with some skepticism, "*spectral action at a distance.*" But later experiments, such as those conducted by physicist Alain Aspect in 1982, confirmed the actual reality of this connection. Nothing, not even the vastness of space, can break it.

In Hinduism, we find parallel thinking in the Upanishads, ancient sacred texts that investigate the essence of being. The Chandogya Upanishad, for example, states.

"All this is Brahman. Verily, Atman is Brahman."

In other words, each fragment of self (Atman) is connected to the whole (Brahman), in an interweaving that knows neither temporal nor spatial boundaries.

An interesting analogy is told in the legend of "*Neti-Neti,*" a spiritual practice that calls for defining what is Atman through negation. "*Not this, not that.*" It is an invitation to discard all illusions of separation and arrive at the common basis of all that exists. This view corresponds to the quantum concept whereby particles, seemingly separate, are actually different states of a single global system.

Then there is the temporal aspect that makes the parallel between Atman and entanglement particularly fascinating. In quantum physics, connections between particles seem to ignore time, demonstrating a synchronicity outside ordinary experience. Similarly, in Indian thought, Atman transcends linear time. The immortal self, present in every being, is not subject to birth or death. It is eternal, as spiritual traditions have described it for millennia.

One need only visit places like Varanasi, along the banks of the Ganges, to intuit this perspective. Here, the cycle of death and rebirth appears still as a natural and indivisible element of daily life. The same concept is reminiscent of the dance of subatomic particles, constantly passing from one state to another, transforming without ever losing their deepest essence.

After all, Atman and entanglement both remind us that, beyond illusions of distance, there is a fundamental link that connects everything. Across time and space, the immortal self is never really alone.

Ego and nothingness:

Hinduism invites us to reflect on the idea of the Atman, the immortal self, which represents the individual and eternal essence, a reflection of the boundless Brahman. This concept finds surprising resonance in the language of modern quantum physics, particularly in the idea of the quantum vacuum.

The quantum vacuum, in physics, is not an absence of everything. It is, rather, a pulsating sea of potential energy, a fundamental field from which any particle or state of matter can emerge. A similar concept is found at the heart of Hindu philosophy: the Atman is perceived not as a separate entity, but as a "core" of potentiality, arising from the mysterious and infinite Brahman.

Niels Bohr, one of the founders of quantum mechanics, once said:

> *"Those who are not amazed at their first experience with quantum physics have not yet understood it."*

The same could be said of those who approach the concept of Atman. The two fields seem to converse: if the quantum vacuum produces all possible particles and states, the immortal self emerges as pure possibility, an idea that transcends time and space as we know them.

A useful comparison is with the "superposition" state of quantum physics, in which a particle can be in multiple states at once. The self, the Atman, seems to share this quality: it is at once individual and universal, personal and cosmic. It is as if each Atman is a droplet that simultaneously contains and reflects the entire Brahman ocean.

From a scientific point of view, the quantum vacuum represents a kind of fertile "nothingness," capable of producing parallel universes or alternative realities according to some interpretations. Similarly, in Hindu philosophy, the Atman transcends the material world, but also constitutes its essential foundation. This view harks back to physicist David Bohm, who in his studies spoke of an "implicit order" hidden beneath visible reality, not far removed from the concept of Brahman.

This convergence of science and spirituality can help us rethink our place in the universe. The idea of Atman and the quantum vacuum show us that "nothingness" is not to be feared: it is a promise, an infinite possibility. Both remind us that out of that void, silent and mysterious, infinite worlds and infinite versions of ourselves can arise.

Maya and the superposition: The reality that the Atman observes.

Hinduism introduces us to the concept of Atman, the immortal self, the deepest core of the individual. It is a divine reflection, a manifestation of Brahman, the universal principle that pervades everything. But if Atman is eternal and unchanging, why does man struggle to see its true nature? This is where Maya comes in, the veil of illusions that clouds reality, similar to the distortion of a mirror that reflects countless images, never showing the authentic one.

In quantum physics, the phenomenon of superposition might offer an illuminating metaphor. Superposition allows particles to exist in multiple states simultaneously, until an observation "collapses" this multiplicity into a single measurable reality. What if what we know as Maya is a kind of cosmic superposition? Reality, as the Danish physicist Niels Bohr suggested, could be not a static entity, but an interweaving of potentialities dependent on observation. Bohr's words:

> *"It is not possible to speak of reality separately from the observer," he said,*

seem to echo the texts of the Upanishads. In them, it is said that those immersed in Maya see the world as fragmented, ignoring that everything is one.

Erwin Schrödinger, a pivotal figure in quantum physics, was deeply influenced by Indian philosophy. In his essay "*What Is Life?*",

Schrödinger writes that the idea of an immortal self not only finds resonance in mysticism, but also in science. He saw in the universe a single collective consciousness, a common theme in the Upanishads.

This vision is not without challenges. What then is real? Is what we observe real, or what exists beyond illusions? This question is central, both in quantum physics and in Indian philosophy. Schrödinger's cat's mind experiment resembles the perennial tension between Maya and Atman: a reality suspended between infinite possibilities that only an act of awareness can reveal.

In both domains, the role of the observer is crucial. The Atman observes, but must recognize that what he sees through the filter of Maya is not the ultimate truth. This path of spiritual awakening brings him closer to universal reality, abandoning the illusion of separation.

Hinduism, with its concepts of Atman and Maia, and quantum physics, with the mystery of superposition, converge in one point: reality is never as it seems, but is what we observe of it. And the observer, in the end, is none other than the self.

Atman and quantum coherence

In the mysteries of quantum physics, a fascinating concept emerges: quantum coherence. This phenomenon describes the ability of a quantum system to maintain a deep connection between its states, creating a kind of invisible harmony beyond the apparent chaos.

The Atman, according to the Upanishads remains intact and unchanging while navigating through the worlds of matter and illusion. The Brihadaranyaka Upanishad strongly proclaims that.

> *"This Self is free from death, free from fear; it is immortal, uncapturable, untouched by pain or suffering."*

Quantum coherence, in its own way, reflects these qualities. In the quantum world, even when particles seem to escape any firm definition, they can maintain a subtle cohesion, like an invisible orchestra working in synergy.

The metaphor of the self as resonance is also echoed in experiments studying the interaction of isolated particles. Erwin Schrödinger, a great physicist, equated the self with the totality of the cosmos, suggesting that the divisions we perceive are mere illusions. When a quantum particle is maintained in a coherent state (for example, in a superconducting crystal at extremely low temperature) it does not lose contact with all other possible states. This quantum "memory" can be compared to the Atman's ability to remain connected to its true nature while living in a human body.

In Benares (modern-day Varanasi), a sacred city on the banks of the Ganges, Indian philosophers have meditated for millennia on the idea that everything that exists is part of an invisible network of spiritual connections. Modern physics, ironically, offers us a mathematical analogy. In 1972, physicist Alain Aspect demonstrated that quantum particles, even separated by great distances, remain mysteriously interconnected through the phenomenon of entanglement. This raises a profound question: if the cosmos itself is quantum coherent, can we see the Atman not as an isolated entity, but as the vibration of a universal reality that transcends space and time?

The Atman is one. The quantum system, which never quite divides, is its mathematical mirror. What Hinduism for millennia has considered eternal and unchanging, science today is beginning to intuit as a fundamental connecting principle. The main challenge remains open: to understand how science and spirituality can share the resonance of a unique truth.

Krishna, Arjuna and Atman: Quantum lessons in the Bhagavad Gita.

The Bhagavad Gita, one of the pivotal texts of Hinduism, is an epic dialogue between Krishna, the divine guide, and Arjuna, a warrior prince. On the battlefield of Kurukshetra, Arjuna is overwhelmed by doubt and moral conflict. Krishna, his charioteer and teacher, instructs him on the deeper aspects of existence. Central to the dialogue is the concept of Atman, the immortal Self, which Krishna describes as eternal, indivisible and beyond death.

In an emblematic passage (Bhagavad Gita, II:12-25), Krishna states:

> *"Never was there a time when I, you or these kings did not exist, nor will there ever be a time when we cease to be."*

This immortal phrase places the Atman beyond time and space, similar to a superposed quantum particle.

In quantum physics, the superposition principle indicates that a particle, such as an electron, can exist simultaneously in multiple states. Only when observed does the particle "collapse" into a specific defined state. This concept offers a striking parallel to the Atman, which Krishna describes as simultaneously unique and universal.

The Atman can be embodied in many roles: warrior, son, king or teacher. Arjuna himself experiences this multiplicity. He is a loving brother, a courageous commander and a spiritual disciple. Krishna explains that these roles are like clothes, worn temporarily by the Atman. The essential identity remains unchanged, as in a quantum system where possibilities coexist to the point of interaction.

For example, Arjuna's duality as warrior and moral man reflects the state of overlap. He hesitates to fight his own relatives, a dilemma that represents the collision between his roles. Krishna resolves the conflict by asking Arjuna to transcend these superficial identities and recognize his eternal Atman.

Krishna introduces a key concept: the universe is "*Lila,*" a cosmic game orchestrated by Brahman, the universal principle from which Atman emanates. As in quantum mechanics where particles dance within a web of probability, so Krishna invites Arjuna to see war not as a material conflict, but as part of a higher order.

Bhagavad Gita, (XVIII:20) explains that an enlightened mind sees "*one Self in all things*." This resonates with the concept of a unified quantum field, where all particles are interconnected, beyond space and time.

Twentieth-century quantum science, developed with the theories of physicists such as Schrödinger and Heisenberg, seems to intercept this message. Erwin Schrödinger himself was deeply influenced by the Upanishads, which develop the same ideas as the Gita.

Atman in the infinite possible.

According to the Vedas, the Atman is unique and unchanging, but at the same time it manifests itself in a multiplicity of individual experiences. This apparent contradiction is reminiscent of the quantum phenomena studied by physicists such as Niels Bohr and Erwin Schrödinger.

In Patanjali's Yoga Sutras, the path to understanding the immortal self is through meditation. Through samadhi, a state of pure concentration, the practitioner can transcend the fragmentation of the egoic self and recognize himself or herself as Atman, a spark of the infinite Brahman. This practice relates to a cosmic metaphor: just as observation changes the behavior of quantum particles, so deep introspection transforms our experience of reality.

An interesting anecdote links this view to a specific moment in modern science. In 1927, at the Solvay Congress in Brussels, Albert Einstein expressed puzzlement about the "indeterminate" nature of quantum physics, saying the famous phrase:

"God does not play dice."

Niels Bohr responded with philosophical advice:

"Stop telling God what to do."

Physics, at that time, seemed to point toward a model of the universe in which reality was not fixed but fluid, potential and complex. Similarly, Atman, according to Hinduism, can only be perceived by going beyond the changing and limited appearances of the material world.

Let us now imagine the microcosm of the Atman and the macrocosm of the quantum universe woven into a single fabric of infinite possibilities. Physicist David Bohm, strongly influenced by Eastern philosophy, described reality as an "implicit order," a hidden web in which every element is interconnected. This view, while scientific, seems to recall the idea of Brahman as the invisible and indivisible basis of all existence.

In the end, the journey through Hinduism and quantum physics shows us that the ancient insights of Eastern sages continue to dialogue with the most advanced scientific theories. In Atman lies not only the secret of personal immortality, but also the key to understanding a universe in which the possible and the impossible are endlessly intertwined.

The return to the quantum self. Atman in the age of modern physics.

The encounter between quantum physics and Eastern philosophy opens new doors to understanding ourselves and the universe. An ancient concept in Hinduism, the Atman, now finds striking parallels with the principles of quantum mechanics. Atman, regarded in the Vedic scriptures as the immortal self and the individual manifestation of Brahman, invites reflection on the meaning of identity and reality.

Quantum physics tells us about superposition, where a particle can be in multiple states at once. This idea resonates deeply with the concept of Atman, which transcends ordinary identity and manifests itself in multiple facets while remaining unique. Just as a particle exists simultaneously in multiple possibilities as long as it is observed, so Atman can be understood as an essence embodying infinite potentialities.

A historical example of this intellectual affinity dates back to the 1970s, when Fritjof Capra, noted author of "*The Tao of Physics*," wondered about the similarities between quantum mechanics and Eastern mystical traditions. Capra observed that:

"subatomic particles are not things, but trends."

Hindu thought, with its emphasis on the impermanence of material things and the centrality of an unchanging self, responds resonantly to these theories.

The iconic site of Varanasi, on the banks of the Ganges, offers a significant cultural connection. For millennia, sages have gathered along the river to discuss the Atman and its connection to the cosmos. Today, those dialogues seem to embrace modern scientific language as well. For example, the double-slit experiment that showed the dual behavior of particles recalls the Vedic teaching that reality is influenced by consciousness.

As science advances, Atman does not cease to be a point of reference. Yogis suggested that by knowing the Atman we overcome the limitations of body and mind. Quantum physics, through concepts such as entanglement or non-locality, invites us to review our idea of separation and reconcile with an interconnected reality.

In these discoveries, ancient and modern, we recognize a universal message: the return to the self, to the Atman, is not only a mystical journey, but also an intellectual challenge in the age of modern physics.

Chapter IX. Karma: the law of cause and effect.

Karma, the law of cause and effect

In Hindu thought, Karma represents a kind of cosmic principle that governs the moral and material universe. Every action, intention or thought generates an effect. This effect, positive or negative, can manifest immediately or cross time, affecting the future, even beyond the individual life. It is not divine retribution, but a universal network of interconnections, like invisible threads that link everything to what has happened and what will happen. This is how Swami Sivananda, master of the 20th century, describes Karma:

> *"The actions of the past shape the present, just as those of today are already shaping the future."*

In other words, the universe dances to the rhythm of our actions, but with the precision of a choreography as old as time itself.

Quantum entanglement and the concept of Karma: invisible entanglements in the cosmos.

The universe moves according to mysterious laws, both in the depths of the human mind and in the meanderings of the subatomic world. In Hindu thought, the concept of Karma defines the universe as a web woven by actions and consequences. Each action produces an effect that manifests in the present or future, tracing a seemingly invisible but real chain of events. In quantum physics, a surprising insight is presented through the phenomenon of entanglement: two particles, once interconnected, remain connected in such a way that a change in the state of one immediately affects the other, regardless of the distance between them.

Quantum entanglement reminds us that the cosmos is interconnected. Particles are not islands, just as human beings are not to Karma. Hinduism has always represented the world as an organic network in which everything is connected. This view is reflected in quantum theory, which tells us of a universe with no clear boundaries. In 1935, Albert Einstein, Boris Podolsky and Nathan Rosen introduced the EPR paradox to discuss entanglement. Yet later events, such as those produced by physicist John Bell in the 1960s and Alain

Aspect in the 1980s, confirmed that these connections are real and fundamental to the structure of the universe.

The concept of Karma works on a similar plane. Every action, no matter how distant in time or mental space, leaves a trace connecting cause and effect. Just as in entanglement, the impact is not immediately visible but is profoundly real. Following this logic, in Hindu philosophy there are no unrelated events. Every current event is the fruit of what has been.

This connection between philosophy and science is no mere coincidence. The Austrian physicist Erwin Schrödinger, father of the equation that bears his name, found inspiration in Indian scriptures. Schrödinger studied the Upanishads, sacred texts of Hinduism, and became fascinated by nondualistic thinking. In his study of entanglement and quantum superposition, Schrödinger imagined the famous "Schrödinger's Cat" to show the unpredictability of the quantum world. But behind this idea was also the Hindu-inspired belief that the separation of object and subject was a mere illusion.

In 1944, in his book "*What is Life?*", Schrödinger wrote:

> *"The mind and its contents are indestructible. Likewise, modern physics shows that the entire universe is energetically interconnected."*

Echoed in these words are the ancient Hindu ideas that all existence, visible or invisible, is part of one indivisible reality, Brahman.

The connection between Karma and quantum physics finds a further meeting point in Heisenberg's indeterminacy. The principle states that both the position and velocity of a particle cannot be known exactly. This indeterminacy places the subatomic world in a constant state of potentiality. Similarly, the concept of Karma does not allow us to predict with certainty the outcome of our actions.

The complexity of Karma:

Karma is not a simple linear cause-and-effect relationship. Actions do not have immediate or easily predictable results because Karma is influenced by an intricate web of factors, such as past actions, intentions, context and individual destiny. This makes it difficult to accurately predict what precise consequence will follow a given action.

Intention and context are as valuable as the action itself: According to many traditions that contemplate Karma, it is not only the action itself that generates consequences, but also the intention that guides it. The same action can produce different results depending on inner motivations or circumstances, making it difficult to make predictions.

Karma is subject to time and personal development:
The effects of actions may manifest immediately, after a long time, or even in a future life (if one considers reincarnation). This varying time frame and the potential influence of one's spiritual growth path prevent a clear and immediate prediction.

Interactions with other people's karma:
People live in an interconnected world, and their actions are constantly intertwined with the karma of others. This means that external factors, such as the actions of others, can also influence the outcome of our needs, further increasing unpredictability.

Reality is not completely deterministic:
Karma operates on universal laws of cause and effect, but it is not completely mechanistic or deterministic. The universe, in many Eastern philosophies, is also governed by mysterious forces and spiritual dynamics that are not easily known or calculated.

In summary, Karma is a complex and profound principle that promotes responsibility and awareness of one's actions, but it is not a tool for predicting the future with certainty. The goal is not so much to control the outcome of actions as to act with awareness and righteousness to foster a positive balance in one's life and the world around him.

We are not isolated spectators but actors.

Entanglement and Karma invite us to reflect on a universe that is not divided. We are not isolated spectators but actors in a great cosmic drama. Every human action, every choice, resonates in the universe like a vibrating string. The ancient Hindu sages sensed this truth centuries before the advent of quantum physics. Today, scientific discoveries only confirm a vision firmly rooted in Eastern thought:

the idea that the cosmos is a living web, unified by deep, interconnected laws.

The uncertain flow of karma.

At the heart of Hindu philosophy, Karma represents one of the fundamental laws of the universe: every action causes an effect, every choice imprints a trace destined to shape the future. This principle of cause and effect is inherently moral and spiritual, reflecting the deep connection between the individual and the cosmic order. But what happens if we view Karma through the lens of quantum physics? The world of the infinitely small offers us perspectives as fascinating as they are unexpected, where concepts such as Heisenberg's uncertainty principle seem to evoke parallels with the complexity and unpredictability of our actions.

The uncertainty principle, formulated by Werner Heisenberg in 1927, is a pillar of quantum mechanics. It states that we cannot know precisely both the position and velocity of a subatomic particle. The more precisely we determine one, the more certainty we lose about the other. This limitation does not result from a lack of technology or human ability, but is an intrinsic property of nature.

In this context, Heisenberg's principle reminds us that the future is never completely determinable. Imagine an experiment with an electron: the moment we observe it, our action (the simple act of measuring it) alters its behavior. This phenomenon, known as "*observer interference*," finds a striking analogy with Karma. Our actions, just like the measurements of physicists, not only affect the present, but create cascading effects that point back to an indefinite and, in part, unpredictable future.

A famous example in quantum physics is the double-slit experiment. When a particle, such as a photon or electron, passes through two slits, it exhibits behaviors that defy common sense: until we observe it, it exists in a superposition state, like a wave of possibilities. But the moment we decide to observe it, the particle "chooses" a specific position or path.

Quantum superposition reminds us of the latent potential of every human action. We make an everyday decision: help a stranger or ignore him or her. This choice might seem trivial, but its consequences can intertwine in unexpected ways with the fabric of future reality, just as the interference of subatomic waves produces unpredictable patterns on a screen. Karma works the same way: every action, no matter how small, produces an effect that spreads beyond our control.

The anecdote of a butterfly flapping its wings in Brazil and causing a hurricane in Texas, known as the "*butterfly effect*" in chaos theory, embodies this same interconnected nature. Karma and quantum mechanics, however distant in language, describe dynamics that we cannot ignore. Both teach us that our choices-as well as every particle in the universe-are part of a larger, more complex system.

An interesting analogy can be made between the behavior of atoms and the karmic cycle of life. Each atom, the fundamental component that makes up matter, is subject to the laws of quantum physics. Its properties-position, energy, behavior-are probabilistic and not completely determinable. The very principles that underpin the structure of matter can remind us that small acts, just like the movements of these atoms, are part of networks of causes and consequences that generate the dynamic universe we inhabit.

This is where physics and philosophy converge: in an interplay of relationships, present choices and future consequences, but without the certainty of being able to calculate every outcome. As Buddha observed:

> *"As the archer straightens an arrow, so the wise man controls his thoughts."*

This control, this awareness, forms the bridge between the law of Karma and the quantum field: being aware of the present moment and the actions we imprint on it means understanding that every choice is a seed planted in the uncertain soil of tomorrow.

In Hinduism's Karma and Heisenberg's quantum uncertainty we find something that transcends mere scientific knowledge or philosophical reflection: an invitation to reflect on the mystery of reality, the role of our actions and the interdependence of all that exists. The idea that the future is never completely predictable, but influential by the choices of the present, unites us all, both as human beings and as explorers of the cosmos.

The dance of Dharma and electrons. Karma and quantum physics.

In the immensity of the Eastern philosophical tradition, Hinduism offers us an extraordinarily profound key to understanding the world: the concept of Karma, the universal law of cause and effect. Every action leaves an imprint, a vibration that propagates over time, affecting the present and the future. But what if we apply this interpretive lens to quantum physics, the field that governs the subtle world of particles? Is it possible to draw a parallel between these two vast dimensions, seemingly so far apart?

The answer lies in the metaphor of cosmic dance, a vision that combines ancient wisdom and modern science. As in the Bhagavad Gita, where Krishna reveals to Arjuna the secrets of Dharma and Karma, the world of subatomic particles reflects the same dynamic and unpredictable energy, an endless dance where creation and destruction alternate.

Werner Heisenberg formulated the famous uncertainty principle. This principle states that it is not possible to simultaneously measure the position and velocity of a particle with absolute precision. In other words, uncertainty dominates in the subatomic realm: every choice of observation affects the result.

This unpredictability, which escapes the rigid laws of classical physics, invokes the very heart of Karma. Just as our actions produce effects that we cannot always predict, particles also behave in ways that seem to defy our ability to control them. When an electron moves, it does not do so by following a fixed trajectory like a billiard ball, but traverses a multitude of possibilities, creating an unstable and fluid reality.

iconic 20th-century experiment known as the *double-slit experiment* demonstrated precisely this "karmic" nature of particles. In this experiment, an electron thrown toward two slits behaves as both a particle and a wave, depending on whether it is observed or not. It is as if the electron "knows" it is being watched and responds accordingly, choosing one of multiple possible journeys. This behaving "*in response*" is reminiscent of the fundamental node of Karma: every action, in this case observation, generates a reaction and influences the future.

To fully understand the value of this correspondence between Karma and quantum physics, we can turn to one of the most powerful images in Hinduism: the dance of Shiva, the Nataraja. Shiva, in his cosmic dancer form, is the keeper of the endless cycle of creation, destruction and regeneration. With one foot he represents the creative force, while with the other he tramples ignorance, destroying the old to make way for the new. His moving arms symbolize the ceaseless rhythm of life, the beating of the drum represents the cosmic Big

Bang, while the fire in his hand corresponds to the destruction necessary for change.

Nataraja is often interpreted as the eternal balancing of material and spiritual energies. But in its rhythm we can also glimpse that of electrons: a continuous motion of uncertainty, a pulsing between creation and destruction, between waves and particles. Electrons, like Shiva, never stand still. Their incessant activity is an echo of the cosmic dance: an expression of the Dharma of the universe.

In the Bhagavad Gita, Krishna says:

> *"You are entitled to the action, but not to the fruits of the action."*

This sentence goes right to the heart of Karma. We are called to perform actions, but we cannot fully control their effects. The electron also moves following infinite possibilities, but which path it will take depends on something that transcends all calculation.

This idea is fascinating because it blurs the boundaries between spirituality and science. Consider the act of throwing a stone into a lake: the waves generated propagate, touch the shores and affect every molecule of water. Karma works in the same way. Every human gesture, no matter how small, creates a resonance. In quantum physics, every particle is subject to a similar "echo," where even a tiny change in one system can alter the whole scenario.

A historical example that encapsulates the idea of Karma and quantum physics is the Manhattan Project. In the creation of the atomic bomb, the motion of a few subatomic particles resulted in one of the most devastating events in human history: Hiroshima and Nagasaki (1945). Political, scientific and military actions were intertwined with quantum physics, demonstrating how actions, in their own small ways, reverberate with enormous karmic consequences.

Karma and quantum physics, when observed together, reveal a profound truth. The universe is a dance. A dance of connections, of possibilities, of causes and effects that intertwine in ways often invisible to human eyes. In both the Bhagavad Gita and Heisenberg's experiments, we learn that reality is neither static nor simple. It is a continuous flux, an intricate web, an endless dance where every action leaves its mark and every particle reminds us that nothing is still.

The echo of Karma. Between human choices and subatomic reactions.

In the world of physics and philosophy, two seemingly distant territories, there are stimulating points of convergence. On the one hand, the concept of Karma in Hinduism describes the universal law of cause and effect, where every action generates consequences, often unpredictable, that reverberate over time. On the other, quantum physics, with Heisenberg's uncertainty principle, shows us a universe in which not everything can be known or determined with absolute precision. Between these two approaches lies the brilliant but troubled J. Robert Oppenheimer, the "father of the atomic bomb," for many the meeting point between Vedic wisdom and the enigmatic world of modern physics.

Oppenheimer, growing up in early twentieth-century New York, was not only an extraordinary scientist. He was also a man who sought profound answers to the nature of existence. During his early studies at Harvard, he discovered the ancient wisdom of the Vedas and immersed himself in the philosophy of Vedanta.

The Bhagavad Gita, one of the Hindu sacred texts, explores many concepts, including that of Karma. For Oppenheimer, Karma and physics seemed to speak the same language: actions that create effects, energies that transform from moment to moment, and unexpected consequences that awaken even after long periods of inactivity.

Karma, at its core, is the cycle of human actions and their consequences. It is not limited to the present: actions taken today can channel effects for generations, influencing future events in profound and unpredictable ways. Vedic texts teach that nothing escapes this web of connections, which calls attention to every choice made throughout life.

Quantum physics, although inscribed in a totally different context, proposes a similar argument. Werner Heisenberg, in 1927, formulated his uncertainty principle, showing that the behavior of subatomic particles cannot be predicted with absolute precision. We cannot simultaneously determine the position and velocity of a particle. Every observation inevitably changes the observed system. This principle generates a world where causality, as we know it in classical physics, dissolves in favor of probability and uncertainty.

The parallels are obvious: Karma and the uncertainty principle remind us that the consequences of actions-whether human or subatomic-move on complex planes that are not always accessible. Nothing can be reduced to a simple cause-and-effect equation.

Oppenheimer's life offers a living example of these connections. His decision to direct the Manhattan Project changed the course of history. But this choice also exposed him to the moral burden of consequences. Like unstoppable Karma, the baleful effects of the atomic bomb echoed through his life, leading him to deep reflections on the role of science and the ethics of human choices.

Many remember how Oppenheimer, in his biography, reflected on the interconnection between individual actions and the collective destiny of humanity. Perhaps, influenced by Vedic thought, he felt he was at the heart of a complex karmic web, an echo of causes and consequences that transcended his own existence. As he said one day:

> *"In the world you will find that everyone is responsible for everything they do. And for everything he avoids doing."*

The Los Alamos desert, where the bomb was built, is today a symbolic place. Here, among cacti and canyons, a human and scientific drama unfolded, where uncertainty reigned supreme. Although mathematical models predicted the bomb's effect, its social, political and ethical consequences remained unpredictable. This is reminiscent of Karma: we cannot always control when, where or how our actions will resonate in the world.

Heisenberg's indeterminacy tells us that at the subatomic level every particle behaves like a potential. We do not know its precise future, but we do know that its interactions will define realities beyond our control. Similarly, Karma functions as a cosmic seed. Every action planted today will sprout tomorrow, but in what way and with what intensity remains a mystery.

As Oppenheimer reflected on these connections, he was constantly living between two worlds: scientific determinism and spiritual depth. In an interview, one of his colleagues recalled that in moments of silence, Oppenheimer often read verses from the Gita, as if seeking answers about uncertainty. Particles followed their own laws, but the Vedic legend of Karma expanded his vision beyond the boundaries of physics.

In an age dominated by technology, reflecting on the concept of Karma and quantum indeterminacy offers us an important lesson. Every action, no matter how small, transforms the world. Similarly, every personal choice, like Oppenheimer's, fits into a web of connections that eludes our complete understanding.

The union of science and spirituality. Karma and probability in the laws of nature.

Karma represents a universal law: every action generates a reaction, which may manifest immediately or at some future time. It is a philosophy that connects by an invisible thread our behavior to the cosmic order, a way to explain how events, large and small, are never random. In all this, quantum physics provides surprising parallels, particularly through Heisenberg's uncertainty principle. This principle describes another form of causality: we cannot predict with absolute certainty the behavior of subatomic particles, only calculate their probability.

The connection between Karma and science is not immediate, but it becomes fascinating when we analyze how both theories accept uncertainty as a fundamental part of reality. Karma and indeterminacy, while belonging to different worlds, prompt us to reflect on the meaning of causality and the consequences of our choices.

The concept of Karma is rooted in the sacred texts of Hinduism, such as the Vedas and the Upanishads, which describe the universe as a web of causes and effects. This moral system emphasizes that every thought, word or action has the power to shape our destiny. There is no randomness in Hinduism: even seemingly insignificant events find an origin in a past action.

A well-known example is the parable of a king named Bharata, mentioned in the Mahabharata. Bharata, after a virtuous life, is reincarnated into a deer because of his attachment to desire. This episode emphasizes how our actions, although microscopic, have long-term consequences. Karma teaches us that every fragment of existence is interconnected, just like quantum particles.

The similarity between Karma and the uncertainty principle emerges in the way they both embrace unpredictability. In quantum physics, the actions of particles do not determine a single certain effect, but a range of future possibilities. Similarly, in Karma, the outcome of our actions is not precisely predictable, even though we know it will generate an effect.

Physicist and philosopher Fritjof Capra points out that Eastern traditions often anticipate scientific insights. Capra writes that the concept of Karma mirrors the dynamic network of relationships described by quantum theory. According to him, both the Hindu sage and the modern physicist share an interconnected view of reality. Nothing happens in isolation; every action produces a reaction, even if we cannot know its exact path.

A popular Indian saying states:

"Every flower that blooms is the result of the rains of a thousand years."

This phrase highlights the importance of every action, even small ones, in the grand scheme of reality. The same concept can be applied to the behavior of subatomic particles. In a famous talk in the 1960s, physicist John Bell suggested an experiment to show that two particles, even at a great distance, can influence each other in unpredictable ways through quantum entanglement. From this phenomenon emerges an idea dear to both Hinduism and quantum physics: everything is interconnected. No event occurs independently, neither in the spiritual cosmos of Karma nor in the quantum microcosm.

Sri Krishna and Heisenberg: How Hinduism explains quantum uncertainty.

The concept of Karma offers a universal ethical message: every action has consequences. Similarly, Werner Heisenberg's uncertainty principle, a cornerstone of quantum physics, emphasizes that every observation inevitably influences the outcome of a system. Although separated by millennia and different cultural contexts, the two concepts present fascinating parallels, revealing a surprising convergence between Eastern philosophy and modern science.

In the Bhagavad Gita, Krishna, the divine incarnation according to Hinduism, explains the law of Karma to Arjuna. One of the most significant passages reads:

"Every action carries with it the seed of its consequences"

Krishna emphasizes that one cannot escape the effects of one's actions. Every human choice leaves an imprint, shaping the individual and collective future. Not only material actions, but also thoughts and intentions influence destiny.

Krishna emphasizes the importance of Dharma (moral duty). Arjuna, the protagonist warrior, must act without attachment to the fruits of his actions. Krishna teaches that the agitation of desire or fear of failure breeds chaos, while action performed with detachment allows harmony, even if the immediate consequences are not always predictable.

The complexity of Karma resembles a great web of cause and effect, a web in which all actions, visible and invisible, are inevitably linked. This sense of interconnectedness can be surprisingly close to Heisenberg's uncertainty principle.

Werner Heisenberg, the father of quantum mechanics, formulated the uncertainty principle in 1927. This principle states that at the subatomic level it is not possible to simultaneously know with absolute exactitude the position and velocity of a particle. The mere act of observation alters its state.

For example, imagine an electron in motion. Measuring its position inevitably implies a change in its velocity. This inherent uncertainty is not a technical error. It is a fundamental property of nature. It implies that the future of a particle-and perhaps of the entire universe-can never be absolutely defined.

Scientists have found that every observation leaves a trace. At the quantum level, simply "looking" at a system creates an impact. It is as if every choice made is never harmless but brings with it a transformation in the world.

At first glance, Karma and the uncertainty principle seem to belong to distant domains: one moral and spiritual, the other physical and scientific. Yet both revolve around a key awareness: every action has consequences, visible or invisible, immediate or deferred.

Krishna states that the world is a web of cause and effect, where nothing arises by chance. Every human thought generates a reaction. Similarly, Heisenberg showed that even a microscopic action, such as observing an electron, creates consequences that no one can ignore. In the physical as in the spiritual world, uncertainty about the future depends not on unpredictability, but on the complexity of the relationships that bind each element of the system.

If Krishna's vision invites moral responsibility, Heisenberg's indeterminacy urges us to intellectual responsibility. Both remind us that free will is not a game without rules. Every choice has an impact on others.

Contemporary culture is faced with the task of reconciling these two seemingly distant worlds. On the one hand, the spiritual teaching of Karma invites us to think before we act. On the other, science teaches us that even the smallest gestures, such as observing a particle, can alter the universe in ways that escape our understanding.

Sri Krishna and Werner Heisenberg, figures distant in space and time, share a message that still resonates today: actions shape the future. Krishna teaches this to Arjuna on the banks of the Yamuna River, inviting him to profound ethics. Heisenberg demonstrates it in the darkness of a laboratory, revealing the mysteries of the atom. Both, in their own ways, point us to a truth: nothing happens without leaving a trace, visible or invisible, in the infinite web of the cosmos.

Predictability and choice in the field of uncertainty.

In Hindu thought, Karma is not just a moral rule, but a universal law. Every action-physical, verbal or mental-produces a consequence, like the stone that causes circles in the water. Hinduism holds that every human act generates "*vasana*" (hidden desires) in consciousness. This influences the future, both in this life and the next. There is no divine judgment in Karma, but a system of balance: you receive what you sow.

Karmic philosophy invites us to understand that we can never control everything. Rather, we can act righteously to generate positive effects on paths that escape our full vision.

Moving from the spiritual universe to the physical world, quantum physics offers us a surprisingly kindred perspective. The uncertainty principle states that we cannot accurately measure, simultaneously, the position and velocity of a subatomic particle. The very act of observing the particle, with an instrument, alters its trajectory.

This phenomenon implies a fundamental reality: at the microscopic level, the future is inherently unpredictable. Yet every observation made by a scientist determines, directly or indirectly, a certain outcome. The quantum universe is not fixed or determined, but a "field of possibilities," just like human life in the concept of Karma.

Many scholars have noted how much similarity there is between Karma and indeterminacy. Both teach us a valuable lesson: Every action-or observation-activates forces that shape the future. In the Karmic view, attention and awareness in daily choices are crucial to one's destiny. In parallel, in quantum physics, observation "collapses" the infinite possibilities of a particle into a single concrete state.

Let us imagine for a moment a pendulum in the real world. In an ordinary situation, we can predict its motion with certainty, knowing force and friction. In the subatomic world, however, that certainty disappears: observing the pendulum would introduce a new variable into the system. Karma works similarly, but on an existential level. An action generates an impulse in the "field" of reality, often with unforeseen effects.

An interesting parallel between Karma and the uncertainty principle also emerges in our use of technologies. Digital platforms, built on algorithms, demonstrate how every online action brings unpredictable results. Clicking on content, for example, contributes to the way data is processed, influencing what is suggested in the future. This phenomenon is reminiscent of Edward Lorenz's butterfly effect: a tiny change in initial conditions-such as the beating of a butterfly's wings-can unleash a storm on the other side of the world.

In the karmic-quantum view, these connections teach us to be more conscious. Every click, like every choice in life, has profound repercussions that we cannot immediately see, but which will shape the future fabric of our society.

Finally, Karma and the uncertainty principle converge in their essence. Both remind us that we live in a complex reality in which our choices and observations have a profound bearing beyond the immediate.

The principle of causality.

In the epic poem of the Mahabharata, the idea of Karma runs through the lives of its protagonists like an invisible thread. One of the most famous episodes concerns King Dhritarashtra and his actions as ruler. Dhritarashtra, blind from birth, literally and figuratively represents the inability to "see" the consequences of his decisions. He does not take a stand to stop his sons, the Kauravas, when they commit injustice against their cousins, the Pandavas. His inaction-which is itself a form of action, as the concept of Karma suggests-leads to the disastrous war of Kurukshetra, an epic battle that not only devastates his kingdom but generates protracted suffering for generations.

Now, let us pause for a moment: how can a decision, or a non-decision, generate such extensive effects, both in time and space? Modern science, through quantum mechanics, offers a fascinating analogy.

German physicist Werner Heisenberg formulated his uncertainty principle in 1927. This principle states that we cannot know precisely both the position and velocity of a subatomic particle at the same time. The universe, at the microscopic level, therefore operates under the sign of the probable, not the certain. This uncertainty does not mean absolute chaos, but suggests an inherent complexity: any interaction at the quantum level can affect other particles, generating results that are not immediately predictable.

Now let us think about Karma. Just as a particle, in its movement, generates unpredictable effects within the quantum field, a human action, according to Hinduism, can also trigger effects that unfold in unimaginable ways. Heisenberg's uncertainty principle becomes, then, a scientific window into the traditional concept of subtle causality: nothing is random, but nothing is entirely determinable.

The Mahabharata describes the effects of actions as strings of a large moving net. Each action vibrates along this net, affecting seemingly distant points. In quantum physics, a parallel concept is quantum entanglement. Two particles can remain entangled even light-years apart: a change in the state of one particle instantaneously affects the other. It is a phenomenon that defies intuition and reminds us how deep the hidden connection is in the world we inhabit.

Take, for example, the character of Karna, often considered the symbol of man struggling with his destiny. Karna, born to Kunti, mother of the Pandavas, but abandoned at birth, spends his life trying to justify his existence. He becomes a great warrior, but remains eternally torn between loyalty and justice. The weight of past decisions-both his own and his mother's-is palpable. Karna is both victim and maker of the complex karmic web that drives the Mahabharata. As in the case of entanglement, his connections with other characters influence the entire narrative, culminating in his tragic downfall.

Another interesting analogy concerns time, both in Karma and in quantum mechanics. In the Mahabharata, the consequences of actions often manifest long after the time they were performed. For example, when Krishna, the wise counselor of the Pandavas, warns Arjuna of the need to fight for truth, he emphasizes that every right or wrong action will come back like a wave. This reflects the idea that, in the quantum world, events do not rigidly obey the linearity of time. A phenomenon at the subatomic level can affect present, past or future events, creating a complex form of causality.

Why does the concept of Karma succeed in such a fascinating dialogue with quantum physics? Both hint at universal interconnectedness. In the physical world as in the spiritual, every action is part of a great web of events. In the Vedic poem, the cosmos is often compared to a fabric of Indra, an endless web of nodes where each jewel reflects all the others. This ancient symbol is a powerful reminder: what we do does not end with us. Just as the movement of one particle can influence the behavior of other particles, every decision we make resonates with the universe.

The Mahabharata and modern science teach us, then, that indeterminacy is not an excuse for inertia. It is an invitation to conscious action. As narrated in the dialogues between Krishna and Arjuna, what matters is not just the action, but the quality of the intention that guides it. Similarly, quantum physics teaches us that, even with uncertainty, every action helps shape the future in ways we cannot always predict.

In conclusion, Karma and the uncertainty principle whisper the same great truth to us: we are responsible for what we put into the web of the universe. Our actions are seeds, sprouting in mysterious but interconnected ways. Looking at these ancient cultural connections through the lens of science does not diminish their power, but rather amplifies them. The web of the cosmos embraces us all, reminding us that, in the end, we are united by the thin but indestructible thread of our actions.

Karma's freedom in an indeterminate universe.

Karma, in the Hindu context, is based on a clear idea: human beings enjoy inherent freedom. They can choose how they act. However, these actions are not without consequences. For example, the Indian philosopher and mystic Swami Vivekananda described Karma as "*the ropes that bind us*," but emphasized that we ourselves weave those ropes through our choices. This freedom is reminiscent of what happens in the quantum world, where a subatomic particle can "choose" a state from multiple possibilities.

Unlike classical physics, which sees the world as a rigidly deterministic machine, quantum mechanics reveals a universe where potential, not certainty, reigns. Even the smallest particles in the universe seem to be endowed with a kind of freedom, since their behaviors cannot be predicted with absolute precision. This uncertainty functions as a bridge to the concept of Karma, which enshrines individual responsibility in determining one's own future. Both systems highlight that there is a dimension in which outcomes are not predetermined, but emerge from a combination of choices and possibilities.

Vedic texts offer an interesting story about causality. The sages of ancient India tell of a man, Ajamil, who, driven by his past actions, leads a life of immorality. However, on his deathbed, Ajamil invokes "Narayana," the name of the god Vishnu, and achieves redemption. This tale illustrates how a single decision can alter the course of fate.

In one respect, it is comparable to the "Schrödinger's cat" experiment, introduced by physicist Erwin Schrödinger in 1935. The famous thought experiment shows how, at the quantum level, a system can exist in several possible states simultaneously, until observation "decides" one of the two states.

In Ajamil's account, just as in quantum physics, a single event-a choice, an observation-is capable of profoundly changing the outcome. Moreover, the idea of "*observer*" in quantum mechanics seems to reflect part of the individual's role in the law of Karma: becoming aware of one's actions and their repercussions.

If we reflect carefully, Karma introduces a moral model based on freedom and responsibility. Heisenberg's indeterminacy, on the other hand, describes a physical universe that is not immutable but "open" to unpredictable dynamics. Both focus on the idea that not everything is fixed in rigid, predetermined patterns. The universe, in both the spiritual and physical realms, is full of possibilities.

The Mahabharata, a great Indian epic, contains an emblematic passage that reads:

"Destiny is not already written, but is the sum of our actions."

This phrase captures the duality of Karma. On the one hand, the individual enjoys the freedom to choose; on the other, those choices build a fabric of relationships that influence the future. This is the same balance we observe in the quantum world.

Niels Bohr, Heisenberg's contemporary and one of the founders of quantum mechanics, argued:

"Predictability is only an illusion of certainty."

This phrase seems to converse directly with Eastern philosophies, which have taught for millennia that freedom lies not in eliminating life's uncertainties, but in recognizing and accepting them, consciously choosing in the present.

The union of Karma and quantum indeterminacy offers a fascinating perspective. On the one hand we find an ancient philosophy exploring moral causality; on the other, a modern science studying the probabilistic basis of nature. Both, however, reject the idea of a completely deterministic universe.

The two disciplines seem to meet at a critical point. Human beings, like subatomic particles, are not condemned to follow a rigid path. The possibility of choice, of action, remains a constant in reality, whether physical or ethical. This link demonstrates that scientific discoveries and spiritual traditions, rather than contradicting each other, can dialogue to offer us a deeper understanding of the mystery of the universe.

In conclusion, Karma reminds us that our every action counts. Similarly, the uncertainty principle emphasizes that, even in the physical world, every choice and every interaction shapes the future. Both concepts celebrate freedom, ultimately making the universe a wonderful abode of possibility.

When physics and philosophy meet.

In the Hindu tradition, Karma is the law of cause and effect. Every action performed by a person leaves an invisible imprint, an energy that will influence future destiny. This force is neither immediate nor linear, but manifests itself over time. A small gesture, a word spoken at a crucial moment, can generate unpredictable consequences that propagate through the lives of individuals and the cosmos.

Patanjali's Yoga Sutra summarizes it this way:

> *"Every action generates a seed, and that seed will produce a fruit, bitter or sweet, in the appropriate season."*

Karma, then, is not a punishment or reward, but an inevitable consequence, rooted in the very fabric of reality.

In the quantum world, an observation made on a particle alters its state. This means that the act of observation is itself an action that alters the future. It is a paradox that reflects an idea surprisingly close to that of Karma: every interaction, no matter how small, has the potential to shape the future in ways that are difficult to predict.

Imagine a subatomic particle that "chooses" a state from among multiple possibilities. Its behavior is governed by a web of probabilities rather than certainties. Similarly, Karma can be seen as this web of possibilities. Human actions do not guarantee a single outcome, but create situations that influence the course of events.

A striking example comes from the world of quantum entanglement. Two particles, even separated by enormous distances, remain connected by a common state. A change on one particle is immediately reflected on the other, regardless of the distance separating them. This phenomenon is reminiscent of the idea of collective Karma : individual actions create waves that propagate through the fabric of the universe, connecting human beings in a web of invisible interdependence.

An interesting Hindu parable tells of a king who, during a hunt, accidentally killed a sacred deer in a forest dedicated to Shiva. Years later, at the point of death, the king remembered that action and realized that the difficulties he encountered in his life could be traced back to that ancient act. This anecdote illustrates how Karma follows a complex logic, often incomprehensible in the immediate moment, but clearly recognizable in hindsight.

Similarly, physicist Niels Bohr, Heisenberg's contemporary, observed that at the quantum level:

"The path we take today we can only interpret tomorrow."

Bohr, ideally, could have recognized a similarity between these seemingly distant worlds.

In India, evidence of this intersection of science and spirituality can be found in places like Varanasi, a temple of wisdom and meditation. Here, ancient teachings on Atman (the self) and Karma cohabit with scholarly conferences on modernity. At *Benares Hindu University*, for example, dialogue between physics and philosophy was encouraged as early as the 1960s.

The bridge between Karma and quantum physics reminds us that there are orders still unknown to human beings. Modern science, like ancient philosophy, explores reality by trying to grasp its invisible threads. Both fields suggest that deep rules, rooted on different scales but equally fundamental, lie behind apparent randomness.

Karma and indeterminacy invite us to reflect on a universal principle: every action has the potential to change the world, both on a cosmic and personal level. As Heisenberg and the sages of Hinduism seem to suggest, we cannot control everything, but we can carefully choose the intention of our actions.

Karma and Schrödinger's cat. Are the choices real?

Imagine that you are faced with a choice. Every action you take today will have consequences, visible and invisible, in the future. This is the central principle of Karma in Hinduism: the law of cause and effect, according to which every action, thought or decision leaves an imprint on the present and what is to come. But what does this have to do with quantum physics? More than you might imagine.

Quantum physics tells us that in the microcosm of subatomic particles nothing is certain until it is observed. This is at the heart of Heisenberg's uncertainty principle: we cannot simultaneously know the position and velocity of a particle with absolute precision. This uncertainty affects not only the microscopic world but also inspires philosophical reflections on the nature of human choices.

According to Hinduism, Karma does not compel: it shapes. The choices we make today determine what we will find tomorrow, but the moment of decision, like the fate of the quantum particle, remains fraught with possibility until it materializes. It could be said that every human action resembles a particle poised between multiple states, ready to define a future when its "observer"-the individual-makes the choice.

To better understand this analogy, let us look at the famous Schrödinger's Cat thought experiment, devised in 1935 by Austrian physicist Erwin Schrödinger. In it, a cat is locked in a box along with a mechanism that can kill it, triggered by the radioactive decay of a particle. Until someone opens the box, the cat is both alive and dead, in a state of quantum superposition.

This experiment invites us to reflect on the simultaneous existence of opposing possibilities until a concrete action collapses them into a single reality. Something similar occurs in Karma. Every action generates a ramification of possibilities. Every thought, every word, every behavior is like the particle in the box: it contains a potential that, when released, turns into a tangible result.

This idea is not new in Eastern philosophy. The ancient Indian poem *Bhagavad Gita* emphasizes the concept of responsibility in choices. In the text, Krishna, the divine guide, explains to Arjuna that every action must be done with awareness, because no one can escape the effects of his or her actions. The modern interpretation of this teaching could be likened to the need to "observe" a choice in order to make it real and bear the burden of consequences.

If we apply this view to Schrödinger's cat experiment, we realize that it is not only the physical observer (the scientist) who decides the cat's fate, but also the level of awareness and responsibility with which one takes action. Similarly, Karma emphasizes that it is the quality of intention in actions that determines which of the infinite possibilities will materialize.

Indian folk wisdom offers a tale that illustrates this dynamic: a peasant, living in a village, finds a bag of gold one day. He could keep it for himself without saying anything, or return it. He hesitates, aware that his decision will affect his future. Like the cat in the box, the choice is not made until he acts. He decides to return the gold, and later discovers that the owner rewards him with half the contents of the bag. This allowed him to live without remorse of conscience. Would the same have happened if he had acted differently? We will never know.

This episode perfectly reflects the intersection of Karma and the quantum superposition paradox. One can imagine that, until the moment of choice, the future remains suspended in multiple states, ready to collapse into a single reality when we act.

This bridge between quantum physics and Indian philosophy is not limited to academics. In 1975, Fritjof Capra, a leading proponent of the dialogue between science and spirituality, related the principles of modern physics to Eastern mystical traditions, including Karma and universal unity. With an open-ended, interdisciplinary approach, he showed that despite centuries of distance, Eastern philosophy and Western science share fundamental insights into the inextricable interconnectedness of all that exists.

Another example comes from the reflections of Carl Jung, who became interested in the concept of synchronicity and the way seemingly random events seem to link through meaning rather than causality. While not directly related to Karma, Jung's idea reminds us that realities and choices do not always manifest through tangible or mechanical causes, but can resonate in a broader field.

The dialogue between quantum physics and the view of Karma shows us a fascinating view of reality: choices are not just simple decisions, but are creative acts. As observers, we are responsible for the collapse of possibilities into one reality. Whether in Schrödinger's cat box or in everyday life, learning to choose with awareness means understanding the power to shape the future. Within this dialogue between East and West, a common awareness emerges: every decision is important precisely because it holds infinity and the future within it.

Karma and the quantum plots of fate.

At the heart of ancient Hindu philosophy, the concept of Karma emerges as an invisible weaver that interweaves human actions with destiny, influencing every moment of the present and future. In a world increasingly dominated by science, this age-old law finds unexpected answers in quantum physics.

In Hindu thought, Karma is comparable to a book written in real time, where each action etches a word, a paragraph, a chapter. Just as a novel cannot be understood by reading a single page, Karma cannot be understood by looking at a single event. It is a collective work, a plot that includes past, present and future.

This symbolic book, in some ways, resembles the behavior of a quantum particle. Consider the double-slit experiment, one of the most famous in the history of physics. When a light wave or particle passes through two slits, it behaves simultaneously as a particle and as a wave, combining two different natures in seemingly contradictory ways. But what is really surprising is that the behavior seems to change when the experiment is observed. The act of "observing" changes the results, just as our choices change the future chapters of the "Book of Karma."

According to Karma, even the most insignificant action has consequences. Similarly, in quantum physics, small causes-often invisible to the human eye-produce immense effects.

We also encounter the idea of entangled systems, known as "quantum entanglement." Two particles, once connected, remain related even at a distance, evoking a philosophical echo of Karma. Even when we are geographically distant, our actions can affect other lives, in mysterious but undeniably powerful ways.

Here we can recall the symbolism of the Ganges, India's sacred river par excellence, which connects the pilgrims in Varanasi with the rest of the country. The river's current, like entanglement, is a metaphor for Karma: it connects souls and destinies, starting from a single source.

Krishna states that no one escapes the cycle of Karma, not even the one who deludes himself into inaction, since even the choice not to act is an action. This is reflected in the quantum paradox: even the apparent "*neutral observer*" changes the course of history. Heisenberg himself sensed this, describing the universe as an eternal process of writing and observation.

The Upanishads investigate Karma in poetic terms. The soul (Atman) is said to be like a "*light in the cave of the heart,*" and our thoughts and actions are the rays that run the world. Today physics describes the universe as a dance of vibrations, of particles bound by uncertainty. Both, physics and philosophy, bring us back to the same conclusion: existence is an intricate web, a book in which every word has a specific weight.

Karma is not meant to be a prison, but an invitation to awareness. Similarly, Heisenberg's indeterminacy is not an insurmountable barrier, but a window showing the complexity of reality. By embracing the mystery, we can find guidance.

Like the ancient rishis, modern physicists wonder who is the author of the "*natural book.*" Schrödinger, influenced by Vedic philosophy, once wrote that;

"The observer and the observed are one."

Even in the laboratory of science or in the quiet of a Himalayan cave, we always return to the same intuition: everything is connected, and nothing exists without mutual influence.

Ultimately, if Karma represents the book of life, Heisenberg's indeterminacy is its pen: we do not know what the next sentence will be, but we are certain that what we write today will influence the pages to come.

From karma to probabilistic waves.

Rooted in the "law of cause and effect," Karma reminds us that every action performed-physical, verbal or mental-plants a seed destined to sprout in the present or future. This universal vision links each individual to the great cosmic web of human and natural relationships. But what happens when we compare this ancient law with the ideas of quantum physics? Remarkably, modern physics, through such principles as Heisenberg's indeterminacy, seems to dance in the footsteps of truths that Indian sages intuited thousands of years ago.

The word "Karma" comes from Sanskrit and literally means "action" or "act." The Upanishads, ancient texts of Hinduism, describe Karma as a cosmic system that records and repays every action, positive or negative. It is not divine punishment, but an impartial principle of balance. Every action-such as choosing to help or ignore someone-creates an energetic trace destined to recur. Like planting a seed in a field, the quality of the harvest will depend on the seed and how it is tended.

A famous parable in Hindu scripture tells of a young prince who, in a fit of rage, shot a bird with an arrow. Years later, during a raid, he himself was struck in the same way in battle. The tale illustrates a basic principle: actions come back, boomerang-like, in ways that sometimes transcend a single lifetime. This karmic dynamic encompasses everything from the movement of the stars to the lives of ordinary people.

The correspondence with quantum mechanics is manifested in Heisenberg's uncertainty principle, formulated in 1927. Werner Heisenberg illustrated how it is not possible to simultaneously know precisely the position and velocity of a subatomic particle. It is a principle that reveals the intricate probabilistic nature of reality. When we observe a particle, we modify its behavior. The effect of observation thus alters its future.

Incredibly, the concept of Karma resonates with this idea. Just as Karma ensures that every action generates a consequence, the Uncertainty Principle tells us that the universe, at the subatomic level, is not deterministic. Particles do not follow fixed trajectories, but probabilities that occur according to specific interactions. As we observe or interact with reality, we modify those probabilities. There is no certainty, only possibility. A karmic seed works the same way: planted today, it will sprout unpredictably, feeding on the surrounding context.

Imagine a still lake. A pebble thrown onto the surface generates concentric waves that propagate. In physics, this phenomenon is often used as a metaphor to describe the wave function: every particle has a probability nature that extends in space and time, like circles on water. Similarly, in Hinduism, every action is a pebble thrown into the lake of existence, and its vibrations spread throughout the universe.

An interesting anecdote comes to us from Mahatma Gandhi's life. During his youth, Gandhi was surprisingly selfish and insecure. Over time, however, he began to internalize the principle of Karma. He began to live by making choices geared toward the collective good, planting seeds of kindness and ahimsa (nonviolence). Those actions not only changed his life but influenced global history, leading millions in the struggle for independence through peaceful means.

Just as in quantum physics, the emitted energy (the karmic seeds) does not disappear; it reappears in other forms. Think also of the double-slit experiment: the particles, by themselves, follow a dual behavior (wave-particle), but observation changes their outcome. This is the essence of Karma: the interaction of our intentions and actions creates different possibilities for the future. The future is uncertain, but it depends on every choice we make.

For centuries, science and spirituality have traveled separate paths. However, ideas such as Karma and quantum probability show that the two worlds overlap in mysterious ways. Mathematician Erwin Schrödinger, one of the fathers of quantum mechanics and a profound connoisseur of Indian philosophies, wrote:

> *"The world is a construction of the mind, and we can never be sure of what is real."*

Schrödinger's reflections, inspired by Eastern traditions, recognize the interconnectedness of all things.

Destiny and causality.

In the ancient philosophical traditions of India, Karma is a principle that spans millennia and generates profound questions about existence and causality. According to Hinduism, every action leaves an imprint, generating effects that manifest in the present or future. This concept, seemingly spiritual and far removed from modern science, finds curious resonances in quantum physics. Among them, the analogy with Heisenberg's uncertainty principle leads to a fascinating dialogue between two seemingly antithetical worlds: East and West.

In essence, Karma holds that every thought, word or deed inevitably generates an effect. The actions performed draw an invisible pattern in the lives of human beings, as a weaver weaves threads that will only be visible with time. This law assumes no moral judgment, but describes an implacable mechanism: what we do produces consequences, both on an individual and universal scale.

Ancient Hindu wisdom sees Karma as that which shapes each individual's destiny. But destiny does not mean rigid predetermination. The human being, in performing conscious actions, is seen as the principal architect of his or her own future. This philosophy opens the door to a surprisingly modern reflection: personal responsibility and the interconnectedness between today's choices and tomorrow's outcomes.

All this finds a clearly parallel echo in one of the most celebrated discoveries of quantum physics: the uncertainty principle.

In other words, in the quantum world, observation itself changes the observed object. For example, the simple act of measuring the position of an electron will inevitably affect its motion. Consequently, subatomic reality presents itself as a fluid and probabilistic realm, where causes do not always lead to certain and determinate effects.

This view challenges the linearity of classical causality, suggesting that there are deeply interconnected aspects of reality, similar to the way Karma describes the influence of our actions on the future.

Quantum physics, like Karma, reminds us that every action-even the smallest-contributes to creating our future. In quantum language, the future of a subatomic system is the result of a probability distributed across multiple possibilities. Similarly, according to Hindu texts, Karma is not an immediate and rigid mechanism, but a process that unfolds over time. Just as in the quantum realm each particle affects the global system, in the realm of Karma each human action affects the universal order.

A tangible example of how uncertainty and karma are intertwined can be drawn precisely from quantum behavior. Imagine a subatomic particle passing through a double slit in a famous experiment: the final outcome depends on how and whether the particle is observed. The mere presence of an observer changes the course of events. In symbolic terms, it is like saying that every action-even just turning attention toward something-alters the fate of the universe.

Similarly, the Buddha, sharing the concept of Karma in a form similar to Hinduism, stated:

"Man is the result of his thoughts."

Even a mental action, therefore, can alter reality. Some modern historians find in this thought a "pre-quantum" suggestion: the idea that intention itself is a form of invisible interaction with the universe.

The connections between Karma and quantum physics remind us that we live in a universe that functions like a huge invisible orchestra. Every note, even the one played in a whisper, contributes to the symphony.

Karma, with its moral and spiritual view of causality, is not far from quantum laws, which teach us how actions-observing, measuring, interacting-alter the course of events. Both of these perspectives invite us to reflect on a fundamental truth: while fate is not completely written, we are at every moment authors of part of our existence. The more we understand this web of interconnections, the more aware we become of the power of our choices, large or small.

On the border between science and spirituality, Karma and quantum physics suggest that nothing is really separate. Actions, thoughts and physical phenomena are intertwined, reminding us that the mystery of life is at least as great as the universe we inhabit.

He who watches shapes the future.

In every action there is a seed that shapes destiny. This ancient view of Hinduism is reflected in the doctrine of Karma, the universal law of cause and effect. According to this age-old philosophy, every action-mental, verbal or physical-generates a reaction, visible or invisible, that affects the present and shapes the future.

Karma is not just a moral principle but a natural law that governs the cosmos, an inscriptible mechanism where the past meets the present to shape future destiny. The extraordinary modernity of this concept emerges even more when one compares it with the discoveries of quantum physics.

The most fascinating relationship between Karma and modern physics is materialized in the Heisenberg Uncertainty Principle, enunciated in 1927 by German physicist Werner Heisenberg. This principle states that the act of observing inevitably changes what is observed. In the subatomic world, particles are in a state of possibility, described by the wave function, until an observer intervenes with his or her measurement. When this happens, the wave function "collapses" and a definite position or energy is manifested. In other words, human observation shapes reality.

This connection appears remarkably akin to Karma. In both contexts, human intervention-be it a physical act or an act of observation-becomes an event that shapes and transforms destiny. The Bhagavad Gita, one of Hinduism's holiest texts, states that every action leaves imprints (samskaras) that generate future events. In a parallel way, quantum physics tells us that the observer actively participates in the creation of reality itself.

Niels Bohr, one of the founding fathers of quantum physics, noted with wonder how observation was a creative act. For many Indian traditions, even at the macroscopic level, every action-every glance, every thought-is a seeding that brings consequences. It is not randomness, but rather a hidden order, a cosmic balance that takes shape only through the conscious intervention of the individual.

If we go back to Heisenberg's or Bohr's laboratory, we can imagine the observer facing the dual nature of an electron, which is both wave and particle. One of the best-known experiments in modern physics, the double-slit experiment, demonstrates this duality. When no one is observing, the electron behaves like a wave and passes through both slits, creating interference. But when an observer steps in to measure, the electron behaves like a particle and passes only through one slit: the act of observation has changed its nature.

In karmic terms, this experiment suggests that reality does not exist as something predetermined, but as a potential field of possibilities. Human beings, through observation and action, are co-creators of the world around them.

An ancient Indian tale illustrates the relationship between conscious action and its effect. It tells of a wise man who planted a banyan tree seed, a tree capable of growing for centuries and spreading deep roots. His seemingly simple action produced an entire forest capable of nourishing and protecting generations of living beings. This is how Karma is described: every action creates a state of future reality, both good and bad.

We can compare this picture with the quantum experiment. The seme-banyan is the potential of the wave function. The forest is the result of observation, of human intervention collapsing the possibilities into a concrete reality.

These connections between Karma and Heisenberg's uncertainty principle are not isolated coincidences. Both of these disciplines, though born millennia apart, share a profound awareness of the importance of man as an active agent in the cosmos. Modern physics, which once sought to reduce the world to an objective entity, has rediscovered the subjectivity of the observer, a concept already central to Indian doctrines.

A cultural bridge stands between Copenhagen-the city where Niels Bohr developed his quantum theory-and the Ganges, where sages of Hinduism meditated for centuries on the interconnectedness of the cosmos. Modern scientific understanding and Vedic philosophy invite us to consider human presence as a phenomenon beyond mechanistic explanation. We are not passive spectators. We are co-creators, bearers of choices, whose effects echo in a constructed universe of possibilities.

Today, as we face global challenges-from climate change to artificial intelligence-this intersection of physics and philosophy invites us to reflect on the importance of awareness. Our daily actions and the observations we choose to make help build the future world. Karma, like Heisenberg's indeterminacy, reminds us of the power we hold in our hands, the infinite potential contained in every choice.

Ancient India, with its philosophy of Karma, and modern physics, with the quantum revolution, leave us with a clear message: everything we do-or choose not to do-leaves an imprint on the fabric of time.

Vibrations, decisions, and waves of possibility.

Werner Heisenberg proposed the uncertainty principle, according to which it is not possible to simultaneously know with absolute precision the position and velocity of a subatomic particle.

In a sense, Karma follows similar rules. Human actions not only produce direct effects, but vibrate in ways that cannot always be predicted. This karmic vibration can be compared to the quantum wave function, which represents all the possibilities of a particle until it is "observed." As the American writer Ralph Waldo Emerson said:

> *"Man sows an act and reaps a habit; he sows a habit and reaps a fate."*

In many Hindu schools of thought, such as the Samkhya and Vedanta, it is believed that every action carries an imprint in the cosmos, a vibration that endures over time. This concept is echoed in the teachings of Paramahansa Yogananda, author of "*Autobiography of a Yogi.*" He writes:

> *"Every action, good or bad, creates a vibratory energy that returns to its source. The human spirit is like a stone thrown into a pond: it creates waves that inevitably reflect back to us."*

Similarly, in quantum physics, a particle does not exist in a single fixed state until it is observed. Instead, it is in a superposition state, a combination of all its possibilities. The moment a choice is made--whether in human actions or in subatomic interactions--the wave function collapses, resulting in a specific outcome.

An Indian anecdote illustrates the idea of Karma. Bharata, an ancient king described in Vedic texts, saved a small, dying fawn one day. This compassionate gesture emotionally bonded Bharata to the creature. In his last moments of life, the king thought of the fawn with such intensity that in the next cycle of reincarnation he found himself living as a deer. This tale highlights that in Karma, as in the quantum world, repercussions do not always follow a straight or predictable line.

Classical physics has accustomed us to reasoning with direct causality: one ball hits another and the second one moves. But at the subatomic level-and, for Hinduism, also karmic-the interactions are complex, interconnected, even surreal. The effect of actions travels like a wave, affecting the cosmos and returning to the source.

Modern science is beginning to recognize the vibratory potential of everything in the universe. Nikola Tesla stated:

> *"If you want to understand the secrets of the universe, think in terms of energy, frequency and vibration."*

Ancient Indian wisdom also views the world as a set of vibrations. "Om," the sacred sound of the universe, is perceived not only as a religious symbol but also as a metaphor for the fundamental vibrations of reality.

Karma could be imagined as a system similar to quantum waves: each decision creates an interference, a propagation in the cosmic ocean of possibilities. Each choice leaves a trace, which persists and interacts with the waves generated by other living beings.

The notion of Karma is not simply a moral justification. It is a tool for understanding the responsibility of our actions and their vibrations. Hinduism teaches that each soul, or Atman, follows a unique path, guided by the consequences of its actions, but also by the possibilities offered by free will.

Modern physics, with indeterminacy governing subatomic observations, reminds us that reality is not rigid. Instead, it suggests a universe of fluid possibilities, materializing through interaction and choice. Every human act, then, could be compared to that single event that interrupts quantum indeterminacy, transforming waves of possibility into concrete effects.

In the "karmic particle" the mysteries of science and spirituality are intertwined. Every action creates vibrations and, at the same time, shapes possibilities. As in quantum mechanics, life does not follow rigid patterns, but vibrations and decisions. Karma is not just a religious principle: it is a cosmic law that invites reflection on individual choices and their impact, today and in the time to come.

So goes an ancient saying from the Upanishads:

"You are that which is your deepest desire. As is your desire, so is your will. As is your will, so are your actions. As is your actions, so is your destiny."

Chapter X. Moksha. Liberation from Samsara.

Moksha: Liberation from the cycle of birth and death.

In the context of Hinduism, *Moksha* represents one of the most profound and significant concepts in the Indian spiritual tradition. It is commonly translated as "liberation" or "emancipation" and refers to the ultimate goal of human life: liberation from the cycle of Samsara, or the continuous wheel of birth, death and rebirth (reincarnation).

Moksha is the liberation of the individual soul (*Atman*) from the material bonds and limitations of the physical world. This state allows the Atman to unite with ultimate reality, which may vary according to the religious philosophical school followed:

In some schools of Hinduism, such as Advaita Vedānta, Moksha is total union with Brahman (the ultimate and absolute reality), recognizing that Atman and Brahman are inherently the same thing.

In dualist schools, such as the Viśiṣṭādvaita or the Dvaita Vedānta, Moksha implies remaining separate from Brahman but living eternally in a condition of divine bliss, in the presence of God.

In any context, Moksha represents the cessation of suffering, the attainment of eternal peace and the realization of the true nature of being.

How to reach Moksha.

The path to Moksha varies according to traditions and practices, but in general it is described through four main paths (*marga*):

Jnana Marga (*Way of Knowledge*). Through self-knowledge and scriptural knowledge, one comes to the realization that Atman is Brahman.

Bhakti Marga (*Way of Devotion*). Complete devotion and love toward a personal deity, such as Krishna, Shiva or Vishnu, helps to free oneself from karmic bonds.

Karma Marga (*Way of Action*). Performing selfless and virtuous actions, without attachment to the fruits of one's deeds, helps purify the self and brings one closer to Moksha.

Raja Marga (*Way of meditation and self-discipline*). Through contemplative practices and yoga, one controls the mind and focuses on self-realization.

Moksha and the Four Goals of Life.

In Hindu philosophy, Moksha is generally considered the highest of the four goals of life (purushartha), which are:
Dharma: The fulfillment of moral and social duties.
Artha: The attainment of material prosperity.
Kama: The pursuit of pleasure and personal fulfillment.
Moksha: Spiritual liberation.
Moksha is the ultimate goal that transcends others, symbolizing the completion of the human spiritual journey.

Moksha, then, is the culmination of the spiritual path in Hinduism: ultimate liberation from the suffering of the material world and union with the divine. This represents a goal of inner revelation and deep peace that many Indian spiritual traditions and practices seek to achieve.

Moksha and quantum decoherence.

In the sacred texts of Hinduism, Moksha represents supreme liberation, a state in which the soul (or Atman) escapes the endless cycle of birth, death and rebirth, known as "*Samsara*." This concept, which spans millennia of philosophical and spiritual thought, can find a curious and fascinating correspondence in a principle of modern physics: quantum decoherence. Both seem to address the same central theme, albeit from completely different perspectives: the transition to a state of freedom through overcoming the constraints of a perceived reality.

In Hinduism, Samsara is often depicted as a wheel that spins ceaselessly. Each life, according to this interpretation, is like a cycle within another cycle. The doctrine of Karma decides fate, linking the actions of one life to the experiences of the next. But Moksha breaks this sequence. To achieve it, an expansion of inner awareness is needed. Practices essential for this purpose include deep meditation, yoga and knowledge of sacred texts such as the Upanishads.

The Upanishads, composed around the 6th century B.C., describe Moksha as the union of Atman with Brahman, the ultimate reality. An emblematic passage in the Chandogya Upanishad reads:

"Tat Tvam Asi" ("That's You").

an exhortation to understand that individual identity is illusory and part of the same cosmic substance. At this point, the ego dissolves, as do the illusions of "self" that bind it.

In quantum physics, the term decoherence describes the moment when a quantum system loses its coherence, i.e., its superposition state, by entering a definite, observable reality. When a quantum entity (such as an electron or photon) interacts with its surroundings, its multiple possible states collapse into one, resulting in a change of state that can be observed.

From a philosophical perspective, this phenomenon is reminiscent of the idea of Samsara: the continuous "collapse" of individual consciousness into limited realities, determined by interactions with the external world and the dynamics of matter. Moksha, from this point of view, could be understood as the overcoming of all external interference, a liberating state in which consciousness is freed from the limitations of matter and the fragmentation of reality.

Meditation: a state of transition.

Meditation, a central practice for achieving Moksha, offers a striking analogy to decoherence. Practitioners of deep meditation often describe a state in which the boundaries between self and external reality dissolve. One of the best-known masters, Adi Shankara (8th century CE), founder of the Advaita Vedanta doctrine, explained that meditation enables direct perception of the unity of Brahman, stripping away sensory illusion. It is a journey beyond the multiplicity of forms, toward undivided awareness.

Treating it as a quantum transition process, one could say that the mind enters a state of "superposition" in which the self experiences a larger reality. The abandonment of external influences allows a transition to unity, an echo of the process in which decoherence is overcome, making way for pure quantum essence before "collapse."

One of the most fascinating parallels is how both concepts, Moksha and decoherence, suggest liberation from a cycle. For a physicist, this cycle is the continuous interaction of particles and observers. For a Hindu philosopher, it is the eternal return of material life. However, both agree on one crucial point: freedom comes by eliminating interference and anchoring in a deeper reality.

When the sages of ancient Indian civilizations wrote about the secrets of Samsara and liberation from the karmic cycle, they could not have known that, millennia later, science would come up with seemingly similar concepts. Yet, the similarities between Moksha and quantum decoherence are a reminder that there are different paths-philosophical or scientific-to investigate the mystery of consciousness and its relationship to the cosmos.

We may never know if spiritual liberation and quantum transition share a common substance, but the fascination of uniting two seemingly so distant worlds through a common language remains: that of the search for truth.

The way to Moksha. Inner ascent and quantum transitions.

Hinduism, with its complex weaving of myths, philosophies and spiritual practices, has always sought to answer the fundamental question of the essence of life and reality. The concept of Moksha-the liberation from Samsara, the endless cycle of birth, death and rebirth-represents the culmination of this quest. In its sacred texts, such as the Bhagavad Gita, Hinduism points the way to break the chains of illusion (*Maya*) and attain the absolute, a theme that, interestingly, finds fascinating parallels in the recent discoveries of quantum physics.

In Hindu thought, Samsara is described as a perpetual round of existences, governed by the law of Karma. Souls-or rather, consciousnesses-repeatedly incarnate, dancing through different forms of life, like actors wearing temporary roles on the cosmic stage.

But beneath this dance, Hinduism tells us that the true Self (Atman) is eternal and unchanging, a spark of the absolute (Brahman). Achieving Moksha means awakening to this truth, moving from the multiplicity of forms to the unity of reality.

This transition can be seen as a transition, a change not only spiritual, but also conceptually akin to phenomena studied today in quantum physics.

Quantum physics tells us of an unfathomable world, of particles that exist in multiple states until an "observer" forces them to collapse into a single reality. This phenomenon, called decoherence, signals a transition: what was previously uncertain and nebulous becomes concrete and determined.

In the context of Hinduism, we can imagine Moksha as a similar passage. The state of the soul trapped in Samsara -- a continuous interplay of possibilities and illusions -- is reminiscent of this indistinct quantum world, in which one is bound to infinite possibilities of existence without a fixed point. In the same way that a quantum particle achieves a single defined state through interaction with its environment, the soul, through meditation and knowledge, frees itself from the illusions of the material world and recognizes its true nature.

An emblematic example of this spiritual journey is offered in the Bhagavad Gita, a pivotal text in Hinduism. In the dialogue between Krishna and Arjuna, Krishna reveals the nature of reality and liberation:

> *"As water does not touch the lotus flower, so sin does not touch those who act without attachment, offering every action to the Divine."*

Here, the "lotus flower" evokes the purity of the liberated soul floating above the murky waters of Samsara.

With the development of inner awareness, Arjuna understands that the path to liberation involves transcending opposites-pleasure and pain, success and failure-to immerse oneself in the knowledge of Atman. This awakening recalls, in modern terms, the transition between undefined potentials (Samsara) and the ultimate recognition of the absolute (Moksha).

Quantum consciousness: a new frontier.

Today, the idea that consciousness plays a fundamental role in the universe is no longer simply the domain of philosophy or religion. Scholars such as Roger Penrose and Stuart Hameroff propose that the human mind, particularly consciousness, may have a quantum basis. According to their theories, phenomena quantum intrinsic to microtubules in brain cells could explain deeper aspects of human experience, such as creativity and self-perception.

This fascinating perspective merges with the Hindu idea that deep meditation and knowing lead to a transition of consciousness. In other words, just as in quantum collapse, the moment when the Self perceives absolute truth is a turning point, a phenomenal leap from the multiplicity of illusions to the clarity of ultimate reality.

The connection between Moksha and quantum decoherence should not be seen as a forced overlap. On the contrary, many scholars see in this parallel a bridge connecting ancient spiritual traditions and modern scientific theories. Indeed, both perspectives remind us of an essential truth: reality is much more complex, and at the same time simpler, than it appears.

In the Varanasi Temple, sacred to Shiva, verses evoking liberation are recited daily:

> *"The world is an illusion, but he who recognizes the divine in himself is already free."*

Thousands of miles away, in theoretical physics laboratories, scientists observe particles and waves, trying to figure out whether reality is made of solid matter or probabilistic puzzles.

And perhaps, in this convergence of science and spirituality, we find a global perspective. The path to self-understanding-whether traced in the verses of the Bhagavad Gita or between the lines of an equation-is the path that leads us to see beyond the veil of illusions. Like the scientist peering into the nature of the atom, and like the mystic looking within himself, both seek the same thing: the truth of the whole.

Transition states in consciousness.

Hinduism describes Moksha as the ultimate liberation. It is the end of the perpetual cycle of birth and death, known as *Samsara*, and represents the attainment of a state of consciousness beyond time, space and matter. These concepts, which have their roots in the Vedas and Upanishads, seem to vibrate in tune with some modern theories of quantum physics. Among them, the phenomenon of quantum decoherence offers an intriguing perspective for exploring the leap from philosophy to science.

In quantum physics, the concept of quantum decoherence describes the process by which a quantum system loses its superposition state. In practice, this means that a particle, which exists in a multitude of possible states (the famous "*Schrödinger's cat*," both alive and dead), is reduced to a single observable reality. This seemingly mysterious transition separates the world of possibilities from the concrete world we experience every day.

Moksha, under this lens, can be seen as a form of "*conscious decoherence*." When an individual realizes the authentic self, or Atman, by freeing himself from the illusion of the phenomenal world (*Maya*), he accesses the ultimate reality: the Brahman, the universal unity. This is an ultimate transition, a "state choice" that grants access to a deeper level of existence. Sri Ramana Maharshi, the enlightened sage of the 20th century, described Moksha as:

"...a merging of individual consciousness into the infinite."

Contemporary theories of the multiverse further broaden the horizon. These hypotheses suggest that parallel realities exist, each delineated by the possible different outcomes of quantum decisions.

Roger Penrose and Stuart Hameroff, with their model of "*orchestrated consciousness*" (Orch-Or) , hypothesize that microtubules within brain cells might function as quantum systems. Their oscillations, subject to decoherence, could generate states of consciousness. In this perspective, Moksha could be seen not only as a release, but as an entry into a mental multiverse.

The Upanishads, ancient wisdom texts of Hinduism, held that the inner self is a whole universe. The Mandukya Upanishad explores four states of consciousness: waking, dreaming, deep sleep and a state of absolute serenity called "*Turiya*," which transcends and unifies all others. Turiya has been described as the access point to Brahman. A parallel could be seen in the quantum transition of a system moving from the fragmentation of superpositions to a coherent singularity.

While the landscapes of physics and philosophy may appear distant, their union opens up extraordinary spaces for contemplation. Anandamayi Ma, a prominent 20th-century Indian spiritual figure, emphasized that the normal mind lives in fragmentation. Only through meditation and knowledge, he said, can one "*see the whole truth*." The same potential for total coherence is what Penrose and Hameroff glimpsed in microtubules: not merely a physical mechanism, but a vehicle for understanding the universe.

Neuroscience now tries to explain phenomena as abstract as consciousness, but the big questions remain. Why do we exist as observers? How can we perceive reality? Deeply human questions that perhaps find their answers in both perspectives: the Vedic and the scientific.

Meditation as a quantum key to liberation.

Moksha is not just an end, but a transcendence of duality, a liberation from the limitations of material and mental existence. To achieve it, meditation turns out to be a central tool, a bridge between the human condition and the realm of that which is eternal. But what actually happens in the mind during meditation? And how is this process intertwined with modern insights from quantum physics?

In Hindu thought, Samsara is seen as a "*cage of the mind.*" The individual, caught in the illusion of time and space, experiences a reality conditioned by opposites: pleasure and pain, success and failure, birth and death. This is the dualistic mind, which separates instead of uniting. Meditative practice is designed to undo this duality and open access to a unified consciousness, Brahman, the indivisible absolute.

According to sacred Hindu texts such as the Upanishads, meditation allows one to shift attention from the ego to deeper awareness. One of the verses of the Mundaka Upanishad states:

> *"The self is not known by the mind, but when the mind calms down, the self shines forth in all its splendor."*

Here the fundamental paradox is outlined: the mind cannot comprehend the self, because it itself is part of the illusion.

Remarkably, quantum physics offers some analogous pictures of this process. Take for example the phenomenon of quantum decoherence. In quantum systems, particles exist in superposition states, where each possibility coexists simultaneously. But when the system interacts with the environment, this superposition "collapses" into a single observable state. Decoherence is that moment when a definite reality emerges, the result of the interaction between the system and the environment.

Physicists have observed that in the absence of this interaction, a quantum system remains in a pure, unfragmented state. It is here that an intriguing metaphor emerges. Meditation can be seen as a process that isolates the mind from the "noise" of the environment, dissolving the dualistic interactions that keep the individual bound to Samsara. By freeing the mind from its fluctuations (which Indian sages call "*vrittis*"), meditation allows consciousness to emerge in a state of unity, much like an isolated quantum system.

The Rishi, the sages of antiquity, were true "scientists of the mind." Through the practice of yoga dhyana (deep meditation), they reported a state of consciousness that transcends the limitations of matter. Patanjali, author of the Yoga Sutras (2nd century B.C.), described meditation as the method of "*ceasing the fluctuations of the mind.*" Upon reaching this inner silence, the practitioner experiences samadhi, the absorption that precedes Moksha.

A classic example is the life of Ramana Maharshi (1879-1950), a modern Indian mystic. Ramana, through meditation and self-inquiry, achieved an experience of unity that he described as "total transcendence of mind." He said:

> *"What we call the destruction of the mind is, in effect, the realization that there has never been a separate mind."*

From a scientific perspective, neuroscience is beginning to detect the potential of meditation to alter brain functioning in tangible ways. Studies of Tibetan monks in deep states of meditation have shown a significant reduction in *beta rhythms* (associated with mental fragmentation) and an increase in gamma activity, considered indicative of a highly coherent mind.

We can consider Moksha not so much as a "place" to be reached, but as a state transformation. It is the liberation from conditioned patterns, a kind of "quantum leap" in consciousness that allows the individual to step out of relative reality and perceive absolute reality. Meditation, as the Hindu schools teach, is the catalyst for this leap.

The image of quantum leaping recalls the behavior of an electron that, in order to move from one orbit to another, suddenly leaps without crossing intermediate distances. Similarly, the meditating practitioner might experience a radical change in perception of reality, a shift from the fragmented view of the dualistic mind to the clarity of a universal consciousness.

In this context, it is worth mentioning the concept of *Indra's Net*, a central image in Vedic texts. Indra's net is an infinite web of gems, each reflecting all the others. This symbol evokes universal interconnectedness, a theme that resonates in both Hinduism and quantum physics. Meditation allows the practitioner to "see" this web, to grasp the unity underlying the apparent fragmentation.

Meditation, in its deepest meaning, can be regarded as an act of decoding, a process of tuning in to a larger reality. It is not just a method of relaxation or a technique for reducing stress. It is, according to Hinduism, a path to immortality, to liberation from Samsara.

Shiva and quantum entanglement.

The universe is born and dies in a breath. In Hindu thought, Shiva is the principle that destroys and transforms, the cosmic dancer who frees matter from its illusions. But how can Shiva, the symbol of the dissolution of illusion, be connected to one of the most mysterious concepts in modern physics, quantum entanglement? To understand this surprising union, one must cross a cultural and scientific bridge: the one that connects Moksha, liberation from Samsara, and the invisible world of subatomic particles.

At the heart of Hinduism, Moksha represents the ultimate spiritual goal: liberation from Samsara, the incessant cycle of birth, death and rebirth. This state is achieved through knowledge, meditation and awareness of the true nature of existence. Samsara, perpetuated by sensory illusions and material ties, is comparable to a complex web of constraints that traps the soul in linear, material time.

Shiva, in his form as Mahadeva, the great destroyer, embodies this liberation. He breaks the bonds of ignorance, destroying maya, illusion. With his dance, called Tandava, Shiva symbolizes the dissolution of earthly bonds and the expansion of consciousness toward the infinite. This spiritual evolution seems to be echoed in a phenomenon of quantum physics: quantum decoherence.

In the quantum world, particles can be simultaneously in different states until an interaction with the environment "breaks" this superposition, causing what is known as decoherence. The particle, once "observed," collapses into a defined state. This process marks a transition from the uncertain to the apparent. However, decoherence also indicates the loss of an invisible link to the previous state.

Can we read Moksha in similar terms? Liberation from Samsara can be seen as an act of spiritual decoherence, where the soul, after living in multiple "states" of existence, abandons the cycle of material illusions to enter a higher dimension. In both cases-spiritual and scientific-there is the breaking of a link with a previous state. The figure of Shiva, destroyer and renewal, then becomes more than a deity: he becomes a cosmic metaphor, crossing cultural and scientific boundaries.

One of the most fascinating phenomena in quantum physics is entanglement. Particles can be "bound" in such a way that, even at enormous distances, a change in the state of one immediately affects the state of the other. This binding transcends the limits of time and space. Entanglement seems to evoke the spiritual idea of Samsara. In the accumulation of past lives and karmic relationships, everything appears connected in a subtle and inextricable web.

This is where Shiva comes in. In many Vedic and Puranic texts, Shiva is described as the one who breaks karmic bonds, destroying any connection that holds the soul in Samsara. In the spiritual world, this "breaking" leads to freedom; in the quantum world, dissolving the entanglement between particles means that each returns to being a free and independent entity. This call between philosophy and science is not accidental. Just as the soul seeks its liberation by breaking karmic bonds, particles dissolve their bonds to enter new states.

Samsara and the physics of consciousness.

How can we interpret a traditionally metaphysical concept through the lens of modern physics? This is where the correspondence between Moksha and quantum decoherence theory comes in, a striking bridge between ancient philosophies and contemporary science.

The ancient Upanishads describe earthly existence as an ocean of continuous agitation, an unstoppable flow generated by illusion (Maya). The Atman (the soul or deep essence) remains trapped, ignoring its unity with Brahman, the universal principle. This condition is likened to a confused state, where multiple possibilities coexist without full recognition of absolute truth.

An interesting parallel is found in quantum physics. When a quantum system exists in a superposition of states, such as an electron being at two points simultaneously, we can imagine a metaphor for Samsara. This state of uncertainty, of undefined alternatives, reflects the illusion in which human beings remain imprisoned, unable to see clearly.

Liberation (Moksha) occurs when the individual overcomes the veil of Maya and recognizes the unity between Atman and Brahman. It is a moment of awakening, described in sacred texts as "*coming out of darkness into light.*"

In quantum physics, a fascinating parallelism is found in decoherence. This phenomenon occurs when a quantum system moves from the superposition state to a definite state. This is the moment when uncertainty ceases, and the system adopts a clarity determined by interaction with the environment. Just like the attainment of Moksha, decoherence also describes a transition to stability and understanding.

Moksha and the observer in quantum physics.

In Hinduism, the observer is the one who, through introspection and meditation, seeks to transcend the illusion of the sensible world to achieve Moksha: liberation from Samsara, the endless cycle of birth and death. In quantum physics, the observer is that mysterious agent who, through the act of observation, determines the collapse of the wave function, transforming potential into reality. This dialogue between two seemingly irreconcilable worlds opens up surprisingly kindred spaces for reflection.

In the Hindu tradition, the concept of Moksha represents the culmination of the spiritual journey. According to sacred texts such as the Upanishads, the ultimate essence of existence resides not in the material world, but in a pure and indivisible state of consciousness. In this state, the individual is freed from the illusion of maya (the veil of appearance) and realizes profound unity with Brahman, the supreme reality . This liberation is not just an "escape" from the pain of the world, but rather an awakening to the true nature of self.

An emblematic passage from the Brihadaranyaka Upanishad reads:

> *"There is neither day nor night, neither being nor non-being,*
> *but only the pure Self, motionless and devoid of duality."*

This state, imagined as a dazzling light that dissolves all shadows, defies the limits of the rational mind.

Liberation in Moksha is thus an act of transcendence. It is the overcoming of the inherent duality of life: light and shadow, subject and object, observer and observed. It is precisely here that a fascinating analogy with quantum physics emerges.

Quantum mechanics introduced a concept in the twentieth century that disrupted traditional science: in subatomic reality, there is no clear separation between the observer and what is observed. Subatomic particles, such as electrons, have no definite position until they are observed. They are in a state of superposition, a mixture of multiple possibilities, represented mathematically by the wave function. However, the moment an observer intervenes, this function "collapses" and only one possibility materializes.

One of the best known experiments on this subject is the famous *double sl*it *experiment*, described as early as 1801 by Thomas Young.

In later years, with the development of quantum physics, it was realized that a particle of light (photon) or matter (electron) behaves like a wave only when it is not observed. The moment a measurement system comes into play, it behaves like a particle. The very act of observation changes physical reality.

Here the parallel with Moksha emerges. In Hinduism, the state of pure consciousness achieved through liberation is similar to the moment when the observer merges with the observed object, transcending duality. Is this not the collapse of all the shadows of individual existence into a single light of infinite awareness?

In this context, the transition of the quantum state from a superposition of multiple possibilities to collapse into a single reality appears as an extraordinary metaphor for the state of Moksha. Liberation from illusion (Maya) is, in a sense, quantum decoherence. Just as a quantum system left to its own devices loses coherence with the world of phenomena, the individual, through meditation, gradually abandons attachment to the world of illusions to collapse into the transcendental unity of Brahman.

Thus, shadow and light--always spiritual metaphors--are no longer forces in opposition, but manifestations of the same ultimate reality. This could be compared to Bohr's principle of complementarity: just as particle and wave are two inseparable aspects of the same entity, so too on the path to Moksha the shadow of suffering and the light of liberation merge into one indivisible state.

Moksha, like quantum collapse, is the point at which multiple possibilities dissolve into unity. It is an act of supreme observation that dissolves the duality between light and shadow, between life and death. At a time when modern science and ancient spirituality seem separated by an unbridgeable gap, these analogies suggest that dialogue is not only possible, but perhaps necessary to glimpse the ultimate truth of existence. And after all, as Hinduism reminds us, light and shadow are but two aspects of the same infinite.

Mental decoherence, Moksha in daily life.

Moksha is not just an otherworldly ideal; it represents a dimension of inner freedom that can be experienced already during life. It is an exit from the chaos of ego, fears and material attachments. But how can something so transcendental find a place in everyday life? And what can quantum physics tell us about this path of personal introspection?

In quantum physics, the concept of decoherence describes the transition of a quantum system from the superposition state, where anything is possible, to an observable, determinate state. This phenomenon marks a crucial moment when the system "collapses" into a definite reality. In philosophical terms, one could interpret decoherence as the loss of potential disorder to achieve present clarity.

Similarly, Moksha implies a type of mental decoherence: the shift from a mind influenced by the cyclic tensions of desire and delusion toward a clearer, more conscious consciousness. In a practical sense, this process happens through small acts of mindfulness, which we can concretely incorporate into our hectic days.

Modern life is full of mental noise, often referred to as "cyclical thoughts." These thoughts recur obsessively, dragging the individual into loops of judgments, regrets and expectations. Meditation, practiced consistently, is a means of breaking these cycles. For example, in a silent village in the foothills of the Himalayas, monks in the Vedanta tradition teach the inhabitants simple gestures: focusing on one's breath in the morning, walking slowly while observing the earth beneath one's feet, or mentally repeating a mantra.

These small acts, repeated daily, are experiments in mental decoherence. They gradually separate consciousness from identification with mental noise and create a detachment from materialistic thoughts. Indian historian and thinker Swami Vivekananda (1863-1902) offers us an enlightening perspective:

> *"God is within you, like the Sun beyond the clouds. Dissolve these clouds, and you will find peace."*

Even these simple efforts of awareness can dissolve fragments of Maya, the illusion that covers our true nature.

If we look at this concept from a scientific perspective, we could compare the small moments of meditation, sacred reading or contemplation to experiments that restore order in a quantum system. The mind, restless and fragmented, comes together. Physicist David Bohm, known for his reflections on the interconnectedness of physics and Eastern philosophy, described the universe as an implicit order, where everything is connected and new states of awareness can be revealed with proper attention. Meditating, according to Bohm, would mean allowing the mind to participate in that which is larger and more harmonious.

The Indian poet Mirabai (1498-1547), who devoted her life to devotion to Krishna, sang of a divine love that freed her from the bonds of the world. Through unceasing prayers, Mirabai found a state of consciousness so pure that no material circumstances could shake her. Her songs are still celebrations of Moksha lived in the present time: a free soul that does not identify with the cycles of suffering.

Like Mirabai, each of us can, while not abandoning daily life, cultivate inner freedom by starting with small moments. A sunset observed without judgment, a moment of complicit silence with a loved one, or a minute of gratitude before bed. These are seeds. They lead, in their own way, to liberation from mental cyclicality.

In the work Bhagavad Gita, the concept of Moksha is described as a state of equanimity, where pleasure and pain, success and failure cancel each other out. Sri Aurobindo, an influential twentieth-century mystic, suggested that Moksha was not a concept exclusive to renunciates but a goal for those living in the world. When a mind observes without reacting, that mind approaches freedom.

How then can we bring Moksha into our modern life? One simple step is to ask ourselves every night:

"Today, in what moment did I feel silence inside me?"

This gesture, as simple as it is profound, is the path to daily mental decoherence. Each small realization sets the stage for Moksha: not a final event, but an ongoing process of inner liberation.

Maya's collapse and return to Brahman.

Maya represents the changing appearance of the phenomenal world, behind which lies an unchanging reality: the Brahman, the Absolute. This idea underlies the concept of Moksha, liberation from Samsara, the cycle of birth and death. Moksha is the moment when Maya dissolves, revealing the true face of existence.

Hindu tradition describes maya as a veil, a web of perceptions that deceives the mind. For example, in the Mandukya Upanishad, the dream is spoken of as a metaphor for everyday reality.

> *"Just as a man who dreams believes the dream is real, so those who live are trapped by Maya and cannot see the truth."*

But what happens when the veil is lifted? The answer is Brahman, the eternal substratum that lingers beyond illusion.

This ancient thought finds surprising resonance in quantum physics. In quantum theory, particles exist in superposed states until they are observed. A photon, for example, behaves as a wave or particle depending on the measurement. It is a world of latent possibilities, a "potential field" reminiscent of Maya. When observation occurs, this uncertainty collapses into a defined reality. This process, called decoherence, recalls the concept of Moksha.

Let's try to imagine: Moksha is the end of confusion. It is the ultimate collapse to a stable reality, similar to what happens when an undefined quantum state becomes an observable reality. Maya, like the superposed states of particles, conceals the ultimate truth of Brahman.

The Katha Upanishad offers a famous tale about illusion and Moksha. Nachiketa, a young Brahmin, meets Yama, the god of death. Nachiketa, seeking truth, rejects all material offerings-wealth, longevity or power-and asks to know what lies beyond the illusion of life and death. Yama teaches him that the ultimate goal is to realize Brahman. This narrative is a powerful metaphor for the journey to Moksha: a path away from the sensory world, breaking free from maya to embrace absolute reality.

The relationship between consciousness and reality has also been studied in science. In 2019, an experiment conducted at the University of Vienna showed that the choice of observation directly influences the behavior of particles. Physicists recorded how the decision of when to measure changed the final result. This phenomenon leaves open questions about the role of mind and consciousness in defining reality.

The parallel with Hindu philosophy is suggestive. Just as Maya shapes sensory experience, so the consciousness of the observer seems to shape quantum behavior. Modern science, with its advanced technologies, thus finds itself exploring a field that the Indian sages had already intuitively described.

Moksha is, in philosophical terms, the collapse of maya. It is the moment when the superimposed truths of appearance vanish and the eternal Brahman is revealed. This "collapse" is a process of inner transformation, which occurs through *jnana*, self-knowledge, and meditation.

Even in quantum physics, "collapse" brings order to chaos. At the macroscopic level, however, this collapse is not immediate. Maya, like quantum states, persists until consciousness peaks.

Breaking the cycle, Moksha and the notion of the arrow of time.

The idea of liberation from the cycle of birth and death, Samsara, is one of the pivotal foundations of Eastern philosophies, particularly Hinduism. This state of transcendence, called Moksha, represents not only an ultimate spiritual goal but also an interesting philosophical challenge to our concept of time. How can a mind immersed in the idea of the linear progression of events understand an idea that transcends temporal sequentiality? A possible bridge for this confrontation is offered by quantum physics, which is capable of suggesting that time, as we conceive it, may be only one of the ways in which we perceive reality.

To understand how these two visions-spiritual and scientific-can dialogue, we must venture into the folds of time, on the one hand through Hindu thought and on the other by deciphering some of the mysteries of the quantum universe.

According to Hindu sacred texts, time in the human condition is experienced as a movement marked by the eternal repetition of the cycle of Samsara. We are born, live, die and are reborn. This uninterrupted succession (Karma) is governed by our actions and intentions, accumulating consequences that accompany us into successive existences.

However, Moksha represents a definitive break. It is not a mere pause or temporary interruption, but a state of freedom in which the very concept of linear time is transcended. Many Hindu texts, such as the Upanishads, describe Moksha as a return to the original essence, that pure awareness unbound by the illusion of separation between past, present and future.

In the world of physics, time occupies an equally mysterious role. In the fundamental equations of quantum mechanics, unlike the macroscopic reality we know, time does not necessarily move in one direction. The fundamental laws governing particles do not distinguish between past and future, suggesting that what we experience as the "arrow of time" may not be a fundamental property of the universe but an illusion created by our observation.

A central concept is so-called quantum decoherence. Decoherence explains how quantum systems, initially in a mixture of possible states (the famous "superposition" of Schrödinger's cat), acquire a visible, concrete reality as they interact with the environment.

From a philosophical perspective, decoherence resembles the transition from Samsara to Moksha. In the cycle of Samsara, individuals live immersed in the illusory web of Maya, convinced that time flows like an arrow. However, Moksha is like detaching oneself from this illusion, stepping out of the limited (and observable) state and returning to a state of unity.

Classical physics and our daily experience make us perceive time as something that advances, with no possibility of reversal. This phenomenon was first described by Ludwig Boltzmann in the late 19th century through the concept of *entropy*. The growth of entropy, which governs the direction of time for macroscopic systems, seems to give our world the idea of irreversible progress toward the future.

In the context of quantum physics, however, many of these certainties vanish. Particles appear "indifferent" with respect to time direction. Situations that seem irreversible on a large scale (such as the breaking of a glass) could, at least in theory, be seen as reversible at the microscopic level.

We can imagine Moksha as a similar change on the scale of consciousness. Liberation from Samsara offers a way out of the perceived arrow of time, breaking the bond that binds human beings to linear causality. It is, in other words, a "coming home," an awakening to a reality that no longer depends on becoming.

The similarities between time-linear overcoming in Moksha and reversibility in quantum states are not pure coincidence. Both require a radical shift in perspective. Both challenge beliefs that the ordinary mind considers obvious.

Moksha and the "Quantum Zeno" Paradox.

The search for absolute truth has always been one of the great questions that unites philosophy and science. Hinduism, with its concept of Moksha, and quantum physics, with its Paradox of the "Quantum Zeno," seem to travel parallel paths in exploring this fundamental question. Both focus on what happens when the cycle is broken: the cycle of birth and death in the case of Samsara, or that of measurement and observation in quantum worlds.

A crucial step in achieving Moksha is to stop observing the world with the eye of the ego. It is like breaking an invisible chain, the one that binds the individual to a partial identity of desires, fears and attachment. For Hinduism, the true self is not the ego, but Brahman, the universal essence.

The "*Quantum Zeno Effect*": the game of time and observation.

On the other side of the world and knowledge, quantum physics struggles with paradoxes that seem to defy common sense. One of these is the *"Quantum Zeno Effect,"* first theorized by Misra and Sudarshan in 1977. According to this principle, constant observation of a quantum system prevents it from changing its state. It is an effect named after Zeno, the Greek philosopher famous for his paradoxes about motion and time.

A classic example to explain the "Quantum Zeno Effect" is radioactive decay: an atomic nucleus, "observed" continuously through measuring instruments, appears never to decay. The very act of observation freezes the system, locking it into its initial state. This paradox seems to suggest that direct attention to a phenomenon, ironically, changes it: stopping it.

And here arises a fascinating dialogue between the East and physics. In Hinduism, the process leading to Moksha requires breaking Samsara by interrupting the egoic observation of life. It is the act of observing-and thus judging, desiring, attaching-that "locks" the soul into the phenomenal world. When you stop observing with ego eyes, you allow your self to transcend. Freeing oneself from the same constant "observation" is what allows the soul to flow toward liberation.

In the *"Quantum Zeno Effect"* something similar happens: continuous observation freezes the quantum system. If we compare the quantum system to human consciousness, egoic observation becomes a constraint, a freeze that prevents change, transformation or, in the case of Moksha, liberation. The decisive action, for both Hinduism and quantum physics, is to interrupt the cycle.

But how does one interrupt egoic observation? Ancient Indian sages-thinkers such as Adi Shankara pointed to meditation as the key. Meditation dampens mental activity and directs attention to a state beyond the ego. It is a process that adds nothing to the mind, but removes that which obstructs access to ultimate reality.

There is a poetic image that goes back to Vedic texts: the Atman (the self) is like a lamp covered with layers of cloth. Each layer represents an illusion of the ego. Through spiritual knowledge and practice, these veils are lifted, allowing the luminous essence to shine through.

If we look at the results of the most modern interpretations of quantum mechanics, we see a striking similarity to the idea of Moksha. On the one hand, quantum physics reveals to us that the universe exists in a web of possibilities, fluctuating between different states until observation "locks" it into a definite reality.

On the other hand, Hindu philosophy states that our deep self is already in a state of unlimited freedom, but remains stuck in Samsara due to constant and rigid observation. To become fluid again, free from the cycle of rebirths, we must stop "locking" our consciousness into an ego-dominated repetitive pattern.

Moksha in loop mechanics.

There is a cycle that governs everything. Birth, death and rebirth are not only at the heart of Hindu spiritual thought, but also an intriguing parallel to the cycles hypothesized by modern physics to explain the reality of the universe. The concept of Samsara - the infinite cosmic cycle of creation, destruction and rebirth - finds a surprising echo in loop quantum gravity theories. This scientific perspective, which is still being developed, imagines the universe in such a way that its "infinite song" resembles the eternal flow of Samsara. But if knowledge and meditation, according to Vedic traditions, can lead to Moksha, what might such a "reset" represent from the perspective of the universe?

Hindu philosophy portrays Samsara as an invisible prison. Every existence, no matter how different, is trapped in this pattern of cause and effect. To break free, according to the teachings of the Vedas and Upanishads, the individual must attain profound knowledge, an ultimate realization that reconnects him or her to Brahman, the nondual Absolute. This is what is called Moksha, the dissolution of individual identity and the end of the cycle.

In the field of cosmology, theories of loop quantum gravity, a fascinating extension of quantum mechanics and general relativity, suggest that the universe itself may be chained in a similar loop. According to this model, the universe would not explode in a single Big Bang and then die in a Big Crunch, as the old cyclic theories assumed. Rather, it would transform, rebounding cyclically. With each bounce, the cosmos returns to an initial condition of extreme density, a kind of "quantum vacuum" from which everything begins again. Each "birth cosmic" brings new configurations of physical laws, new universes distinct from the previous ones. Here, the concept of Moksha connects to a possibility: is there an instant, a limit, at which this cycle could be interrupted? Is it possible to conceive of liberation even for a universe?

If we apply Moksha logic to the universe, the answer may lie in a central idea of quantum physics: decoherence. In the subatomic world, particles exist in multiple, overlapping states until an observation or interaction with the external environment reduces this uncertainty, collapsing the particles into a definite state. This process is similar to what happens with a consciousness emerging from confusion in meditation: only by concentrating can clarity be achieved.

In the universe, each cycle or rebound corresponds to a new decoherence. But there are those who speculate that, as with human Moksha, there could be a "stopping" of the cycle in the cosmos. It could happen at a point where quantum gravity can no longer sustain cyclic restructuring. As scientist Carlo Rovelli, a leading theorist of loop quantum gravity, puts it

"Time may cease to exist."

What if the universe could reach a level where it no longer needed to transform itself? Here we could imagine Moksha as a leap outside of time itself, a condition in which there is no longer a need to repeat the past.

In Hinduism, the metaphor of the cycle is ubiquitous. Scriptures such as the Bhagavad Gita describe the universe as an immense wheel. But it is in the Upanishads that the technical and profound meaning of Moksha emerges. For example, the Isha Upanishad teaches that:

"Whoever sees all beings in his own Self and his own Self in all beings will no longer feel fear."

Fear, then, is what holds one back in Samsara. Similarly, in physics, uncertainty about the nature of time and space seems to be the key that keeps the universe locked in its cyclical motion.

Indian philosopher Adi Shankara, one of the greatest thinkers of Advaita Vedanta, likened the universe to a flame that, while constantly changing shape, remains the same at its core. In modern cosmological reasoning, something similar has been described by cosmologists such as Martin Bojowald, who studies the so-called "*Big Bounce*." According to his work, the universe never really comes from nothing: rather, it is a continuous transformation. But what happens when this transformation stops?

Ancient thinkers and modern scientists have many things in common. Both attempt to explain the why and how of existence. Both try to solve the riddle of the cycle of birth and death, through different tools: deep meditation for Hinduism, mathematical formulas and observations for physics. The bridge between Eastern philosophy and science is not always guaranteed, but offers novel insights.

If Moksha is that which liberates from eternal return, then the universe itself could find a moment of liberation. What could take place? An eternal return to emptiness, to "nonbeing," as described in quantum physics? Or a leap to a new state of reality, inconceivable for now, such as cosmic enlightenment?

Questions like these may never be answered. Yet, questioning allows the dialogue between East and West to continue, between the velvety words of ancient sages and the precise calculation of scientists. As in every cycle, from the joining of the parts, new possibilities emerge. And perhaps, from the very act of asking the question, it is already possible to come a little closer to that liberation that transcends all.

The illusion of mind and the collapse of the wave function.

Hinduism sees mental delusion as the main obstacle to liberation. The obsession with fragmented, ego-identified thinking is associated with an intangible prison; only the dissolution of this mental dynamic can lead to truth.

On the other side of the world, quantum physics offers an intriguing concept: the collapse of the wave function. In quantum mechanics, the wave function describes a probabilistic set of possible states for a particle. When an observer enters, this function "collapses," and the particle assumes a defined position or state. The act of observation, therefore, appears as a determinant of reality.

The juxtaposition between Moksha and quantum collapse comes naturally precisely because both concepts revolve around the idea of dissolution. In Hinduism, Moksha dissolves mental illusion, leading to ultimate truth. In quantum physics, the role of the observer seems to dissolve the probabilistic uncertainty of the wave function, creating a concrete statement in reality.

David Bohm, developed a theory known as "implicit order," suggesting that visible reality was only a projection of a deeper, invisible, interconnected dimension. Bohm himself said he was fascinated by Vedic philosophy and "Om," the primordial sound that, according to Hinduism, encapsulates the principle of the universe. Bohm emphasized how quantum principles could align with these spiritual concepts, suggesting a masterful dance between the manifest and the implicit.

The mind, according to Hinduism, functions just like the wave function: it creates endless possibilities and probabilistic scenarios through imagination and fear. Only by dissolving these projections does the practitioner become liberated and access a stable, indivisible reality called Brahman. Physicist John Wheeler, for his part, proposed the "*participatory anthropic principle*": the universe exists because it is consciously observed, a concept that seems to echo the Hindu idea that consciousness (in this case, Atman) is inseparable from ultimate reality.

Here arises a curious similarity: just as mysticism invites us to look beyond the illusions of the mind, quantum physics requires us to understand that matter, too, is elusive in its objectivity. We stand at the crossroads between two different languages that, surprisingly, speak of the same fundamental truth: the world is a game of illusions, influenced by what cannot be seen or measured directly.

Hinduism offers the scientific debate a surprisingly timely perspective. Richard Feynman, speaking ironically about quantum mechanics, said:

"If you think you get it, then it means you don't get it."

The same phrase, perhaps, could be applied to the mysteries of the mind and the search for Moksha.

More and more researchers are combining meditation and neuroscientific studies to explore the relationship between deep meditative states and altered levels of consciousness. Indeed, it is not surprising that monasteries and laboratories are beginning to openly dialogue. In particular, in Dharamsala, home of the Dalai Lama, scientists and Buddhist monks regularly discuss quantum physics and spirituality, looking for points of connection.

Moksha, understood as overcoming mental illusion, resonates deeply with the collapse of the wave function, where reality emerges from conscious intervention. Spiritual liberation and scientific inquiry find themselves, though following different paths, in the same field of reflection. Perhaps this dialogue, still ongoing, will never be fully resolved. But what remains certain is that both offer us a perspective: that the boundary between the real and the illusory may be thinner than we dare to imagine.

Chapter XI. Dharma: Duty and Ethics.

Ethics in Hinduism.

Ethics is a central pivot in Hinduism. At the heart of this age-old tradition is the concept of Dharma. This Sanskrit term encapsulates the idea of a universal order that governs the cosmos and human life. It guides each individual to live in balance with himself, with others and with the universe.

Dharma as a universal structure.

Ethics in Hinduism is not just a set of rules to be followed, but an entire worldview, rooted in the concept of Dharma. This Sanskrit term is vast and complex, difficult to translate into a single word. At its core is the idea of a universal order that maintains the cohesion of every aspect of existence: the cosmos, human life, nature and social relations. Dharma is not static. On the contrary, it is dynamic and specific to each individual as it takes into consideration the personal, social and spiritual context of each human being.

The term Dharma expresses the natural law that permeates the universe. Hindu philosophy describes it as the principle responsible for order and balance. However, it is not limited to cosmology; Dharma also guides every person's ethical behavior. In the Indian epic poem of the Mahabharata (in its versions dating from the 4th century B.C.-4th century A.D.), the sage Bhishma states that:

"the Dharma sustains all that exists: it is the soul of actions."

This conception not only connects human beings to the divine, but also offers a moral code that guides every individual choice, maintaining the interconnectedness of all entities in the world.

In Hinduism, however, ethics is not limited to personal behavior. On the contrary, it is intertwined with the macrocosm. By following the Dharma, human beings participate in universal harmony. This concept is evoked in one of the fundamental philosophical texts of Hinduism, the Bhagavad Gita (dated between the fifth and second centuries B.C.), where Krishna, a divine figure, explains to Arjuna the importance of following one's SvaDharma, or personal Dharma. Krishna states:

> *"It is better to fail by following your own Dharma than to succeed by following someone else's."*

Personal Dharma is essential: each person must live in accordance with his or her role and responsibilities, thus contributing to the overall balance.

A central and universal principle of Hindu ethics is *ahimsa*, nonviolence. This value emerges in many sacred texts, but it achieves special resonance in the Vedas and the Upanishads Ahimsa represents not only the absence of physical violence, but also the rejection of all thoughts and actions that harm other living beings. Hinduism holds that evil inflicted on others causes a misalignment with the universal Dharma.

A modern representative of ahimsa was Mahatma Gandhi. Educated in the principles of Hinduism from childhood, Gandhi transformed ahimsa into an instrument of ethical and political revolution. His struggle for India's independence from British rule was based on the idea that nonviolence was not a form of passivity, but an act of deep connection with truth, or *satya*. Gandhi explained that:

> *"ahimsa is the attribute of the soul: to live ahimsa is to live in tune with the Dharma."*

Hindu ethics transcends the individual and focuses on interconnectedness. One of the most famous metaphors to describe this reality is the "*net of Indra*," found in the texts of Vedanta and Mahayana Buddhism. This cosmic web is described as an infinite fabric of jewels, each of which reflects all the others. Each action of an individual affects the entire system, creating a ripple effect that runs through the universe. This image metaphorically represents the fact that following one's Dharma is not an isolated act, but contributes to the well-being of all.

A practical example of this view is the system of purushartha, the four goals of human life, which include Dharma (right action), artha (wealth and sustenance), kama (desire and pleasure) and Moksha (spiritual liberation). Living in balance among these goals not only generates a harmonious existence, but also ensures respect for universal interconnectedness.

Hindu ethics has influenced various disciplines, even outside religious contexts. In parallel with quantum physics, where each part of the universe is connected to a larger whole, the concept of Dharma in Hinduism is presented as a deeply systemic idea: every individual action affects the whole. A universal message emerges, inviting not only personal responsibility but also a collective vision.

Ultimately, Hindu ethics, centered on Dharma and ahimsa, suggests a model oriented toward cooperation, harmony and respect for life in all its forms. This system, rooted in a tradition thousands of years old, still reminds us today that our choices affect not only our personal destiny, but the balance of everything around us. As Mahatma Gandhi said:

"Be the change you want to see in the world."

In Hinduism, this transformation begins by following one's Dharma.

Dharma in daily life.

The idea of Dharma is one of the oldest and most fascinating cornerstones of Hindu ethics. Dharma is a complex term to translate. Dharma could be defined as the harmony that governs the cosmos, human relationships, and individual behavior. It is not a rigid law, nor a set of intangible rules, but a fluid principle, adaptable to circumstances, age, gender and social role. The elasticity of Dharma makes it an extraordinarily contemporary concept, capable of inspiring conscious action even in today's life.

In Hinduism, there is no one-size-fits-all morality. Dharma changes according to the context and the individual. For example, a teacher's Dharma is to educate with patience and wisdom, while a doctor's Dharma involves caring for patients with empathy and dedication. But beyond specific roles, what all forms of Dharma have in common is moral responsibility toward the common good and interconnectedness among all living beings.

According to the Bhagavad Gita, one of the holiest texts in the Hindu tradition,

"It is better to follow one's own Dharma, even imperfectly, than the Dharma of others, even perfectly."

This fragment emphasizes a fundamental truth: each person must act in accordance with his or her own nature and position in life, without imitating the path of others. Knowing one's Dharma is a process that requires introspection and awareness.

Dharma often manifests itself in the most intimate and everyday relationships, such as family relationships. Take the case of a father and his children. The Manu Smriti, an ancient Hindu ethical text, depicts paternal Dharma with great elegance:

"The father who educates his children according to the Dharma fulfills his divine duty."

This means not only providing materially, but also inspiring children to live with integrity and discernment.

At the same time, the Dharma of children toward their parents includes respect, care and gratitude, especially in old age. Indian culture still celebrates the value of mutual care in the family. An emblematic example is the holiday of Pitru Paksha, a time of year dedicated to ancestors during which descendants offer prayers and food to honor their memory. The holiday symbolizes the Dharma of gratitude, which is expressed as much toward those who are alive as it is toward those who are gone.

Another illustration of Dharma is found in the Ramayana, one of the great epic poems of Hinduism. In the tale, Prince Rama represents the very essence of Dharma. Rama unhesitatingly accepts exile in the forest for fourteen years, even though he knew he was entitled to the throne of Ayodhya. His decision stems from a desire to honor his father's promise to his half-brother Bharata. In this account, Rama's Dharma emphasizes the importance of loyalty and respect for family, even when it involves sacrificing personal ambitions.

While rooted in Indian culture, the Dharma contains a truth that transcends geographical and religious boundaries. It recalls the idea of global responsibility that many feel today regarding the climate crisis. Just as the Dharma requires individuals to act in harmony with their environment, so the current need to preserve the planet reflects a collective Dharma, requiring awareness and collaborative action.

Mahatma Gandhi himself, when he spoke of nonviolence (*ahimsa*) as a universal principle, implicitly linked it to Dharma:

"Our greatest responsibility is to all living things."

Gandhi believed that Dharma was inseparable from duty to truth and compassion.

In modern daily life, Dharma becomes a tool for dealing with the complexities of the world. For example, in the workplace, following one's Dharma means acting ethically, treating colleagues and co-workers with respect, and avoiding shortcuts that violate integrity. In personal relationships, Dharma can translate into prioritizing authentic communication and mutual understanding.

As a practical cue, the Hindu tradition proposes the concept of sankalpa, or conscious intentionality. Before taking an action, one should stop and reflect, "*Is this decision in harmony with my Dharma? Will it contribute to the well-being of myself, others and the world around me*?"

Dharma, in its essence, is not a distant spiritual abstraction. It is a living principle, adaptable and implementable in every area of life. It teaches us the importance of acting with awareness and responsibility in the unique context of each existence. From the father raising his children to the individual questioning his own environmental impact, Dharma guides toward a life centered on interconnectedness and ethics. In Hinduism, as in everyday reality, Dharma is not just a rule, but an invitation to live in balance with oneself and the world.

From the personal to the collective.

Ethics in Hinduism develops as a journey from the personal to the collective, from individual inwardness to universal harmony. Its deep roots find expression in the conception that all humanity, and even the cosmos, is an interconnected whole. This perspective is summed up in the famous Sanskrit maxim, "*Vasudhaiva Kutumbakam*," meaning "*The world is one family.*" This principle, handed down in sacred texts such as the Upanishads and Puranas, reflects an ideal of global empathy that calls for seeing every living being as a member of the same community. Hindu ethics is thus not limited to duties to the family or village, but extends to the whole of creation, inviting universal responsibility.

Among the philosophical-religious works of Hinduism, the Bhagavad Gita, part of the Mahabharata, is a key text for understanding Dharma ethics. In the dialogue between Krishna and Arjuna, Krishna stresses the need to act without selfishness:

> *"Do your duty without expecting rewards. Act for a higher good."*

This principle, often referred to as "Karma yoga," encourages the individual to harmonize his or her own interests with those of the collective. Ethical commitment is not limited to personal good, but aims at collective welfare.

For example, a farmer sowing a field should not work with only his own harvest in mind, but with the contribution that farming makes to the livelihood of the community. Similarly, a warrior, like Arjuna, should fight not out of vanity or hatred, but to restore justice. In this sense, the message of the Gita nurtures an ethic based on self-transcendence, a momentum beyond the ego toward the fulfillment of a universal duty.

A vivid example of how this ethic translated into history is the life of Ashoka the Great, one of India's most celebrated emperors. Ashoka ruled the Indian subcontinent in the third century B.C., expanding his empire with violent military campaigns. However, the devastation caused by the Battle of Kalinga, which resulted in tens of thousands of deaths, profoundly transformed his worldview. The pain and suffering he saw in the survivors led him to renounce wars of conquest and follow the ethical principles of Dharma.

Ashoka adopted the teachings *of ahimsa* (nonviolence) and undertook a series of reforms aimed at collective welfare. He had his edicts engraved on pillars scattered throughout the empire. These pillars, such as the famous Pillar of Sarnath, promoted values such as religious tolerance, respect for all forms of life, and social justice. Ashoka also sent emissaries to spread peace and Dharma beyond India's borders, seeking to create a civilization based on universal ethics. His reign remains a historical symbol of how a personal ethic can be transformed into a commitment to the collective good.

Ethics as the connection between the individual and the universe.

In Hinduism, the interconnection between the individual and the universe is fundamental. The daily practice of Dharma, understood as "right behavior," governs every aspect of life. It is not a rigid code, but a set of flexible principles that help balance personal needs with those of the whole community. On a cosmic level, it is believed that every action (Karma) has a consequence, not only on the life of the individual, but on the entire cycle of existence. Therefore, acting ethically means contributing to universal harmony.

The environment and living beings are also an integral part of this ethic. Many Hindu texts, including the Vedas, highlight respect for nature as a sacred duty. Natural elements such as the Ganges River, trees, and animals are revered not only for their spiritual value, but for their role in the web of life. This respect for the environment relates to the concept of interdependence: to harm one part is to compromise the whole.

The ethics of Hinduism invites one to see beyond the boundaries of one's individuality. It moves from family responsibility to a broader vision that includes society, humanity and the entire universe.

Ethics in spiritual practice

In Hinduism, ethics represents more than just a set of rules to guide daily actions. It is a broader view, encompassing the entire existential path, including spiritual growth. At the heart of this perspective is the concept of Dharma. This Sanskrit term is rich in meaning: it can be translated as "duty," "universal order," or "moral law." Dharma is not static, but varies according to context, individual and historical moment, guiding each person in finding a balance between personal ethics and cosmic harmony.

A universal foundation for understanding the ethics of Hinduism emerges in Patanjali's Yoga Sutras. These ancient texts (written probably between the second century B.C. and the fourth century A.D.) provide a practical guide for those seeking union with the absolute. Patanjali introduces yamas and niyamas, two categories of ethical precepts that represent the first step on the path of yoga.

The yamas concern relations with the outside world. Among them, the cardinal principle is ahimsa, non-violence. This concept is extends far beyond simply avoiding physical harm. It also concerns words, thoughts and intentions. Being nonviolent means living with respect for all forms of life, rejecting all behavior based on hatred and aggression.

Niyamas, on the other hand, are regulations that guide inner discipline. Among them we find satya, truth. Speaking the truth, according to Hinduism, is not an end in itself. Truth is indispensable for maintaining harmony because it generates clarity and trust both within ourselves and in our relationships with others.

These principles, while deeply rooted in the philosophical and spiritual context of India, possess universal force. They are timeless tools for the individual, who can use them to face the daily challenges of modern life.

One of the most distinguished exponents who embodied the ethical principles of Hinduism was Mahatma Gandhi. Born on October 2, 1869, in Porbandar, India, Gandhi transformed ahimsa into an unprecedented philosophy of life and political strategy. For him, nonviolence was not mere passivity but disruptive action capable of shaking the pillars of British colonial oppression. Gandhi often declared:

> *"Ahimsa is the attribute of the soul. It cannot be practiced by those who lack courage."*

For him, nonviolence was an active force, requiring total dedication to truth and immense inner strength. For example, in 1930 he led the famous Salt March, an act of civil disobedience made powerful by its nonviolent nature. Thousands of individuals followed Gandhi on the path to Dandi, proving that ethics could inspire an entire people to resist without hatred.

Hinduism also emphasizes the interconnection between individual and universal ethics. According to this view, every action taken by the individual has repercussions on the entire cosmos. This principle emerges clearly in the concept of Karma, according to which every choice leaves an imprint in the fabric of the universe, generating effects that affect the future.

Dharma in daily life. Family ethics and social responsibility.

Dharma, one of the pivotal concepts in Hindu ethics, represents universal order and a sense of individual duty. In everyday life, it translates into moral guidance that governs family relationships and social responsibilities, adapting to context and circumstances.

In families, Dharma acts as a common thread that promotes harmony and mutual responsibility. The principle "*matru- devo-bhava, pitru-devo-bhava*" invites children to regard their parents as a divine manifestation, expressing filial respect and devotion. An emblematic example of this principle is narrated in the Ramayana: Rama's obedience to his father Dasharatha's command. Despite difficulties, Rama accepts exile to honor his father's wishes, embodying the ideal of a devoted son.

On the parents' side, Dharma includes the obligation to protect, educate and prepare children to live righteous lives aligned with ethical principles. This exchange of duties between parents and children strengthens the family bond and lays the foundation for building a society based on respect and responsibility.

According to Hindu tradition, an individual's life is divided into stages (*ashrama*). The Grihastha Ashrama, the stage devoted to family and social life, emphasizes the importance of duties to the community as well as to one's family. The head of the family (*Grihastha*) is responsible for supporting others economically and morally, sharing wealth and resources with those in need, in accordance with the tradition of *dana* (charity). This responsibility is not a burden but an opportunity to contribute to the harmony of the universal order.

Dharma is not limited to family boundaries, but extends to collective challenges. In modernizing, the concept has found new applications, such as commitment to environmental ethics. The principle of Vasudhaiva Kutumbakam ("*The world is one family*") sums up this global vision. As in a family all members are protected, the Dharma encourages a sustainable and compassionate approach to nature and living beings. This is reflected in environmental initiatives and attempts to balance economic progress with respect for the planet.

One example is the current climate justice movements in India, often inspired by ancient scriptures and respect for the planet as a sacred entity. This link between traditional ethics and modern challenges highlights the perpetual value of Dharma as a moral compass.

In the era of climate crises and rising social tensions, the Dharma offers a perspective that balances individual rights with collective duties. Its emphasis on interconnectedness among living beings, as well as between humans and nature, could inspire a global code of ethics. The principle of living in balance with oneself, others and the planet can become an answer to the challenges of the 21st century.

Dharma and interconnectedness. Ethics between spirituality and science.

The idea of interconnectedness is a central theme in Hinduism, a millennia-old philosophy that not only analyzes the relationship between human beings and the universe, but also guides ethical behavior through the concept of Dharma. In the contemporary world, this perspective finds surprising new correspondences with discoveries in quantum physics, offering a bridge between spirituality and science.

In Hinduism, Dharma represents the moral and cosmic order that each individual is called upon to uphold through his or her actions. It is a dynamic force that holds the universe together, indicating the "right way to live" according to individual, social and natural context. The cardinal idea is that every action has an effect on everything else, a law of interdependence that guides human behavior toward harmony with the world.

For example, the Bhagavad Gita, one of the fundamental texts of Hinduism, emphasizes that living beings do not exist in isolation, but are part of a larger whole. In Chapter III, Krishna teaches Arjuna that man should act according to his Dharma not out of personal ambition but for the collective good, respecting and cooperating with the forces of nature.

This principle of interconnectedness is found not only in spiritual speculation but also in everyday practices. A practical example is the ancient belief in *ahimsa* (nonviolence), which affects not only human relationships but all forms of life, emphasizing the importance of recognizing the inherent value of every living being. This respect for universal interconnectedness guides much of the ethical choices in Indian culture.

Many centuries later, modern physics has discovered scientific concepts that seem, in fascinating ways, to echo Hindu spiritual insights. One of the highlights is the perspective of quantum physics, where the interconnected nature of the universe emerges powerfully.

Fritjof Capra, in his famous essay "The Tao of Physics," explores in detail this complementarity between Eastern thought and the concepts of modern science. Capra states that, just as Eastern philosophies teach, there are no isolated entities in the field of quantum physics. Physical phenomena arise and develop through the network of their relationships. For example, the scientist explains how subatomic particles do not have meaning as separate entities, but only as a function of their interaction with other particles, creating a "cosmic network" of interdependencies.

The parallel with Dharma is obvious: just as each particle acts within a larger whole, so every human action, according to Hinduism, affects the universal balance. It is a profound reminder of personal responsibility, not only to other living beings but also to the planet itself.

One of the most powerful examples of the application of Dharma in a modern context is the Chipko movement, which began in India in the 1970s. In this initiative, led mostly by women from Himalayan villages, local communities put into practice an ancient principle of interconnectedness and respect for nature. The symbolic gesture of hugging trees ("chipko" in Hindi means "*to embrace, to cling*") represented their attempt to protect forests from speculators, reminding the world that environmental destruction has direct consequences for human life.

In the Chipko movement, we find the beating heart of the Dharma: protecting that which sustains life to maintain overall balance. The women who led the movement, such as Gaura Devi, embodied this ethical vision, showing the world how an ancient philosophy can translate into concrete, transformative action. The gesture was not just symbolic, and the movement led to legislative changes that stopped wild logging in many areas.

Universal interconnectedness, detected by both spiritual insights and quantum physics, offers innovative answers to contemporary challenges. The crises we face today-from climate change to biodiversity loss-highlight the failure of a vision dominated by fragmentation. Hinduism, with its insistence on Dharma and recognition of the web of interdependencies, suggests an ethic that may be surprisingly relevant today.

As Capra reminds us, the dialogue between modern science and Eastern spirituality not only provides fascinating metaphors, but can contribute to a transformation of our worldview. A vision that recognizes the urgency of acting collectively, aware that each individual gesture has an impact on the whole.

At a time when ethics and sustainability are global priorities, the integration of Hindu principles and quantum science offers common ground for working out a new paradigm. The Dharma, with its emphasis on responsibility to the whole, invites us to rethink our relationship with the planet and with each other. It is an invitation to rediscover harmony, not only through meditation and a personal spirituality, but also through laws, policies and concrete actions.

Hinduism, with its profound respect for interconnectedness, proves to be an invaluable guide to addressing the moral challenges of our time, reminding us that every atom, like every human gesture, is linked to an infinite web of relationships. It is a message of responsibility and hope, and a call to recognize that, in the fabric of the universe, every thread counts.

Dharma and Quantum Entanglement.

In the philosophy of Hinduism, Dharma represents individual duty in the context of personal, social and spiritual circumstances. It expresses the unique role of each being in the cosmic order. Arjuna, for example, is a kshatriya, a warrior. His dharma is to fight for justice, even if personal emotions cause him to flee. Krishna teaches him that his behavior is not limited to immediate effects. Every action, every decision operates like a wave in the great ocean of existence, touching other lives and altering their course.

This viewpoint is echoed in recent developments in quantum physics. The principle of entanglement states that two particles, once "entangled," remain connected regardless of distance. A change in one particle is instantaneously reflected in the other, creating a perfect match. The invisible web that binds two particles appears to be a microscopic reflection of Dharma: an individual's choices vibrate throughout the universe, influencing a collective order.

In the 1980s, physicist Alain Aspect demonstrated that, against all classical logic, particles can influence each other instantaneously, violating the "separation" predicted by relativity. Entanglement, verified several times in the following years, reveals an interconnected reality . The universe, instead of being a series of separate events, functions as a large "fabric" in which everything is connected.

In Hinduism, the concept of Karma parallels that of Dharma. Karma and Dharma are threads that weave the carpet of life. Every action leaves an imprint that is reflected back in time and space. Similarly, entanglement reminds us that we do not exist as separate islands, but as parts of an indissoluble reality: the behavior of one reflects on the fate of another, just as in human relationships, ecological systems or, indeed, between subatomic particles.

Arjuna represents the eternal human dilemma. To act or not to act? His dialogue with Krishna tells us about the need to accept a role in the grand scheme of life. The philosophical importance of this choice is reflected in physics through the so-called unified field, theorized by physicists such as Erwin Schrödinger and Werner Heisenberg, both inspired by Indian philosophy. Schrödinger, an admirer of the Upanishads, perceived the universe as one indivisible reality, where the individual was not separate from the whole.

Dharma is not a static concept. It changes according to time, circumstances and context. It is subjective and universal at the same time. Quantum entanglement offers a different but complementary picture: each event, though unique, is joined at a larger scale. There is no action without reaction, no particle without connection.

If we think about our lives, its challenges continually pose questions similar to those Arjuna asks himself on the field of Kurukshetra. What is our duty? How do we operate without harming others and at the same time evolve as human beings? Science and philosophy seem to point us in a common direction: to live in awareness of our impact, knowing that every action is reflected in universal harmony.

The Dharma invites us to see life not as a sum of random events, but as a network of interconnected responsibilities. Quantum entanglement offers us a modern simile: the universe is interdependent, just as the Upanishads teach.

While quantum physics unlocks the secrets of the microcosm, the sacred texts of Hinduism continue to explore the infinity of the spiritual macrocosm. Both remind us that life is an adventure of balance. Every choice, big or small, is a strand that connects the individual to the whole.

The Kurukshetra Field and Heisenberg physics:

On the vast expanse of the field of Kurukshetra, Arjuna, the hero of the Bhagavad Gita, finds himself in a moment of profound crisis. Faced with his duty as a warrior, he hesitates. He has to fight against friends and relatives. In this context, Krishna, embodiment of the divine, urges him to follow his Dharma: his duty, ethical and spiritual. But Arjuna is blocked by uncertainty. This timeless scene vibrates with questions that echo in the most advanced reflections of modern physics.

The parallelism with Werner Heisenberg's uncertainty principle, formulated in 1927, illuminates this connection between ancient philosophical thought and contemporary science. Heisenberg stated that, at the subatomic level, it is not possible to simultaneously know with absolute precision the position and velocity of a particle. This uncertainty generates a universe of possibilities and probabilities, where every choice - or observation - affects the system under consideration.

Arjuna, like every human being, finds himself in an existential condition dominated by the multiplicity of possibilities. His dilemma reflects the uncertainty of the quantum world. On the field of Kurukshetra, each possible decision negates another. To fight or abandon the battle becomes a moral bifurcation that mirrors, in some ways, the behavior of a particle. By measuring a position, the physicist loses information about speed. By choosing one action, Arjuna renounces all other avenues.

Quantum physics teaches us that every event is interconnected, just as every atom influences other atoms in the fabric of the cosmos. Similarly, the Kurukshetra camp, in the epic narrative, becomes symbolic of a complex web of familial, political and cosmic relationships. Arjuna's decision affects not only his personal fate, but that of all humanity.

Krishna is a key example of this interconnectedness. As a guide, he does not impose a solution on Arjuna. Instead, he leads him to a higher understanding of reality: one in which individual choices, while uncertain, bend to the universal fabric of Dharma, the cosmic law. Similarly, in quantum physics, a particle never exists "by itself." Every change in its state affects the whole system.

Arjuna's crisis and Heisenberg's uncertainty principle share a powerful truth: humanity must live with uncertainty. Arjuna cannot know with certainty what the outcome of his actions will be. The physicist cannot say exactly where an electron will be. However, both Arjuna and scientists face this uncertainty with distinct but complementary tools.

Krishna teaches Arjuna to trust in the universal law of Dharma, which guides every living being toward its own unique purpose. Modern physics, on the other hand, addresses uncertainty with statistical probability and mathematical language. Both, however, teach us a crucial fact: freedom does not come from certainty, but from the ability to act in the limbo of possibilities.

The field of Kurukshetra becomes a metaphor for life itself, a place where every choice has far-reaching consequences. Likewise, the uncertainty principle reminds us that every observation makes the world a little less mysterious, but never fully knowable.

In a sense, Heisenberg himself seemed to resonate with this view. In his writings, he noted how Eastern philosophy, including the concepts of Dharma and impermanence, offered a striking affinity with the findings of modern physics. As he said:

"Man is always simultaneously actor and observer in the universal drama."

The Battle of Kurukshetra and Heisenberg's uncertainty principle invite us to rethink the meaning of our decisions. Both tell us that there is no easy way through the uncertainty of the world. However, it is precisely in this uncertainty that the wonder of our existence manifests itself. Physics and philosophy converge in suggesting to us that, whether Arjuna before his bow or a scientist before a microscope, the very act of choosing transforms the present and the future.

Dharma as an ordering force. The dance of particle interaction.

he Sanskrit word "Dharma" comes from the root "*dhri*," meaning "to uphold." This universal principle supports the structure of the universe and governs the harmony between beings and nature. Surprisingly, a parallel idea also emerges in quantum physics : the intricate interactions between particles. These relationships, though chaotic, generate an order that sustains the entire fabric of reality.

To understand the Dharma, Hinduism offers a powerful symbolism: Shiva Nataraja, the Lord of Dance. Standing on a circle of flames, Shiva moves in a perpetual dance, a continuous cycle of creation, preservation and destruction. This profound image is not only a religious icon, but a representation of the cosmos. Each gesture of Shiva regulates what exists, destroying what is obsolete to create space for the new.

In many Tamil temples in South India, such as the famous Chidambaram Temple, the statue of Nataraja embodies this vision. Here dance is more than an aesthetic act: it is a metaphysical principle. Devotees see Shiva as the very rhythm of the cosmos, the pulse of union between chaos and order.

Quantum physics allows us to glimpse a reality that reflects the cosmic dance of Shiva. At the subatomic level, particles move incessantly, appearing and disappearing in the quantum vacuum. This frenetic dynamics might seem random. However, interactions between particles follow precise rules, discovered through the equations of quantum field theory.

For example, the behavior of photons (the particles of light) when they move through a double slit has unraveled the mystery of "decoherence." The interaction of observers with the particles seems to establish order. Just like the Dharma that governs the actions of individuals, observation creates a balance in the physical world.

Just as the Dharma assigns a unique and interdependent role to each being, subatomic particles do not exist in isolation. The interaction principle states that each particle influences the field in which it is located. This quantum correspondence echoes the conception of Dharma as a principle that harmonizes the actions of each individual in the social and cosmic universe.

The metaphor of cosmic dance can be extended to natural laws. In the apparent chaos of particle collisions or ethical choices of human beings, an invisible order emerges. This "hidden order" is the manifesting Dharma. For Hinduism, it is eternal and cyclical. For physicists, it is the fundamental grid of reality, described mathematically.

The dance of Shiva and the subatomic universe, then, are not just symbologies or theories, but different languages for expressing the same truth. At a deep level, Hinduism and quantum physics teach us that the universe is a web of relationships. Nothing exists alone. Every action, every interaction, every particle affects what happens around it. In the chaos lies the Dharma: the underlying structure that holds everything together.

The Dharma of the Photon. A lesson in ethics from the wave-particle.

Wave-particle duality, first described by French physicist Louis de Broglie in 1924, shows how light and other particles behave in different ways depending on circumstances. This seemingly contradictory behavior offers fascinating insights into exploring concepts as old as the Dharma of Hinduism.

The double-slit experiment, conducted by Thomas Young in 1801 and refined in the following decades, remains a milestone in quantum physics. Imagine a beam of light passing through two thin slits on a surface. When no one "observes" the passage, the light behaves as a wave, creating an interference pattern on the screen behind it. However, when we introduce a device to observe the passage of individual particles of light (photons), they behave like particles, tracing precise trajectories through one of the slits.

This change in behavior is not random. The photon seems to adapt to the context. The photon's "choice" depends on the situation: when observed, it acts as a particle; when let loose, it manifests as a wave.

This is reminiscent of the concept of Dharma in Hinduism, where individual duty changes according to circumstances.

The most fascinating aspect of the wave-particle duality is its ability to adapt. In Hinduism, Dharma implies a balance between the self and the other, between the microcosm and the macrocosm. The photon, in its behavior, reminds us that change is not simply a reaction, but a meaningful adaptation to the surrounding reality.

In human terms, the Dharma teaches us to live situational ethics. A practical example is that of Gandhi. During the struggle for Indian independence, Gandhi rejected violent extremism not because it was unequivocally wrong, but because nonviolence reflected the Dharma of the Indian people in the historical and cultural context of that time. Like the photon, Gandhi acted in harmony with the conditions around him.

Even in science, physicists have adapted. Niels Bohr, father of quantum theory, embraced the contradictions of quantum physics as an opportunity to rethink the nature of reality. This spirit of adaptation seems to echo the philosophical concept of Dharma.

The wave-particle duality reminds us that the universe does not fit a binary logic but an "interdependent" logic. A similar example emerges in the Buddhist concept of Pratītyasamutpāda (interdependence), which suggests that everything exists in relation to others. Quantum physics, with its phenomena such as entanglement, shows a similar aspect: each particle is connected to all others.

Fritjof Capra explores precisely these connections. He points out how modern scientific discoveries often have affinities with the mystical insights of Eastern religions. The wave-particle duality, according to Capra, reflects the same kind of "holistic-contextual" understanding so dear to Eastern sages.

Photon behavior, while a scientific phenomenon, offers us a moral and spiritual lesson. Living our Dharma does not mean abiding by granitic rules, but finding balance and harmony in the context in which we find ourselves. As the photon teaches us, we too can be waves at one moment and particles at another.

As in the double-slit experiment, our choices are shaped by the interaction between our being and the world around us. In this dynamic interchange, ancient Eastern philosophies and quantum physics meet, revealing a universe where science and spirituality merge in the search for meaning.

Quantum relation theory.

In an ancient chant of the Rig Veda, we read:

"This universe is like a large network. Each node is connected to the other."

This metaphor, dating back more than 3,000 years, seems to foreshadow a fundamental concept being studied in quantum physics today: universal interconnectedness.

Hinduism, with its vast set of sacred and philosophical texts, explores the interdependence of reality through the concept of Dharma. Dharma represents the duty, ethics and harmony that each individual must maintain with the cosmos. Each action of a person is never isolated. It influences the surrounding world, contributing to the complexity of the universe. A similar idea emerges in modern physics, where interactions between subatomic particles build a fabric of relationships that shapes reality itself.

The concept of Dharma is rooted in the principle of cosmic harmony. In the Bhagavad Gita, another famous Hindu sacred text, Krishna reminds Arjuna that each person has a unique duty, a role to play in the vast theater of life. To ignore one's Dharma is to break the universal balance.

In the world of quantum physics, we find a fascinating parallel in the theory of decoherence and entanglement. According to quantum mechanics, each particle is connected to others through intricate relationships. Observation or action on one particle can have effects on other parts of the system, even at enormous distances. There is, therefore, no such thing as an isolated or neutral particle. Similarly, in Hinduism, each person is seen as an inextricable part of a universal web, where Karma-the principle of cause and effect-traces paths similar to those analyzed by physicists.

One of the scholars who has best explored the connection between Eastern philosophy and modern science is Fritjof Capra. He states that the insights of the Vedic sages surprisingly reflect modern cosmology. Capra, a physicist and popularizer, points to the profound similarity between the poetic language of the ancient Vedic texts and the scientific theories of contemporary physics. The network described in the ancient verses is not very different from the idea of a quantum field united by interactions and correlations.

For example, the process of quantum entanglement, in which two particles remain linked even when separated by incredible distances, recalls the Vedic view that all things, living or not, share a common divine spark, originating from the primordial sound of "Om."

A practical example of how well the concept of interconnectedness was embedded in Vedic culture can be found in the workings of traditional Indian temples. These places were not only spaces of worship, but symbolically represented the entire cosmos. Each part of the temple, from the foundation to the top, embodied an aspect of the universe. The "*garbha griha*," the inner chamber, with its concentrated energy, was reminiscent of the indivisible core of reality, analogous to what physicists today study in particle physics.

Another interesting insight relates to the philosophy of the great rishi Yajnavalkya, who lived around the 9th century B.C. Yajnavalkya was one of the first thinkers to question the unity of all things, stating in the Brihadaranyaka Upanishad that what exists in the universe is reflected in human beings and vice versa. In modern terms, it is the holistic concept of the "unified field" of physics, according to which all particles and forces are manifestations of a single fundamental origin.

Underlying it all, the indivisible Dharma and the behavior of subatomic particles remind us that the meaning of existence, both spiritually and physically, is based on a principle of relationship. Hinduism has celebrated this holistic view for millennia, encapsulated in the symbolism of "*bindu*," the central point of the universe.

For physicists, the quantum field represents the fundamental level of reality, where particles "come and go" in a continuous whisper of interactions. For Vedic mystics, Dharma guides every element in the universe toward a cosmic dance of order and harmony.

The connection between Eastern philosophies and quantum physics returns us to a worldview in which science is nothing more than a continuation of mysticism with different tools.

Modern physics continues this millennia-old conversation, offering new metaphors for reality and bringing us closer, perhaps, to the same universal truth that the rishis intuited: everything is connected, and every node in the network contributes to the grand cosmic design.

Dharma and the Multiverse: Infinite Choices and Possibilities between Ethics and Science.

Every choice counts. This is not just a moral lesson, but a principle that interweaves ancient philosophy and modern physics. In

Hinduism, Dharma is at the core of this idea: the duty and ethics that govern every aspect of life, from the most everyday actions to the most existential decisions. But what if every single choice created a new world? This is the point of contact between Dharma and the quantum theory of the multiverse proposed by Hugh Everett III, a visionary physicist in the 1950s.

The concept of Dharma is deeply intertwined with Indian culture. It is the individual and collective code of ethics, the bond that holds the world together in its cosmic order. Each person has a specific Dharma-a unique path shaped by his or her nature and historical and social context. This principle is reflected in the great Indian epic of the Mahabharata, where we find one of the most emblematic stories related to ethical choices and their consequences.

Prince Yudhishthira, leader of the Pandavas, is famous for his unwavering adherence to the Dharma. However, this virtue confronts him with harrowing dilemmas during the Kurukshetra War. One of the most famous moments is his decision not to lie in order to win, even though he knew it would cause immense suffering.

A crucial moment in the moral dilemma occurs when Yudhishthira is asked to tell a half-truth to contribute to the victory of the Pandavas. Underlying this decision is the strategy designed by Krishna to defeat Dronacharya, one of the most powerful warriors in the Kaurava ranks. Krishna knows that Dronacharya, the preceptor of the Pandavas and Kauravas, is invincible when fully focused in battle, but also that his attachment to his son Ashwatthama could be exploited. The strategy involves making Dronacharya believe that Ashwatthama is dead, causing him to lose the will to fight and thus allowing the Pandavas to overwhelm him.

Yudhishthira, who is known for never telling a lie, finds himself in a harrowing situation: if he chooses to tell the truth, the outcome will be the prolongation of the battle and the loss of many lives; if he agrees to tell a half-truth, he compromises his moral integrity but could save countless people by ending the war more quickly. In the end, Yudhishthira utters the ambiguous phrase, "*Ashwatthama is dead,*" adding under his breath, "*Ashwatthama is an elephant*" This deception leads Dronacharya to lay down his arms, believing that his beloved son has fallen, and to be killed.

What would have happened if Yudhishthira had chosen differently? Perhaps, the world itself would have followed a different course of events.

Hugh Everett's "many worlds" theory offers a strikingly similar perspective. According to this interpretation of quantum mechanics, whenever an event can take different directions, the universe "bifurcates." Thus, there are infinite parallel universes, each the result of alternative choices.

Imagine Yudhishthira in this context. If the multiverse exists, then there would be a world in which the prince chooses to lie and another in which he does not. In each universe, the consequences of that choice shape reality in a unique way. Again in the Mahabharata, this concept is reinforced by the notion of Karma: every action, good or bad, leaves an indelible imprint on the cosmic balance. Analyzed through the quantum lens, Karma becomes a kind of "source code" that guides the evolution of worlds in the multiple branches of the multiverse.

The connection between Dharma and multiverse is not only poetic but also practical. Quantum mechanics suggests that the behavior of subatomic particles is influenced by a kind of "potential possibility." For example, the famous double-slit experiment shows that a particle can exist in many different states until an observation fixes one. But if we take this phenomenon back to a macroscopic scale, what happens to our decisions?

Conscious behavior, dictated by Dharma, can be seen as a tool for navigating the infinite possibilities that the multiverse offers. Each choice reshapes the fabric of reality, tying in with the profound ethical responsibility that comes with human freedom.

When Yudhishthira walks the fine line between truth and falsehood, he also embodies the tension between freedom and responsibility. Although the multiverse suggests that each choice creates a new world, this does not reduce the value of ethics. On the contrary, it amplifies the importance of Dharma as supreme guidance. The famous phrase attributed to the Mahabharata,

"The Dharma protects those who protect it."

It thus takes on a deeper meaning. Respecting one's moral order is not just a virtue, but a creative act that shapes the universe itself.

The parallels between Hinduism and quantum physics remind us of the universality of the human quest for meaning. On the one hand, ancient Indian rishis meditated on the intertwining of ethics and reality. On the other, modern physicists like Everett and Niels Bohr explored quantum uncertainty and its cosmic implications. Both camps converge on a surprising truth: choices matter, and every action leaves a trace.

In today's accelerated world, the Dharma invites us to reflect on our decisions. Every small act has the power to influence the "quantum fabric" of reality, opening up possibilities beyond our immediate perceptions.

Reflecting on Dharma in the context of the multiverse is not just a mental exercise, but a profound challenge to live mindfully. Just as Yudhishthira chose Dharma in the epic war of the Mahabharata, we too must choose-every day-when faced with the dilemmas of modernity. We do not know how many realities will arise from our decisions, but we can be sure that the thread of ethics will run through them all.

Neutrino oscillation and Dharma flowing.

The neutrino oscillation, a phenomenon discovered in the 20th century, might seem far removed from the ancient Vedic texts, but perhaps it is not at all. Connecting this subtle quantum dance to the notion of Dharma, the duty and ethics of every individual, a profound reality emerges: life and its purpose are dynamic, just like the universe.

Neutrinos, nearly massless particles, silently traverse the universe. Despite their elusiveness, they are fundamental to understanding some of the most secret dynamics of existence. One of their most fascinating properties is oscillation: as they travel, neutrinos change "identity." They change from one flavor (electronic, muonic or tauonic) to another, in a process that seems to contradict the idea of fixity or permanence. It is like observing someone constantly changing role and perspective depending on the situation.

This transformation is not random. It is influenced by their environment and the physical laws that govern the cosmos. Such behavior can be seen as a mirror image of the concept of Dharma in Hinduism, which changes in response to the needs of life, time and context.

In Sanskrit, Dharma represents duty, justice and cosmic harmony. However, it is not a rigid law. Krishna, one of the central figures of Hinduism and the protagonist of the Mahabharata, exemplifies how Dharma is not static. In the Bhagavad Gita, Krishna begins as a peacemaker, trying to prevent war between the Pandavas and the

Kauravas. But when peace efforts fail, Krishna abandons the role of mediator and becomes a master strategist and warrior guide.

This transformation of Krishna is not inconsistency or weakness. It is adaptation to context. Like the oscillation of the neutrino, his behavior shows that the Dharma, to be true, must respond to the surrounding reality. If Krishna had persisted in his vision of peace even when it was now impossible, he would have failed in his task.

Neutrino oscillations and flexible Dharma point to a shared principle: the deep interconnection between events and constant change. In quantum physics, each particle is never isolated. Every action, even the smallest, affects the whole system. This principle is in line with the Hindu idea of Brahman, the universal indivisible reality.

Vedic scriptures thousands of years ago stated that every being is interconnected in a divine network called Rta, the cosmic order. Recent experiments with neutrinos at research facilities such as Gran Sasso in Italy or Super-Kamiokande in Japan seem to breathe new life into this ancient notion. When neutrinos change state, they do so in harmony with the laws of the universe. Similarly, when human beings act in accordance with the Dharma, they come into harmony with a larger order.

Krishna taught that Dharma should not be seen as a rigid set of rules, but as an art. A warrior's duty in peacetime is to protect and negotiate. But in times of war, it is to fight justly. This principle applies not only to kings, but to every individual. Just as neutrinos dance between different states, our Dharma can and should swing, following the rhythm of life.

Hinduism reminds us that change is inevitable. Like the neutrino, we cannot always remain in the same state. Our role in family, society and the world constantly fluctuates. Resisting change is not only futile, but contrary to the very laws of the universe.

The neutrino oscillation and the concept of Dharma remind us that life is not about static certainties. Neither science nor spirituality finds definitive answers in stationary things. Everything evolves, everything connects. Just as quantum physics has taught us to see the universe as a constantly moving web, Hinduism invites us to see Dharma as a flowing, changing and adapting river.

In the Bhagavad Gita, Krishna urges Arjuna not to fear change but to embrace it wisely. This age-old lesson, now enlightened by modern science, also accompanies us today. The duty of the individual is to flow with time, just like neutrinos, transforming and responding to the needs of existence. Nothing is static, and in this movement lies the very essence of life.

"Aligned" in Chaos.

Every gesture, thought and choice, according to Hinduism, contributes to a deeper connection with one's purpose and the balance of the universe. In an age dominated by science and quantum exploration, we can find surprising parallels between Dharma and the deep mechanisms of the subatomic world.

Quantum physics reveals to us an invisible landscape of complex interactions. Subatomic particles do not act in isolation. Every event on the quantum plane-an absorbed photon, an emitted electron-has an impact on the fabric of reality. This phenomenon, known as quantum coherence, is what allows particles to maintain an intrinsic order despite apparent chaos. Interference, such as an interaction with the external environment, disrupts this order and causes what physicists call "decoherence."

Hinduism suggests something similar with Dharma. Each individual is part of a larger system, an ordered macrocosm.

A practical contemporary example of quantum coherence is found in quantum computer technology. These devices exploit quantum coexistence to perform computations at a speed unthinkable for classical computers. Each particle, like a quantum bit (qubit), must remain "aligned" within the system to preserve a functional order. Here, however, a danger emerges: even the slightest disturbance is enough to disrupt connections. Interestingly, in Hindu philosophy, alignment with the Dharma requires such vigilance. Any choice, any deviation, can lead to "decoherence" in life, that is, disharmony with one's purpose.

Let us take the tale from the Mahabharata, an ancient Indian epic poem. One of the heroes, Yudhishthira, the righteous king, is known for his dedication to the Dharma. In a famous episode, Yudhishthira faces a crucial question: sacrifice his principles to save his family or remain true to the Dharma? He chose the Dharma, even at the cost of enormous suffering. Physicists would say he preserved consistency, remaining "aligned" despite the chaos.

The parallel with modern science invites us to reflect. On the one hand, quantum physics studies the behavior of particles, the language of the invisible universe. On the other, the concept of Dharma invites us to live ethically and consciously. Both suggest to us that order arises from an invisible but omnipresent discipline.

Today, Dharmic order could be interpreted as a guide for dealing with the chaos of modern life. Ethical decisions, like quantum interactions, may seem insignificant individually, but collectively they shape the world. A quote from Nobel laureate Niels Bohr, a deep thinker of quantum mysteries, fits perfectly with this idea:

"In the realm of physics, every part depends on the whole."

Hinduism would have said the same, but in more poetic language:

"Tat Tvam Asi"-"You are That."

Staying aligned in chaos is a challenge. But Dharma, like quantum coherence, shows us the way: accept our unique purpose, act with awareness and find harmony in the apparent disorder.

Dharma complementarity.

Bohr's principle of complementarity, proposed in 1927, introduces the idea that two seemingly opposite aspects of reality can coexist and complement each other. In the context of quantum physics, a photon, for example, can exist as both a wave and a particle, depending on how it is observed. This duality does not imply contradiction, but instead suggests that reality is more complex than it appears.

Similarly, Dharma moves between two interconnected truths: personal duty (called *SvaDharma*) and universal duty (*Sanatana Dharma*). The former refers to individual obligations, related to the context in which a person lives, his family, his profession, and his "position" in the world. The second, on the other hand, represents the universal principles that govern the cosmic order. Like wave and particle, these two dimensions do not exclude each other but complement each other.

Dharma complementarity can also be related to another aspect of quantum physics: interconnectedness. In quantum mechanics, particles are connected through invisible laws, such as quantum entanglement. When two particles interact, their fate remains intertwined, regardless of distance. Every action on one particle has an effect on the other.

This principle reflects a central aspect of Dharma. According to the Hindu view, every individual action inevitably affects the entire cosmic fabric. A famous episode from the Rigveda, one of India's oldest texts, represents this idea. The story tells of a sage who, in cutting down a tree to build a shelter, is confronted by a crow that lives in that very tree. The sage realizes that his act, however seemingly justified, takes life away from another being. This leads him to reflect on how any act, no matter how insignificant, can alter delicate universal balances.

Philosopher Niels Bohr, interviewed in the 1950s, emphasized how the concept of complementarity was applicable not only to physics, but also to psychology, biology and even philosophy. Without ever explicitly mentioning Dharma, Bohr recognized in the ancient Eastern traditions a profound intuition of balance between opposites. Indeed, Hinduism has always incorporated dichotomies into its worldview: light and shadow, life and death, creation and destruction.

A parallel can also be drawn with modern approaches to sustainability. In an age dominated by the urgency of balancing economic progress and environmental preservation, the Dharma offers a model for navigating between these two demands. Each individual, as a particle in a physical system, must consider how his or her personal actions contribute to the universal order.

The Dharma concept of complementarity is not just arcane Eastern knowledge, but a powerful lesson for the present. We live in an increasingly interconnected world, where every gesture can have global repercussions, such as climate change or economic crises. This requires us to balance personal and collective responsibilities, making choices that reflect both truths: our individual lives and our place in the universe.

Non-locality and the global Dharma. Influences beyond space and time.

When Albert Einstein sarcastically defined the phenomenon of "non-locality" as a "*spectral action at a distance*," he probably did not imagine how crucial that definition would prove to be in redefining our understanding of the cosmos. In quantum physics, non-locality represents one of the most fascinating enigmas: two particles, separated by cosmic distances, can affect each other instantaneously, as if space and time were an illusion rather than a barrier.

Surprising parallels with Eastern philosophies, particularly Hinduism, emerge from this perspective. The concept of Dharma, which we might translate as "duty," "moral law," or "ethics," expands to a universal dimension that transcends the simple everyday. Just as in quantum non-locality, every individual action, no matter how seemingly isolated, generates consequences that reverberate far beyond immediate boundaries in time and space.

According to Hinduism, Dharma is not only the moral guidance of the individual in his existential context, but is a law that governs the entire cosmic order. This principle manifests itself not only in personal acts, but also in intersubjective, social and cosmic relationships. The idea that the individual is an integral and responsible part of an interconnected network of events invokes precisely the concept of non-locality: nothing happens in isolation. There are no actions unrelated to the overall system of existence.

In this sense, Hindu philosophy invites us to see "duty" not as an individual constraint, but as a global, shared duty that takes into account the long-term repercussions of each gesture. Gandhi, one of the most emblematic figures of global ethics, embodies precisely this principle. The Mahatma argued:

"You must be the change you want to see in the world."

This seemingly simple phrase holds a powerful parallel to non-locality: one right action, rooted in the Dharmic, can reverberate infinitely, affecting realities far away in space and time.

Experiments on non-locality, such as Bell's famous theorem of the 1960s, have shown that quantum reality works through invisible connections. Two particles that have interacted once remain "linked," regardless of the distance between them. Similarly, Indian thought teaches us that every human thought, word or action weaves invisible threads of connection that guide collective destiny.

Imagine, for example, a historical event such as the *Salt March* of 1930. Guided by his Dharma, Gandhi undertook a seemingly limited but symbolically immense act: a walk to protest the British colonial monopoly on salt. A simple local action, rooted in a specific context, turned into a message that crossed India's borders and inspired national liberation movements around the world. "Locality," at that moment, was transcended: one man's gesture generated a global-ethical rebound.

The concept of global Dharma, inspired by non-locality, urges us to reconsider our responsibilities. If every event is connected, as both Indian philosophy and quantum physics suggest, then our "duty" is not reduced to the personal sphere. It expands to the protection of humanity, the planet and future generations. This calls for an ecological and social ethic that transcends national boundaries, cultural limitations and time prejudices. Exactly like Gandhi, who saw his duty not only to India but to a united humanity.

We could summarize this vision with a universal image: the beating of a butterfly's wings. The idea, made famous by deterministic chaos, gains new light in the interweaving of Dharma and non-locality. A flap of wings can indeed unleash a hurricane on the other side of the world. An act of kindness can transform a life far away. An act of violence can trigger global conflicts.

The parallelism between non-locality and Dharma translates into a call for awareness. In today's context marked by global interdependence, global Dharma may be our only chance to address challenges such as climate change, economic inequality, and geopolitical tensions. Eastern philosophies and modern physics, in their own ways, tell us the same thing: we are all connected. And with this connection comes a responsibility that knows no boundaries, either in space or time.

Dharma and the superposition principle.

Dharma represents the principle of duty and personal ethics. But how does this ancient view connect to modern quantum physics? Surprisingly, it connects through the principle of superposition. Just as a particle can be in multiple states simultaneously, Dharma emphasizes the need to recognize the multiple options in life and act within one's role.

In Hinduism, Dharma means much more than "religious duty." It is the moral compass that guides every individual in daily behavior. It is often depicted as the pillar of the universe, the principle that governs the interactions between people, nature and the cosmos. This philosophical idea is intertwined with the notion that every action, even the smallest, affects the universal balance.

Dharma is both personal and collective. Siddhartha Gautama, the Buddha, is an example of this. Born in the kingdom of Kapilavastu in the 6th century B.C., Siddhartha was faced with a profound dilemma: to follow the royal Dharma as prince and ruler, or to take the spiritual path to seek enlightenment. Eventually, he left the luxury of the court and chose the path of meditation and renunciation, showing the world how personal action can reveal universal harmony.

Quantum physics describes reality as a mosaic of possibilities. One of its most fascinating concepts is that of superposition, which states that a particle can be in several states at the same time, until an observation is made to determine its position. We can imagine a particle as a being suspended between "doing" and "not doing," between multiple potential trajectories.

Applying this lens to Dharma philosophy, the choice of action is presented as an overlay of possibilities. Each decision creates a new line of reality. As in physics, choices are not neutral. They influence the fabric of life and personal and collective destiny.

The parallel between Dharma and physics becomes evident when we consider modern life. Every day we are faced with ethical choices that appear as overlapping possibilities. Take, for example, environmental activist Vandana Shiva, who was deeply influenced by Hindu concepts. A physics graduate, Shiva decided to abandon her scientific career to devote herself to safeguarding the planet and the rights of farming communities in India. In her case, too, her choice of personal Dharma impacted the world around her, weaving a larger web of collective impact.

Let us return to Siddhartha Gautama for a deeper lesson. Legend has it that before attaining enlightenment, he meditated under a Bodhi tree in Bodh Gaya. During this meditation, he faced not only temptations sent by the demon Mara but also an internal battle. Conflicting thoughts held him back: continue living as an ascetic or give up and return to royal ease. This phase can be seen as his "*quantum superposition*." Siddhartha chose his Dharma, transcending both material and ascetic life.

This choice had universal consequences. Siddhartha became the Buddha and inaugurated a philosophy that still resonates today. His example teaches us that possibilities are not infinite only in theory. They become reality through meaningful action.

Both Dharma and quantum physics highlight a holistic view of the universe. When we choose ethically, we live in harmony with this interconnectedness. Just as observing a particle affects its state, our actions shape reality and the future.

When imbalance begets evolution.

In ancient Eastern philosophies, the concept of Dharma represents the universal duty, the inherent ethics that every individual must follow to maintain harmony in the cosmos. In a phrase, it is the calling of every living person in the grand scheme of existence. However, what happens when this balance is broken? Is it really always "*immoral*" to deviate from the predestined path? Here a fascinating idea arises: imbalance may be necessary for an evolution.

In quantum physics, a parallel can be found in the behavior of particles out of equilibrium. When a system leaves its steady state, new dynamics emerge that often lead to more complex and adaptive states. This concept-that of an "*order emerging out of chaos*"-reverberates deeply in mythological tales and Vedic texts related to the dynamic Dharma view.

An emblematic figure embodying "broken Dharma" is Parashurama, the sixth avatar of Vishnu. According to the texts, Parashurama was destined to be a priest (Brahmana), a man of peace and knowledge. However, his personal experiences prompted him to act against his traditional role. When a dynasty of warrior kings (kshatriya) abused their power, sowing injustice, Parashurama took up arms to restore order.

This act breaks his original Dharma and transforms him into a warrior, an ambivalent identity that mixes the wisdom of the priest with the strength of the fighter. Parashurama is said to have swept away entire generations of oppressors with his justified anger. Yet, he did not act out of personal revenge. He acted for the sake of cosmic balance, performing a "*dynamic Dharma*," which abandons immobility to adapt to the moment.

This myth raises profound questions: is it still Dharma if it requires an act of destruction? Is it possible to see the breach of duty not as a failure, but as a way to evolve to a higher state?

To understand these ideas in the light of modernity, we can look to the world of quantum physics. In a quantum system, the concept of disequilibrium is not always synonymous with disordered chaos. For example, phenomena such as phase transitions-the transition of frozen water to a liquid or vapor state-represent the breaking of an equilibrium state to create something new.

Similar dynamics are observed in the study of particles. Experiments with the Bose-Einstein condensate, carried out in the 1990s and early 2000s, show that systems, when pushed out of equilibrium state, create new forms of order previously unattainable. Nature itself seems to follow a paradigm similar to the broken Dharma described by Hinduism: stability is sacrificed for evolutionary progress.

This concept of imbalance as a creative tool also appears in another central figure in Hinduism, Shiva. Known as the "Destroyer," Shiva does not represent a final end, but the eternal cycle of transformation. With his Tandava, the cosmic dance that shakes existence, Shiva destroys what is stagnant to make way for the new. This parallel between the philosophical concept and quantum dynamism is particularly intriguing: both disciplines recognize that change is necessary to progress.

In addition, Shiva and Parashurama are connected by a significant episode described in the Shiva Purana. Parashurama, filled with pride in his destructive strength, challenges Shiva, but is defeated by the latter. Shiva, after humiliating him, is said to restore his composure, reminding him that destruction without wisdom only brings chaos. This moment symbolizes the need to find balance-even in imbalance-to remain true to the Dharma.

In its deepest concept, Dharma is not a rigid law. It is a principle that requires discernment. A step into the unknown -- into imbalance -- may seem like a mistake, but it can also become the engine of progress. In the Bhagavad Gita, Krishna explains to Arjuna that sometimes breaking one's role is the only way to fulfill a greater purpose.

In the modern sciences, complex systems show the same evolution. Chaos may appear as a fracture of order, but it is often an opportunity to generate structures never seen before. In galaxies, particles and human lives, disorder is almost a call to new possibilities.

Whether it is Parashurama's anger, Shiva's dance, or out-of-balance quantum dynamics, the message is clear: imbalance is not always a threat. It can become a creative force. Broken Dharma does not destroy the cosmos, but transforms it, guiding it toward deeper evolution.

Balance, then, should not be sought as a permanent condition, but as a starting point. Only by accepting imbalance can we explore the hidden potential of what is still possible to become.

Goddess Saraswati and quantum flux,

In Indian culture, Saraswati represents wisdom, art, and the continuous flow of knowledge. Goddess of knowledge and the arts, Saraswati is also the guardian of speech, language and creativity. Her image is associated with the flow of a river, a symbol of movement, knowledge and constant adaptation. This key figure in Hinduism offers a fascinating cue to reflect on concepts from quantum physics, such as the balanced system and dynamic flow of energies that affect reality.

In Hindu tradition, Saraswati is not simply a goddess, but a symbol of "*applied Dharma.*" Dharma also includes the concept of universal order. Saraswati embodies the spiritual duty to learn, adapt and flow with the events of life, just as a river adapts to the terrain it flows through.

This is a concept closely related to balanced quantum systems, where equilibrium is not static, but dynamic. In the subatomic world, particles interact with each other in a fine fabric of energy exchanges. Each event causes a change in the system, but the overall order is maintained precisely because of the continuous adaptation of its components.

n interesting parallel can be drawn between Saraswati and the principle of "superposition" in quantum physics. A particle, or rather a quantum system, exists in many states simultaneously until an observation establishes its final state. This process depends on previous events, creating a "flow" of cause and effect that is reflected in the reality we perceive.

Saraswati represents a continuous flow of knowledge, a constant adaptation to situations. In the Dharma context, this means acquiring new information and applying it to maintain harmony in a larger system. A traditional anecdote tells that Saraswati, in her divine aspect, chose to reside in books and musical instruments. The message is clear: knowledge and art should be used to improve life, with the same dynamic spirit we see in quantum interactions.

The Rig Veda, one of the oldest texts in Hinduism, celebrates Saraswati as "*she who gives life to ideas through language.*" This description relates to the concept of creative flow, which modern psychology today calls the "*flow state.*" This state occurs when a person is so immersed in an activity that he or she achieves a perfect balance between challenge and competence. It is not just a subjective experience: a study done in 2021 by neuroscientist Mihaly Csikszentmihalyi shows how the human brain, in a state of flow, achieves a neurochemical harmony similar to a balanced quantum system.

From another perspective, Saraswati reminds us that balancing is never static, but is a smooth accompaniment of change. This is the heart of the Dharma: living mindfully, adapting without renouncing universal truth.

Quantum physics describes reality as a web of events and interactions. Nothing exists in isolation, and each particle influences all others, even at impressive distances. This phenomenon, known as "quantum entanglement," closely resembles the Hindu idea of universal interconnectedness. Saraswati, with its meaning of flow of knowledge, embodies this relational vision.

A practical example of this philosophy can be taken from the story of Rabindranath Tagore, the great Indian poet and thinker. In 1926, Tagore met Albert Einstein to discuss precisely the relationship between science and spirituality. Tagore argued that the cosmos could not be separated from human consciousness, a viewpoint that reflects both Dharma and quantum insights into universal interdependence.

The figure of Saraswati offers a rich and inspiring metaphor for interpreting the principles of quantum physics. Just as a balanced quantum system must continuously adapt, human beings are also called to follow the Dharma of knowledge: to learn, adapt and flow with change. Saraswati teaches us that wisdom is not static, but is a continuous process of recognition, adaptation and application.

In the modern world, where scientific and spiritual knowledge sometimes seem distant, Saraswati represents a bridge between these two worlds. His message invites us to seek balance, not in stillness, but in continuous movement. Through Dharma and the flow of knowledge, we can find our place in the vast fabric of the universe.

Dharma and wave function collapse....

The East and the West have for centuries seemed to speak different languages, especially when it comes to philosophy and science. Yet, at certain times, these two directions of thought meet with surprising harmony. One such point of contact is the Indian concept of Dharma and the collapse of the wave function into quantum physics. Two worlds that, at least on the surface, could not seem more distant, come closer together when it comes to choices and the infinite possibilities that each choice brings with it.

Dharma is the moral, social and spiritual duty that every individual has in his or her life path. In the Indian epic, the Mahabharata, Dharma represents an essential guide, an ethical compass.

Quantum physics, with its revolutionary principles, tells us that the world is not deterministic. Before a measurement, subatomic particles have no definite position or state. They are in what physicists call "*superposition*": a multitude of potential coexisting states. Only at the moment one observes or interacts with the system does the wave function collapse, and only one of those states becomes real.

It is a subtle but fundamental process: reality, at the quantum level, is shaped by an act of choice.

The idea of wave function collapse can be surprisingly juxtaposed with Dharma. Every choice we make in life is like a quantum observation. Until we choose, unlimited potentials open up before us. Each option represents a possibility, a road not yet traveled. But the moment we act according to our Dharma, just as happens in the measurement of a particle, we define reality. We "collapse" the infinite into a single concreteness.

Let us take Mahatma Gandhi's life as an example. Gandhi could have been many things: a wealthy lawyer, a quiet father, an ordinary man. Instead, he chose his Dharma: nonviolence and fighting for India's freedom. In that moment, endless other possibilities faded away, and Gandhi became what he chose to be: a symbol of peace and justice.

At the quantum level, this is what happens whenever a particle, interacting with the world, takes "form." Dharmic choice, then, is not just moral or practical; it is a creative force.

In the quantum world, the seeker actively participates in the creation of reality simply by observing it. Similarly, Hinduism suggests that each individual is an inseparable part of Brahman, the cosmic whole.

This interconnectedness directly relates to Dharma. Each individual decision influences not only the individual's present, but also the overall course of existence. It is like a pebble thrown into a pond: the waves generated propagate endlessly, shaping destiny.

Despite the differences between philosophy and science, both worlds prompt us to reflect on the deeper meaning of our choices.

Richard Feynman, one of the most celebrated physicists of the 20th century, said that:

"The act of observing alters what is observed."

This principle of quantum mechanics echoes the age-old wisdom of Hinduism regarding Karma and Dharma.

In one's life, choosing according to one's Dharma is not only an ethical act. It is a way of shaping the chaos of possibilities. Just as the collapse of the wave function defines the state of a particle, following one's Dharma defines our existence.

Choices, whether subatomic or human, are fundamental. In physics as in spirituality, nothing happens without conscious intervention. Ancient Eastern philosophies, such as Hinduism, and modern theories, such as quantum physics, offer us tools to understand this universal truth: infinite potential can only become real through action.

Dharma, therefore, is not an abstract concept. It is the way we translate the possible into reality. Recognizing one's Dharma is not just an ethical imperative, but a profound creative act. And this, perhaps, is the greatest teaching that Eastern philosophies and quantum physics share.

The purpose of the universe: cycles and quantum physics.

More than just a moral or ethical duty, Dharma reflects the harmony of a cosmic order, a balance that holds together the personal microcosm of each individual and the universal macrocosm. We are all fragments of a larger design, where every action carries weight and every choice is inscribed in an infinite web.

Quantum physics, paradoxically, seems to speak the same language. Scientists observe that in the subatomic world a single event never exists "by itself." The principle of entanglement, for example, reveals that particles can remain connected regardless of physical distance: a change in one immediately affects the other. Similarly, the notion of Dharma in Vedic texts implies that what happens in one part of the universe has repercussions everywhere. It is not difficult, then, to find a parallel between the natural laws of physics and the ancient mystical conceptions of India.

According to the Puranas, sacred texts in the Hindu tradition, time is not linear. Traveling forward, it would return to the starting point. This concept, known as Kalachakra (*the Wheel of Time*), describes an endless sequence of creative and destructive cycles. In Hindu cosmology, the universe is born, grows, dissolves and is reborn. Every event, even the smallest, contributes to maintaining the cyclical order.

Modern theories seem to agree. Theoretical physics suggests that the universe may operate according to a complex order, where space and time are intertwined. String theory, for example, postulates that the universe is composed of infinitesimal, string-like vibrations that pulse along dimensions beyond our perception. Each string has its own "duty," its own unique rhythm, which contributes to the collective symphony.

This scientific narrative recalls the concept of cosmic Dharma: each element, each particle, acts according to a precise function that is repeated over time. The universe, thus, finds balance in a seemingly chaotic but remarkably coordinated order.

Entanglement and the unity of things

In 1944, the distinguished Austrian physicist Erwin Schrödinger, known for the famous "cat" experiment, defined entanglement by saying:

"It is not a something, but the something."

For Schrödinger, the mysterious connections between particles represented a fundamental truth of the universe. If we transfer this idea to the field of Eastern philosophies, we find direct assonances with the Vedantic doctrine of unity (Advaita): everything that exists is connected, in an indivisible field of energy and consciousness.

Each action contributes to a larger mosaic, as each particle affects the quantum fabric of the universe.

A symbolic example of this connection is found in the life of Jagadish Chandra Bose. The Bengali scientist and philosopher, who studied vibrations in plants in the early twentieth century, demonstrated that even the simplest living organisms respond to universal patterns. For Bose, there was no fundamental distinction between the human, plant or mechanical realms: everything follows shared dharmic dynamics.

In the present context, the implications of these ideas are profound. To study the universe is not simply to discover how the physical world works, but to understand our role in it. Drawing parallels between the Dharma and modern physics can offer us not only a unifying vision, but a guide to living with greater awareness.

Like particles vibrating in synchrony in string theories, so in the Dharma conception every human action is a wave propagating in the cosmic universe. Time, as the ancient Puranas suggest, is not an arrow but a circle. And we, tiny particles in this endless cycle, have the task of harmonizing with the great rhythm of the universe.

Quantum ethics.

The discoveries of quantum physics not only revolutionize our understanding of reality, but also invite us to reflect on profound ethical issues. Ethical reflection that dialogues with science is not only responsible for guiding the use of technology, but also for redefining the role of human beings in the universe. Hinduism, with its holistic and interconnected view of reality, offers valuable insights to address these challenges and to guide technological progress toward a compassionate and responsible vision.

Ethics and interconnectedness. When Hinduism meets quantum physics.

The discoveries of quantum physics have thrown open the door to a new understanding of reality, revealing a universe where everything is deeply interconnected. It is not just about formulas and particles, but also about fundamental questions about our existence and values. How we choose to live and act in a world based on interconnectedness forces us to reflect on our ethics. And in this very dialogue between science and spirituality, Hinduism can offer us a rich perspective.

The quantum universe, revealed in the 20th century by scientists such as Niels Bohr and Werner Heisenberg, seemed to challenge every rationalistic certainty constructed by the West. As physics discovered that subatomic particles were not independent entities, but related systems strongly influenced by the observer, a striking similarity emerged with the ancient worldviews of Hinduism. The latter had for millennia described the cosmos as a dynamic network of relationships, well symbolized by the image of *Indra's Web*. Through this metaphor, Hindu sages were already inviting people centuries ago to see reality as an infinite web, where each node reflects all the others, in perfect balance.

But what does this mean for ethics? Quantum physics, like Hinduism, offers us an uncomfortable but liberating truth: we are all interconnected, not only with human beings, but with the entire universe. Ignoring this reality is not only irrational but also dangerous.

Taking a cue from these insights, many scholars and cultural exponents have begun to explore the foundations of a "*quantum ethics*" that can respond to the challenges of our time.

According to the Bhagavad Gita, one of the founding texts of Hindu philosophy, every individual action has an impact on the cosmic order. This principle, reflected in the law of Karma, teaches us that every behavior, even the most invisible, contributes to shaping shared reality. Is this not the same message we get from the uncertainty principle or the phenomenon of quantum entanglement? Science, through quantum physics, shows us that no particle-and therefore no being-really exists or acts in isolation.

An important historical example of this convergence of science, spirituality and ethical responsibility is the thought of Erwin Schrödinger, one of the fathers of quantum physics. The scientist was deeply influenced by Hinduism. In his philosophical reflections, Schrödinger adopted the Indian view of Advaita Vedanta, according to which the entire universe is one indivisible essence.

This thought has an immense ethical scope. If everything is One, if every human being, animal and particle is part of the same reality, how can we justify behavior that ignores this interconnectedness? Destroying the environment, exploiting others, acting for pure personal profit appears not only immoral, but also contrary to the true nature of reality.

In the 21st century, the idea of a quantum ethic based on interconnectedness is proving more urgent than ever. The environmental, social and technological crises we face are the result of a fragmented pattern of thinking that does not take into account the global consequences of our local actions. The ethical teachings of Hinduism can help us direct scientific progress toward an idea of universal well-being.

David Bohm, Hinduism and quantum physics: a unified view of reality.

One of the most fascinating intellects of the 20th century, David Bohm (1917-1992), an American physicist and visionary thinker, tried throughout his life to solve a problem that has plagued humanity for centuries: the fragmentation of reality. For Bohm, this fragmented perception was not just the product of the human mind, but an ingrained cultural pattern to such an extent that it even influenced modern science. In particular, quantum physics and Eastern philosophy became fundamental tools for him to imagine a new approach to understanding the cosmos.

Bohm, a student of Robert Oppenheimer and colleague of Albert Einstein, was not just a physicist. He had the rare ability to think beyond the boundaries of academic science, exploring terrain that also embraced spirituality and philosophy. In the 1950s, as he developed a new interpretation of quantum physics, he began to think deeply about concepts that mysteriously echoed the traditions of the East, especially those of Hinduism.

The starting point of his journey was standard quantum theory, which at the time emphasized the so-called "*wave-particle dualism*": light, and in general any subatomic particle, could behave either as a continuous wave or as a discrete particle depending on the conditions of the experiment. Bohm found this fragmentary view unsatisfactory. He perceived it as a reflection of a Western mind obsessed with separation: man and nature, matter and spirit, individual and society.

To counter this fragmentation, Bohm introduced his famous "*theory of hidden variables*" in 1952 with the idea of "*implicit totality.*" For Bohm, there was an underlying, invisible order that connected everything in the universe. At the macroscopic level, things appear to us to be separate and independent, but at a deep level they are connected by an indissoluble, fluid, continuous reality. This concept is strikingly reminiscent of that of Hinduism: the idea that everything is a manifestation of a single entity, Brahman, the absolute reality underlying the phenomenal world.

Hinduism, among the major philosophical traditions of the East, has reflected on the theme of interconnectedness for millennia. According to Vedic texts and Upanishads, everything we perceive is maya, an illusion, a fragmented representation of something greater, eternal and indivisible: the Brahman. In this sense, Bohm's insight into the "implicit all" is not only similar but almost parallel to the doctrine of Vedanta.

Bohm did not consider himself a mystic or a religious man. However, he was deeply impressed by the Eastern idea of a dynamic, holistic universe, where spirit and matter are not separate entities but different expressions of the same reality. In one of his famous dialogues with Krishnamurti, an Indian philosopher and celebrated spiritual teacher, Bohm had this to say:

"Western thought has broken the world into fragments because it has forgotten that the observer and the observed are one."

Bohm believed that science could find common ground with these teachings, helping to overcome the existential alienation of humanity.

But how to explain "implicit order" in concrete terms? Bohm used a simple image inspired by physics and optical games. He spoke of "*two vortices that seem separate but belong to a single river*." Imagine a hologram: each of its parts contains the information of the whole. Similarly, according to Bohm, each fragment of observable reality contains within itself the "wholeness" of the universe, but in a hidden, implied form.

As an example, Bohm told of an ancient experiment conducted by naive pioneer scientists. Near his home in London, Bohm showed visitors a transparent tube filled with two unmixed liquids, such as oil and glycerin. A drop of ink inserted in the center fragmented as the tube was turned. By turning in the opposite direction, however, the drop would magically reassemble. This simple experiment served to show how a seemingly "scattered" structure is not really lost: each of its fragments is still connected to the whole and can be rebuilt if the process can be reversed.

What makes Bohm such an attractive figure is not only the elegance of his theories, but also his ability to build bridges between seemingly opposite worlds. Bohm demonstrated that quantum physics, with its bizarre implications, could dialogue with millennia-old spiritual philosophies such as Indian Vedanta without falling into dogmatic mysticism.

Today, his ideas continue to spark interest. Researchers, philosophers and even artists find inspiration in Bohm's holographic universe. His work invites us to rethink reality as a fluid, interconnected whole, where everything is intrinsically connected to everything else.

Bohm leaves us with a great lesson: human progress cannot come from fragmentation, but from rediscovering wholeness. Today, more than ever, his words resonate powerfully in a world torn apart by social, political and economic divisions. Perhaps it is time to remember that, as the ancient wisdom of Hinduism says, we are all drops of the same ocean. And that "Brahman" that the Vedic sages sought with faith and intuition, Bohm sought with mathematics and passion. One way or another, the universal message remains the same: everything is One.

Dharma and quantum physics. An ethical way to govern the future.

Never before, in the midst of a technological revolution that challenges our understanding of reality, is humanity faced with a crucial question: how to govern new discoveries without losing ourselves in moral anarchy? Quantum physics, with its paradoxes and application power, gives us a glimpse of immense opportunities but also frightening risks. To find an appropriate ethical compass, some contemporary thinkers are looking to the East, particularly to Hinduism's concept of Dharma, the cosmic and ethical order that governs the world.

Dharma, in Hindu philosophy and religion, represents the set of natural and moral laws that ensure the harmony of the universe. It is a universal principle, impersonal and intrinsic to all things. To understand this concept, it is useful to think of the metaphor of "cosmic rhythm": a balance that embraces nature, human actions and the transcendent.

Curiously, quantum physics also thrives on equilibrium. Quantum systems, in their infinite possibilities, find "coherence" only through the interaction of seemingly separate entities. Quantum states exist in superposition until an observation-an interaction-"connects" them to a definite reality. This underscores a principle of interconnectedness that resonates with the idea of Dharma: nothing is isolated; every action reverberates throughout the whole system, whether it is the universe or a society.

In the Bhagavad Gita, one of the foundational texts of Hinduism, Krishna explains to Arjuna that adherence to one's Dharma-that is, one's ethical and cosmic duty-is essential, even when it involves difficult decisions. This teaching, adapted to contemporary dilemmas, could offer valuable guidance for dealing with the challenges of quantum physics and its applications, such as quantum computing, artificial intelligence and the manipulation of subatomic particles.

History has already warned us about the dangers of advanced science without an ethical soul. During World War II, the Manhattan Project generated the first atomic bomb. Scientists like Robert Oppenheimer, one of the key figures in the discovery, found themselves confronted with the legacy of their work in unprecedented moral torment. After the bomb explosion at Hiroshima, Oppenheimer quoted a passage from the Bhagavad Gita: "*Now I have become Death, the destroyer of worlds.*" This sentence, uttered with deep anguish, reflects the weight of knowledge employed without a clear ethical compass.

The Oppenheimer case teaches that scientific progress needs a "higher principle" to guide it. Dharma, understood as respect for universal harmony, could represent that principle.

Today, technology based on quantum physics has similar, if not greater, potential than nuclear power. Quantum computers, for example, promise to solve problems unsolvable by traditional computers, but they could also break down current security systems, making current cryptographic technologies obsolete. What will be the price of access to such power?

The ancient masters of Hinduism would have answered this question by invoking the concept of Ahimsa, nonviolence. In the Dharma perspective, every technological decision must adhere to two basic principles: not to cause harm and to contribute to the collective welfare. Similarly, the architecture of artificial intelligence-one of the fields most influenced by quantum physics-should be designed to serve humanity, not to dominate or destroy it.

A central element of the Dharma is the recognition of the web of life, an idea reflected on the scientific scene today in quantum entanglement. This phenomenon shows that two particles, even at enormous distances, remain mysteriously connected: actions on one particle immediately affect the other. A similar principle, in the Hindu view, applies to the relationship between human beings, nature and the universe: what we do to the world we ultimately do to ourselves.

A practical example of this ethical principle can be found in the field of technological sustainability. Today's dependence on rare resources, used in the construction of quantum computers, raises obvious questions: how can we balance their use with respect for the planet? The answer may lie in the Dharma systemic approach: no resource can be exploited while ignoring the interconnectedness of the environment, society and the future.

The journey from Dharma to quantum physics is not just an intellectual exercise, but an urgent challenge to govern the future. Advanced technologies, if not guided by an ethical compass, can amplify the human capacity to destroy, but also to create. The lessons of Hinduism, which promote balance and respect for cosmic interconnectedness, offer a valuable agenda for contemporary science.

Today, we can choose whether to follow a Manhattan Project-like path or to listen to the ancient wisdom of Dharma, integrating knowledge and morality. As Gandhi wrote, deeply inspired by the concept of Dharma:

> *"Science without humanity is one of the seven causes of man's decadence."*

The future ultimately depends on our commitment to balance the power of knowledge with respect for universal balance.

Technology and power: the need for quantum ethics.

Ideas from quantum physics are rapidly redefining the global technological landscape. Advances in quantum computing, ultrafast cryptography and particle entanglement-based communication systems promise extraordinary innovations. However, behind the enthusiasm for these breakthroughs lie potentially devastating risks. Like any powerful technology, quantum technology risks being used for ends that could increase inequality, reinforce oppressive systems or give rise to new and unpredictable forms of conflict. In this context, the need for an ethical soul has never been more urgent.

In 1945, at the end of World War II, the world watched in dismay as Hiroshima and Nagasaki were destroyed. The use of atomic bombs-the result of the discoveries of nuclear physics-represents the most dramatic example of how scientific knowledge can be used without appropriate moral limits. Over 200,000 people died as a result of the explosions and the long-term consequences, such as radiation. This episode teaches us that the absence of sound ethics in guiding the use of science can lead to incalculable tragedies.

Quantum technologies, while not weapons in the traditional sense, can reflect this same dynamic if left unchecked. For example, quantum computers could override global digital security systems in a matter of seconds. Current encryption, the basis of online privacy and financial transactions, would lose its effectiveness, opening the door to a new era of global vulnerability. At the same time, entangled particle-based communications could grant unprecedented control to governments or corporations, enabling massive surveillance or centralizing power in insidious ways.

To understand how science can be guided toward constructive rather than destructive ends, we can refer to the concept of ahimsa, a fundamental tenet of Hinduism. Ahimsa, generally translated as "nonviolence," is not limited to the rejection of physical violence. It expresses a broader ethic that includes respect for all life, balance with the natural world, and compassion toward every sentient being. Gandhi, during his lifetime, turned this principle into a practical philosophy, using ahimsa to peacefully combat colonial oppression and inspire social justice movements around the world.

Applied to scientific progress, ahimsa invites us to ask: Does this technology respect life? Does it produce more harmony than conflict? Does it serve humanity or privilege the few at the expense of the many? This kind of ethical reflection is not only theoretical, but fundamental to addressing the implications of quantum technologies.

For example, envisioning an entanglement-based communication system that respects the ethics of ahimsa would mean investing in the dissemination of information for global benefit, protecting privacy, ensuring fair access and preventing any abuse of power. Similarly, in the field of medical research, quantum computers could revolutionize the discovery of advanced cures for hitherto incurable diseases. This would be compatible with ahimsa ethics, as it would significantly improve human well-being.

Ahimsa is not the only ethical value applicable to science, but its interconnected view of the world fits perfectly with the principles of quantum physics. In the quantum realm, the concept of entanglement-that mysterious phenomenon whereby two distant particles remain connected, to the point that action on one instantaneously affects the other-reminds us that everything in the universe is connected. This principle, when extrapolated to the ethical plane, implies that every action has consequences that are reflected far beyond the individual or single event.

For this reason, philosophers and scientists today face a crucial responsibility. The scientific community must adopt an approach that takes into account the long-term consequences of its technological discoveries. In addition, governments and academic institutions must commit to creating policies that balance innovation with principles of justice and sustainability.

Hinduism, with the wisdom of its millennia-old teachings, reminds us that technological progress without ethics does not lead to humanity's evolution, but to its potential self-destruction. Therefore, the concept of ahimsa could be an enlightened guide for navigating the still uncharted waters of the quantum revolution.

In the 21st century, science has the extraordinary power to transform the world at a previously unimaginable speed. With quantum technologies just around the corner, we must learn from the lessons of the past. The ahimsa principle offers us a way to balance technological innovation and moral responsibility. Just as Gandhi argued that the end can never justify the means, we must today ask ourselves what end we want our discoveries to serve. This is not just about technological progress, but the future of the entire civilization.

The role of compassionate vision.

In an increasingly interconnected world, reflection on the relationship between science and ethics becomes central. David Bohm, a 20th-century physicist and philosopher, is a key figure in this discussion. Bohm, whose studies of quantum mechanics have profoundly influenced the modern view of reality, was convinced that the universe consisted not of separate entities but of an "implicit order." This order suggests a unified fabric in which each element is connected to the others.

Bohm also developed this vision through his dialogues with Indian philosopher and mystic Jiddu Krishnamurti. The two explored topics such as consciousness, the nature of perception and the meaning of compassion. Bohm believed that a fragmented society, lacking a sense of harmony, was the victim of a "disconnection" from deeper reality. Compassion, for Bohm, was not only a moral value but also a necessary component for understanding universal interconnectedness.

The idea of compassion is also central to Hinduism. In the Bhagavad Gita, an ancient sacred text, we find the concept of *Karma yoga*: selfless action for the common good. Krishna, the deity who guides Prince Arjuna, teaches that all action should be undertaken without attachment to personal fruits, but with thoughts toward universal welfare. This deeply ethical approach translates into practice the idea that we are all parts of a single reality.

Quantum mechanics suggests that, at the subatomic level, boundaries we perceive as solid dissolve. Seemingly separate particles behave as one in dynamics known as quantum entanglement. Bohm interpreted this phenomenon as a scientific demonstration of the interconnectedness of all things. This perspective surprisingly aligns with the views of Hinduism, which for millennia has affirmed the unity of creation, expressed in the concept of Brahman.

Compassion then becomes an ethical response to this interconnectedness. If every element in the universe is connected, our every action affects the whole. This insight prompts one to see compassion not only as an individual virtue, but as a global responsibility. As Bohm pointed out, true social evolution cannot happen through competition, but only through cooperation and mutual respect.

A significant example interweaving science, philosophy and compassion is the aforementioned dialogue between David Bohm and Jiddu Krishnamurti. This exchange, which took place in the 1970s and 1980s, showed a rare encounter between modern scientific thought and Eastern philosophies. In those conversations, Bohm recognized that science, lacking an ethical and compassionate vision, is in danger of becoming destructive.

An interesting parallel is found in Indian and Tibetan monasteries, where monks have integrated meditation practice with the concept of universal compassion. The Dalai Lama himself has often cited quantum physics as proof of Buddhist doctrines of universal connectedness. Although from different traditions, Hinduism and Buddhism share a similar ethical vision: personal well-being cannot be separated from collective well-being.

Even in the modern environmental movement we can observe traces of this vision. Indian activist Vandana Shiva, for example, combined the struggle for biodiversity conservation with a profound ethic inspired by Hinduism. Shiva emphasized that the destruction of natural resources is not only an ecological problem, but also an ethical one. According to her, compassion for the earth and all living things is essential to ensure our own survival.

Bringing science and ethics together is no easy task, but thinkers like Bohm and texts like the Bhagavad Gita offer valuable guidance. The compassionate vision, understood as a recognition of universal interconnectedness, could revolutionize the way we deal with global problems, from the climate crisis to social inequalities. At the same time, this vision invites us to reflect on our individual role in the larger system of existence.

Technological progress and compassion. A new vision from the Bhagavad Gita.

Technological progress is undoubtedly one of the hallmarks of our age. Revolutionary applications such as those offered by quantum physics are changing the way we conceive of reality. However, these extraordinary possibilities are accompanied by unprecedented ethical dilemmas. History teaches us that every new discovery brings with it a responsibility: the choices made can lead to positive outcomes for humanity or to immense catastrophes, as in the case of the use of atomic energy. In this context of progress and responsibility, the Bhagavad Gita offers a universal key.

In the Bhagavad Gita, Arjuna is faced with a moral crisis on the battlefield of Kurukshetra. As a symbol of humanity poised between action and responsibility, Arjuna questions whether fighting is right in a context where every choice risks causing suffering. Krishna, the divine guide and voice of wisdom, teaches him the value of action imbued with compassion and detachment. This is not an escape from responsibility, but an invitation to act in a spirit of service, fulfilling one's Dharma, or moral duty, without attachment to the fruits of actions.

Today, we face similar dilemmas. Quantum technologies promise innovations that once belonged only to the realm of science fiction: impenetrable cryptography, molecular simulation to cure diseases, the possibility of revolutionizing the energy sector. However, these same technologies could be used destructively. Power, as the lessons of the Bhagavad Gita teach, is never neutral. It requires ethics deeply rooted in altruistic actions.

Krishna, in the Gita, introduces the concept of Karma Yoga: a path that urges one to act in the world with a spirit of service and compassion, without giving in to selfishness or greed. This message is extraordinarily relevant today. When we translate this principle to our present, we can see how quantum technologies can be tools for serving the common good, rather than becoming means of oppression or destruction.

One example is the use of quantum supercomputers to address global problems such as climate change. With their extraordinary computational power, these tools could optimize renewable energy production, prevent desertification, or aid in water resource management. Yet the same technological power could be used to develop advanced weapons or further concentrate resources in the hands of a few, exacerbating social inequalities. Ethics must therefore guide every step of progress.

An interesting parallel is found in the choices made in modern history. When the atomic bomb was created, in the context of the Manhattan Project during World War II, physicists like Robert Oppenheimer were faced with questions capable of tormenting the conscience.

Quantum physics is not only a field of practical application; it is also a revolution in the way we perceive the world. The concept of entanglement, for example, demonstrates how two particles can remain interconnected even at great distances. This idea presents a striking analogy with Eastern philosophies, especially Hindu thought, which sees interconnectedness as one of the fundamental truths of existence. Krishna himself points out that every action, no matter how small, will inevitably affect the whole system.

If we accept this view, it becomes evident that every technological development must consider its global impact. Every choice, from the use of an algorithm to the distribution of resources, has repercussions that can change the fate of the planet. Progress, therefore, cannot be a privilege reserved for the few. It must be a means of alleviating human suffering while respecting all living beings and the environment, a principle echoed in Vedic texts.

One practical example demonstrating the meeting of technological progress and compassion is the use of quantum technologies in medicine. Quantum computers are already revolutionizing medical research because of their ability to simulate complex molecules and open up new avenues for the treatment of diseases hitherto thought incurable. To think of these applications as bringing compassion is to focus on the highest goal of progress: saving lives.

In the energy sector, too, the promise is enormous. Advances in quantum computing could create revolutionary materials for storing solar energy or creating more efficient batteries, helping to reduce our dependence on fossil fuels. But, as Krishna teaches us, these breakthroughs must be guided by a clear ethical vision.

The Bhagavad Gita reminds us that progress without compassion is sterile. Krishna teaches that action should always be undertaken selflessly, with a view not to personal benefit but to the common good. As we face global challenges and endless possibilities offered by quantum physics, these principles remain as relevant as ever.

It is time to adopt a vision of compassion and interconnectedness, making ethics the beating heart of our decision-making. Only then can technological progress truly serve humanity, realizing what the Gita calls the "universal Dharma": living and acting with the collective good in mind.

Conclusion: science and compassion, a possible future.

Hinduism and quantum physics travel on seemingly distant paths, but on closer inspection, they seem to converge in one central point: the understanding of interconnectedness. In the Bhagavad Gita, one of Hinduism's most celebrated sacred texts, we find words that seem to speak the language of modern science today:

> *"He who sees the One in the apparent multiplicity and the multiplicity in the One is indeed a sage."*

Similarly, quantum physics has shown us that reality is not fragmented into separate compartments, but is a deeply interwoven system in which each particle influences the others through unseen links.

The idea that everything is closely connected is an ancient concept in Indian philosophy. The word "Brahman" describes the unique and indivisible essence of the universe. Every living being, every action, every sound is an expression of that universal energy that permeates everything. This, translated into more modern terms, finds an amazing echo in the quantum web: a set of invisible relationships that binds particles light-years apart. One of the most famous demonstrations of this interdependence is quantum entanglement, first observed experimentally in 1982 by physicists Alain Aspect and his colleagues, and now confirmed by subsequent findings.

The question, then, is no longer whether we are interconnected, but how we choose to act within that connection. And it is here that the great ethical challenge before science and philosophy emerges.

According to the principle of Karma, every action, thought or word generates a wave that propagates through the universe, inevitably returning to its starting point. For those even superficially familiar with this spiritual concept, it is easy to see parallels with quantum laws. If in the quantum world a localized action on one particle can affect another even at a distance, in Indian ethical thought every effect, no matter how small, takes on a cosmic scope.

This principle forces us to think about the consequences of our choices. There are no isolated or irrelevant actions. The environmental crises of the 21st century, for example, demonstrate how collectively our connection to a living planet has global consequences. Deforestation, pollution and climate change are not problems of "other places." The suffering of a felled forest is reflected in a climate that affects everyone. Movements like Chipko, born in India in 1973 to protest indiscriminate logging, embody this awareness spiritual and scientific. In the Himalayan mountains, groups of women literally hugged trees to protect them from devastating felling. It was a symbolic gesture that spoke of interconnectedness, mercy and shared responsibility.

Today, however, the great challenge is not only to understand these connections, but to apply them ethically to chart a sustainable future.

Science today has reached extraordinary frontiers. Technologies such as artificial intelligence, genetic engineering and even the manipulation of quantum matter offer us incredible opportunities, but also immense dangers. The atomic destruction of Hiroshima and Nagasaki-a project born out of enthusiasm for quantum physics-has reminded us how fragile the line between progress and destruction is.

Only a universal ethic, based on compassion and awareness of interdependence, can prevent this from happening again. Hinduism often speaks of ahimsa, the philosophy of nonviolence, as the starting point for a harmonious humanity. All great spiritual leaders, from Gandhi to the figure of Mother Teresa, have embodied in their actions that compassion which transcends cultures and religions, which responds to the call of those who suffer, wherever they may be.

This awareness is also emerging in science. The Dalai Lama frequently collaborated with scientists, from biologist Francisco Varela to physicist David Bohm, to find common ground between scientific and spiritual knowledge. Bohm, in particular, believed that there was no separation between science and spirituality. In his book "*Wholeness and the Implicate Order*," Bohm proposed that reality was an indivisible whole, a "great sea of energy," echoing views typical of the Vedānta.

Looking ahead, we have a responsibility to choose. Today we know how much one gesture can affect everything. If we regard the planet as a living organism, our actions must be directed toward the good of the whole system.

This does not mean giving up technology or science, but creating an ethical science, guided by compassion. On the other hand, as Hinduism teaches, we can never separate ourselves from others. Even the quantum principle explains to us that no particle, nor any human being, is truly isolated.

Chapter XII. Integration of spirituality and science.

Toward a new paradigm.

In the heart of sacred India, along the banks of the Ganges, Hindu spirituality is intertwined with ancient questions about the universe and reality. For thousands of years, Hindu masters and philosophers have reflected on the relationship between human beings and the infinite, seeking answers beyond visible matter. Today, this millennia-old spiritual speculation meets curiously with the world of quantum physics, a discipline born only in the 20th century. This encounter propels us toward a new paradigm, a revolutionary perspective that redefines our understanding of ourselves and reality.

Quantum physics, with its counterintuitive laws, has challenged traditional models of Western science. Concepts such as wave-particle dualism, entanglement and wave function collapse challenge classical notions of space, time and causality. Remarkably, these concepts recall some of the fundamental insights of Hinduism.

Hinduism, through its sacred texts such as the Vedas and the Upanishads, describes the cosmos not as a fixed entity but as a dynamic dance. For example, Shiva, in Hindu tradition, is often depicted as the *Nataraja, the Lord of Dance*, who creates, preserves and destroys the universe in an eternal cycle. Modern cosmology, for its part, speaks of the universe as an energy field in constant motion and transformation. This parallelism is not just poetic: physicists such as Fritjof Capra have juxtaposed the holistic view of Eastern philosophies with the language of quantum physics.

One of the central themes of quantum physics is the role of the observer. The double-slit experiment showed that the behavior of a particle changes if the particle is observed, highlighting an interconnection between consciousness and matter. This phenomenon, still debated today, seems to reflect the Hindu concept of Maya, that is, the idea that material reality is nothing more than an illusion conditioned by the perception of our senses.

Hinduism does not regard phenomenal reality as objective, but as a subjective reality projected by the mind and senses. Spiritual liberation, or Moksha, consists precisely in transcending this illusory perception and recognizing the unity of the self (Atman) with the absolute (Brahman). Similarly, some quantum physicists explore the possibility that consciousness plays a fundamental role in shaping reality, suggesting that matter and energy may not exist independently of observation.

The fusion of science and spirituality is not alien to human history. Over the centuries, great thinkers and mystics have sought to overcome the barriers between empirical knowledge and transcendental intuition. In the 20th century, Erwin Schrödinger, one of the fathers of quantum mechanics, became interested in Eastern philosophy. Schrödinger studied the Upanishads intensively and argued that modern physics, in a sense, converged with Hindu monism.

Even J. Robert Oppenheimer, the father of the atomic bomb, read the Vedas and quoted the Bhagavad Gita while witnessing the first atomic test in the New Mexico desert. The juxtaposition of science and spirituality may seem paradoxical, but these men saw in both disciplines a search for truth, albeit with different tools.

The new emerging paradigm proposes an expansion of perspective. Hinduism invites us to see consciousness as an integral part of the universe, not as an accidental byproduct of biology. Quantum physics, for its part, reminds us that reality is far more nuanced and interconnected than it appears. Together, these two views offer us a profound message: consciousness and matter are not separable; rather, they are two sides of the same coin.

According to physicist David Bohm, another pioneer of quantum thinking,

> *"Reality is an indivisible whole, in which mind and matter cannot be seen as separate elements."*

This view is echoed in the Hinduism of the Vedic texts, where the individual is not separate from the universe, but is like a drop that contains the whole ocean.

In recent years, scientists and philosophers have begun collaborating to explore the boundaries between modern physics and ancient spiritual traditions. Centers such as the *Institute of Noetic Sciences* in California promote studies that combine neuroscience, cosmology and contemplative practices. At the same time, in India, the *Vedic Science Institutes* advocate the importance of integrating ancient knowledge with modern discoveries.

On a personal level, the union of quantum physics and Hinduism can inspire a deep exploration of self and reality. As the American philosopher Alan Watts put it:

> *"Spiritual renewal is not about finding new ideas but about seeing the unity behind everything that exists."*

Perhaps, the way to understand the universe inevitably passes through understanding our own spirit.

In this journey toward the new paradigm, we are called to work for unity. On the one hand, the rigorous language of physics offers us models and data. On the other, the age-old wisdom of Hinduism invites us to reflect on the meaning of existence. United, we can glimpse a future in which science and spirituality are no longer separate worlds, but points in the same orbit: a common exploration of the universal mystery.

New ways in research and spirituality.

The frontiers of science and those of spirituality are increasingly embracing each other in unexpected ways. In particular, in Hinduism, one of humanity's oldest philosophies, surprising concepts are emerging that seem to be in dialogue with quantum physics and other scientific disciplines such as bioethics and psychology. This dialogue may represent one of the most fascinating epistemological entanglements of our time, offering new ways to understand the world and our role in it.

One of the central teachings of Hinduism is the idea of the unity of all existence. In the Vedas, the oldest sacred texts of this tradition, it is stated that everything is permeated by Brahman, the universal principle that gives life and connects everything. This concept draws striking parallels with quantum physics, where the unified nature of existence emerges through phenomena such as quantum entanglement.

Entanglement, intuited as early as the 1930s thanks to the studies of Albert Einstein, Boris Podolsky and Nathan Rosen (EPR Paradox), describes a property whereby two particles, separated even by immense distances, remain connected in ways that surpass any classical explanation. The discovery was confirmed in 1982 and then in 2022, when three scientists-Alain Aspect, John F. Clauser and Anton Zeilinger-were awarded the Nobel Prize for demonstrating how quantum behavior is beyond the limits imposed by local realism.

Similarly, Vedic texts suggest that all living beings and matter share a single underlying reality. As the Indian philosopher Adi Shankara stated in the 8th century, the material world (Maya) is an illusion, and what really exists is the indivisible unity of being. This philosophical conception seems to anticipate a quantum view of the universe, in which diversity and separation are ultimately manifestations of a deeper harmony.

The concept of Dharma, which in Hinduism represents the moral and cosmic order, offers relevant insights for modern bioethical reflection. Today, complex issues such as genetic engineering, artificial intelligence and cloning raise increasingly urgent ethical questions. How do we guide our scientific choices without compromising the balance of the natural world and society?

According to Hinduism, Dharma is the guide for maintaining universal balance and protecting the interconnectedness of all living beings. This could be applied to modern bioethics through a principle of collective responsibility. Nishit Kumar Soni, a contemporary physicist and philosopher, has argued that integrating spiritual values into scientific decisions could help avoid the environmental and social devastation caused by technology developed without moral limits.

Recently, the dialogue between psychology and quantum physics has intensified due to concepts inspired by Eastern philosophies. Carl Gustav Jung, one of the fathers of modern psychology, took a deep interest in Hinduism and the concept of synchronicity, or the simultaneous manifestation of events that, while not causally connected, display symbolic or archetypal meanings. Jung collaborated with physicist Wolfgang Pauli, one of the pioneers of quantum theory, exploring the ways in which mind and matter seem to interact through coherent, non-random patterns.

These studies make a fundamental concept of Hinduism more acceptable today: the mind possesses creative power. In Vedic texts, consciousness (Chit) is seen as the principle that shapes reality. Some modern scientists, such as Amit Goswami, theoretical physicist and author of "*The Self-Conscious Universe*," argue that consciousness is not a byproduct of matter, but the very basis of existence, anticipating a vision in which spirituality and contemporary physics meet.

In addition to philosophical and conceptual parallels, real fields of interdisciplinary research are emerging. Neurotheology, for example, explores how the brain reacts during spiritual practices such as meditation, which is deeply rooted in the Hindu tradition. Neuroscientific studies have shown that regular meditation can change the structure of the brain, improving the ability to cope with stress and increasing the sense of connection with others.

Some researchers see in these findings the possibility of developing scientific tools to explore higher states of awareness. Richard Davidson, a neuroscientist at the University of Wisconsin, has conducted studies of Tibetan monks who, during deep meditation, show unprecedented levels of gamma waves in the brain. These states, termed "*transcendental unity*," seem to confirm what ancient Indian scriptures describe as *Samadhi*, or a state of supreme consciousness in which the individual would dissolve into the universal.

The encounter between Indian philosophy and quantum physics is not just an intellectual endeavor, but a true exploration of new human possibilities. Ancient wisdom can help science remember that its ultimate goal is not just to advance technology, but to understand the nature of existence and improve the quality of life. As David Bohm, theoretical physicist and collaborator of Einstein, said:

"The universe is not a separate entity, but an indivisible whole, a totality in which each part contains the whole."

In this spirit, perhaps, the future of science will rediscover a deeper dimension, in which the unity between spirit and matter will become the focus of human knowledge.

Bibliography for further study.

This list provides a solid framework for exploring the complexities between Eastern philosophies and quantum physics, allowing for a logical and comprehensive exploration.

Almost all of the books mentioned have been translated into various languages, and can be found through the author's name.

Adams Fred	The Quantum Connection: Bridging Science and Spirituality. New York: Harmony Publishers, 2002.
Amaldi Ugo	La fisica di tutti i giorni. Dai quanti ai buchi neri # Rizzoli, 2019.
Aranya Swami Vivekananda	The Complete Works of Swami Vivekananda (in più volumi) Advaita Ashrama, 1989.
Aurobindo Sri	The Life Divine. Pondicherry: Sri Aurobindo Ashram, 1940.
Baggott Jim	Beyond Measure: Modern Physics, Philosophy, and the Meaning of Quantum Theory
Balagangadhara S. N.	The Heathen in His Blindness: Asia, the West, and the Dynamic of Religion. Leiden: Brill, 1994.
Baruah Bijoy H.	Hindu Dharma and Modern Science. New Delhi: Inter-India Publications, 1985.
Bergé Marc	La quête de l'unité: Méditation spirituelle et physique quantique # Editions Albin Michel, 2021.
Bohm David	Wholeness and the Implicate Order # Routledge, 1980.

Bohm David	Science, Order, and Creativity # Routledge, 1987.
Bohm David, Krishnamurti Jiddu	The Ending of Time: Where Philosophy and Physics Meet
Capra Fritjof	The Turning Point: Science, Society, and the Rising Culture # Bantam Books, 1982.
Capra Fritjof	La rete della vita. Una nuova interpretazione scientifica dei sistemi viventi
Capra Fritjof	The Tao of Physics: An Exploration of the Parallels Between Modern Physics and Eastern Mysticism. Berkeley: Shambhala, 1975.
Chakrabarti Arindam	Classical Indian Philosophy of Mind: The Nyāya Dualist Tradition. Abingdon: Routledge, 1998.
Chatterjee Satischandra	An Introduction to Indian Philosophy. Calcutta: University of Calcutta, 1950.
Chattopadhyaya Debiprasad	Lokayata: A Study in Ancient Indian Materialism # People's Publishing House, 1959.
Chopra Deepak	Quantum Healing: Exploring the Frontiers of Mind/Body Medicine # Bantam, 1989.
Cohen Andrew	Embracing Reality: The Direct Path to Liberation. Nevada City: Moksha Press, 2001.
Dalai Lama XIV (Tenzin Gyatso)	The Universe in a Single Atom: The Convergence of Science and Spirituality
Dasgupta Surendranath	A History of Indian Philosophy (in cinque volumi) # Cambridge University Press, 1922-1955.
Deussen Paul	The Philosophy of the Upanishads # Dover Publications, 1906.
Deutsch David	The Fabric of Reality: The Science of Parallel Universes - And Its Implications # Penguin, 1998.
Dyczkowski Mark	The Doctrine of Vibration: An Analysis of the Doctrines and Practices of Kashmir Shaivism
Einstein Albert, Infeld Leopold	The Evolution of Physics # Cambridge University Press, 1938.
Eliade Mircea	Yoga. Immortalità e libertà # BUR Biblioteca Univ. Rizzoli/Martinez Roca Ediciones, 1951/1999.
Esfeld Michael	Physics and Philosophy. Oxford: Oxford University Press, 2004.
Feuerstein Georg	Yoga Tradition: Its History, Literature, Philosophy and Practice # Hohm Press, 2001.
Feuerstein Georg	The Philosophy of Classical Yoga. Rochester: Inner Traditions, 1980.
Goswami Amit	Quantum Mechanics and Consciousness # Hampton Roads Publishing, 1995.

Goswami Amit	The Self-Aware Universe: How Consciousness Creates the Material World. New York: Putnam, 1993.
Guénon René	L'uomo e il suo divenire secondo il Vêdânta # Adelphi, 1925.
Gupta Bina	The Disinterested Witness: A Feminist Perspective on Indian Aesthetics and Ethics. Evanston: Northwestern University Press, 2000.
Heisenberg Werner	Physics and Philosophy: The Revolution in Modern Science. New York: HarperCollins, 1958.
Iyer K. Satchidananda Murty	Indian Philosophy of Advaita # Motilal Banarsidass, 1982.
Jacobs Alan	The Bhagavad Gita: A Poetic Transcreation # Watkins Publishing, 2005.
Jones Richard H.	Science and Mysticism: Exploring the New Paradigm # Bucknell University Press, 1986.
Joshi Kireet	Philosophy and Yoga of Sri Aurobindo and Other Essays. New Delhi: Indian Council of Philosophical Research, 2005.
Kak Subhash	The Architecture of Knowledge: Quantum Mechanics, Neuroscience, Computers, and Consciousness. New Delhi: Center for Indic Studies, 2004.
Kaku Michio	Physics of the Impossible: A Scientific Exploration into the World of Phasers, Force Fields, Teleportation, and Time Travel # Doubleday, 2008.
Kapra Ajit	Vedanta and Quantum Physics: A Synthesis of Science and Spirituality # Eastern Light Publications, 2012.
Karmarkar Vishwanath	Indian Thought and Quantum Physics: An Exploration # Lotus Press, 2011.
Karnath Ananda	Brahman, Atman, and Quantum Theory: Revisiting Indian Metaphysics # Jain Books, 2020.
Katz Jerry	One: Essential Writings on Nonduality # Sentient Publications, 2007.
Krishna Daya	Indian Philosophy: A Counter Perspective # Oxford University Press, 1991.
Krishnamurti Jiddu	You Are the World # Krishnamurti Foundation, 1972.
Kumar Vijay	Quantum Physics and Vedanta: Scientific Linkage of Quantum Physics and Ancient Hindu Philosophy. New Delhi: Harper, 2012.

Laszlo Ervin	Science and the Akashic Field: An Integral Theory of Everything. Rochester: Inner Traditions, 2004.
Lipner Julius J.	Brahman: A Comparative Theology # Routledge, 1999.
Mahadevan T. M. P.	Invitation to Indian Philosophy. Delhi: Chand & Co., 1974.
Majumdar Ramesh Chandra	The History and Culture of the Indian People (in undici volumi) # Bharatiya Vidya Bhavan, 1951-1977.
Malhotra Rajiv	The Audacity of Self-Identity: Hindus and the Global Order. New York: HarperCollins India, 2013.
Mallick Sanbandhu	The Quantum Vision of Indian Metaphysics # Wisdom Publications, 2015.
Matsoukas T.	Quantum World and the Mystic East # Omega Print, 2009.
Mehta Rohit	Philosophy of Patanjali # Indian Institute of World Culture, 1956.
Menon Sangeetha	Consciousness and Creativity in Indian Performing Arts. New Delhi: DK Printworld, 2008.
Mitchell John D.	Physics Meets Mysticism: Insight into the Advaitan Universe # HarperOne, 2008.
Mitra Kanji Lal	India's Philosophy: Integrating Science and Spirituality # Aryan Books International, 1995.
Mukhopadhyay Ramesh	Quantum Holism and Process Thought. Kolkata: Center of Process Philosophy, 2000.
Muller Gabriel	Hindu Cosmology and Quantum Physics: Bridging Ancient and Modern Knowledge
Murthy K. Krishna	Metaphysics in Indian Philosophy. Delhi: Bharatiya Vidya Bhavan, 1981.
Nagarjuna	Mūlamadhyamakakārikā: The Fundamental Wisdom of the Middle Way. Trans. Jay Garfield. New York: Oxford University Press, 1995.
Nagendra Swami Sivananda	Vedanta Simplified # Divine Life Society, 1960.
Nikhilananda Swami	The Upanishads: A New Translation # HarperOne, 1949.
Oppenheim G.	Brahman and Quantum: Philosophical Reflections on Unity in Diversity # Eastern Publications, 2004.
Oppenheimer J. Robert	Science and the Common Understanding. New York: Simon and Schuster, 1954.
Pande Govind Chandra	The Meaning and Process of Culture. Varanasi: Banaras Hindu University, 1961.

Panikkar Raimundo	The Intercultural Challenge of Hindu Philosophy # Orbis Books, 1993.
Penrose Roger	La strada che porta alla realtà. I segreti dell'universo, della vita e della mente # Rizzoli, 2004.
Prabhupada A. C. Bhaktivedanta Swami	Bhagavad-gita As It Is # Bhaktivedanta Book Trust, 1972.
Radhakrishnan S	Indian Philosophy. London: Allen & Unwin, 1927.
Raju P. T.	The Philosophical Traditions of India. Delhi: Motilal Banarsidass, 1971.
Rosen Steven	Krishna's Song: A New Look at the Bhagavad Gita. Santa Barbara: Praeger, 2007.
Sharma Arvind	Classical Hindu Thought: An Introduction. New Delhi: Oxford University Press, 2000.
Shastri Hari Prasad	Philosophy of Gorakhnath with Goraksha-Vacana-Sangraha. Delhi: Motilal Banarsidass, 1998.
Sri Nisargadatta Maharaj	I Am That. Bombay: Chetana Publications, 1973.
Swami Vivekananda	Complete Works of Swami Vivekananda. Kolkata: Advaita Ashrama, 1897.
Tagore Rabindranath	Sadhana: The Realisation of Life. New York: Macmillan, 1913.
Thakur Upendranath	The Aesthetic Face of Being: Indian Art Through the Ages. Delhi: Heritage, 1975.
Tolle Eckhart	The Power of Now: A Guide to Spiritual Enlightenment. Novato: New World Library, 1997.
Upadhyay Ramachandra Dattatraya Ranade	A Constructive Survey of Upanishadic Philosophy. Poona: Oriental Book Agency, 1926.
Vatsyayan Kapila	The Square and the Circle of the Indian Arts. New Delhi: Abhinav Publications, 1997.
Venkatraman Bharati	Quantum Yoga: The Theory of Relationships Between Modern Physics and Yoga. Chennai: Yogi Sciences International, 2014.

Bruno Del Medico, blogger, writer, and editor, specializes in popularizing topics related to current social issues and the new frontiers of science. He is the author of many publications, including a specialized series on quantum physics and metaphysics.

Hinduism Brahman, Atman Karma Moksha Dharma, quantum physics, quantum entanglement, quantum consciousness, spirituality meditation, Tao Zen Buddhism.

#David Bohm, philosophy of science, metaphysics, intelligent universe, soul survival, immortal soul, afterlife, NDE, quantum mechanics, elementary particles, photons, quantum collapse, particle wave, quantum superposition, Carl Jung. synchronicity, collective unconscious.

www.ingramcontent.com/pod-product-compliance
Lightning Source LLC
Chambersburg PA
CBHW041301210926
51847CB00010B/98
* 9 7 8 8 8 3 2 2 7 3 4 9 6 *